THE
BEST
OF
MAKE:

Volume 2

THE BEST OF MAKE:

Volume 2

65 Projects and Skill Builders
from the Pages of Make:

The Editors of Make:

MAKER MEDIA™
SAN FRANCISCO, CA

The Best of Make: Volume 2
65 Projects and Skill Builders from the Pages of Make:
By the Editors of Make:

Printed in Canada.

Published by Maker Media, Inc.,
1160 Battery Street East, Suite 125,
San Francisco, California 94111

Maker Media books may be purchased for educational, business, or sales promotional use. Online editions are also available for most titles (http://safaribooksonline.com). For more information, contact our corporate/institutional sales department: 800-998-9938 or corporate@oreilly.com.

Publisher: Brian Jepson
Editor: Roger Stewart
Production Editor: Happenstance Type-O-Rama
Proofreader: Happenstance Type-O-Rama
Interior Production: Happenstance Type-O-Rama
Cover Designer: Brian Jepson

Special thanks to Craig Couden
See our webpage at makezine.com/go/bom**2**

September 2015: First Edition

Revision History for the First Edition
2015-09-15: First Release

See http://oreilly.com/catalog/errata.csp?isbn=9781457186899 for release details.
Make:, Maker Shed, and Maker Faire are registered trademarks of Maker Media, Inc. The Maker Media logo is a trademark of Maker Media, Inc.

978-1-680-45032-3
[TCP]

CONTENTS

We're All Makers >>>>>>

A quizzical little magazine called *Make:* appeared in bookstores and on newsstands in 2005, stuffed with geeky DIY projects like "Kite Aerial Photography" and a hacked VCR Cat Feeder. And I do mean "little"—its small format was eye-catching, more like a paperback book than a traditional magazine. Suffice it to say, *Make:* got noticed by a new generation of readers itching to take technology into their own hands and get building things. I hopped aboard as a humble copyeditor, and it's been my great privilege to ride the wave ever since—surfing along at the forefront of a burgeoning Maker Movement. We published *The Best of Make:* in November 2007, collecting 75 beloved projects from our first 10 issues.

Then things got really busy. The *Make:* website took off like a rocket, posting maker news and projects every day. Our "Weekend Projects" video series broke out on YouTube. *Craft:* magazine inspired a new audience to transform traditional crafts. Maker Faire began in 2006 as a Bay Area DIY festival and erupted into a global phenomenon, with Faires in New York, London, Shenzhen—more than 100 cities in 2015. We launched our *Make:* books imprint, as well as the Maker Shed store to purvey the finest in DIY books, kits, and tools. Two years ago, Maker Media, Inc., spun off from O'Reilly Media. We moved our digs to San Francisco and relaunched the magazine in a full-size format aimed at a wider audience, kicking off our most exciting chapter yet.

Next thing we know, *Make:* is 10 years old, with editions published in China, Japan, and Germany, and as of this writing we're coming up on our 50th issue. It's time we served up another slice: *The Best of Make: Volume 2*—65 great projects and skill builders for all levels.

Change accelerates, but the essence of *Make:* is eternal—people just love making things and learning new skills.

While we delight in chronicling the brilliant makers and powerful new tools that are driving the Maker Movement, to my mind it's the DIY projects and tutorials that remain at the heart of *Make:* magazine. We'll always teach you to solder new circuits and MacGyver new gadgets from old—it's what we do.

Have the projects changed over the years? Absolutely. Technology waits for no one. Laser cutters are everywhere now. CNC routers and personal 3D printers have evolved from quirky prototypes—remember the plywood MakerBot on the cover of *Make:* Volume 21?—into smart, reliable tools for new kinds of making, like fashionable, flexible 3D printed "Cyberpunk Spikes" or an ear-splitting "CNC Air Raid Siren."

And drones—radio-controlled aircraft with the robotic brains to level themselves, fly acrobatically, and even navigate autonomously—went from faraway headlines to the familiar multirotor copters (like you'll find in "The HandyCopter UAV" and "Build Your First Tricopter") that everyone's nephew is building and flying today.

Maybe most significant, the rise of inexpensive microcomputers you can embed in almost any project—from Arduino microcontrollers to credit-card-sized Linux computers like the Raspberry Pi—has made possible new kinds of inventions, like a "Million Color HSL Flashlight" or a "Raspberry Pirate Radio."

And if Maker Faire has taught us anything, it's the enormous variety of people who identify themselves as makers. We've discovered new communities of DIY innovators, in traditional arts and crafts reshaped by digital design, in wearable electronics ("Luminous Lowtops"), and in modernist cuisine, ranging from new cooking technology ("Sous Vide Immersion Cooker") to the art and science of fermented foods ("Three-Day Kimchi").

The projects in this book cover a very wide range of topics, for every skill level—from robots and rockets to making soap—because we see all makers as part of the Maker Movement. I'm most satisfied with Make: when it's a variety show like Maker Faire: high tech meets arts and crafts, garage engineering, backyard science, a big portion of family fun, peppered with delightfully unclassifiable projects.

Why would you build "The Most Useless Machine"? How could you not?

As I scan the projects in this book, I'm overwhelmed by happy memories of the entire Make: editorial and creative teams, anchored over the years by Dale Dougherty, Mark Frauenfelder, Paul Spinrad, Shawn Connally, Goli Mohammadi, Gareth Branwyn, Mike Senese, Jason Babler, and our new chief, Rafe Needleman. I have them to thank for my sense of what Make: is and what it can become.

But really I'd like to thank the makers whose words you'll read in this book. For sharing their projects and workshops, their tips and tricks, and their genuine joy in making things and showing you how to make them too. It's that spirit of glee and generosity that moves the Maker Movement. You'll find it in abundance in these pages. It's a wellspring you can drink from again and again.

—KEITH HAMMOND, Projects Editor, Make:

P.S. What will Make: be ten years from now? You tell me! E-mail me your delightful projects at khammond@makermedia.com.

THE MAKER'S BILL OF RIGHTS

If you can't open it, you don't own it.

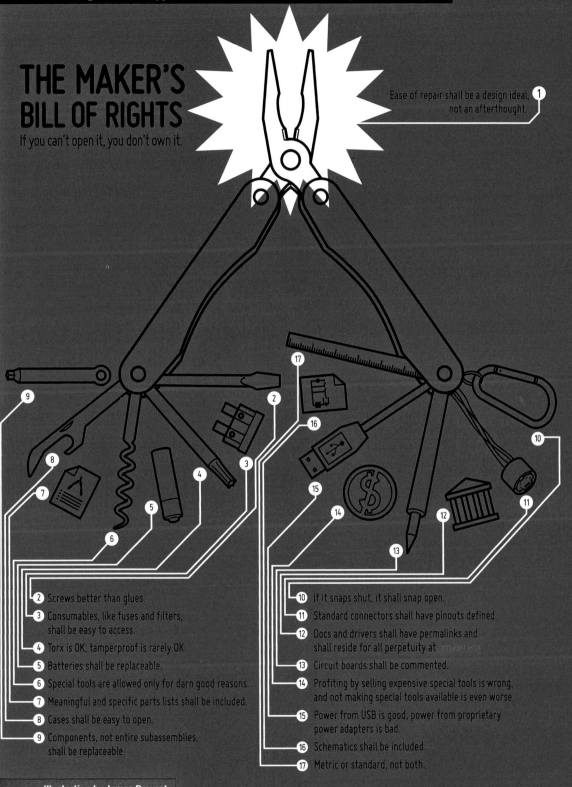

1 Ease of repair shall be a design ideal, not an afterthought.

2 Screws better than glues.

3 Consumables, like fuses and filters, shall be easy to access.

4 Torx is OK; tamperproof is rarely OK.

5 Batteries shall be replaceable.

6 Special tools are allowed only for darn good reasons.

7 Meaningful and specific parts lists shall be included.

8 Cases shall be easy to open.

9 Components, not entire subassemblies, shall be replaceable.

10 If it snaps shut, it shall snap open.

11 Standard connectors shall have pinouts defined.

12 Docs and drivers shall have permalinks and shall reside for all perpetuity at archive.org.

13 Circuit boards shall be commented.

14 Profiting by selling expensive special tools is wrong, and not making special tools available is even worse.

15 Power from USB is good; power from proprietary power adapters is bad.

16 Schematics shall be included.

17 Metric or standard, not both.

Illustration by James Provost

Right to Repair

Fight for your right to truly own your things.

Written by Kyle Wiens ■ Illustration by Jim Burke

IF MY PHONE WERE A PERSON, IT WOULD BE THE BIONIC WOMAN.
Its body has been broken and rebuilt more times than I can count. Its brain has been modified, tinkered with, and improved.

In the past three years, my iPhone 4S has been jailbroken and wired into a home automation system. Its Apple-approved glass back panel has been replaced with a transparent one. It's been water-drenched, dismantled, and completely cleaned. Twice. Thanks to an app from the free-as-in-speech Cydia store, I'm tracking my battery's performance in ways Apple won't allow. And I've pried up and replaced that battery over and over again.

It's the phone that will not die — at least not if I have anything to say about it.

Ten years ago, I started iFixit, the world's free online repair manual. Our goal is to teach everyone how to fix the stuff they own — whether it's laptops, snowboards, toys, or clothes. And we're not alone. iFixit is part of a global network of fixers trying to make the stuff we own last forever.

On the surface, fixers and makers are cut from different cloth. Makers put things together; fixers take them apart. One creates new gizmos; the other rebuilds existing ones. But I've always thought that, under the skin, they're incredibly similar — two different sides of the same coin.

We are, all of us, tinkerers. We're motivated by the same ideals: an inexhaustible curiosity, an appreciation for things done by hand, a sentimental attachment to the smell of wood shavings, and a never-ending pursuit of understanding the things around us.

As tinkerers, we become more than just consumers. We are participants in the things we make, own, and fix. But over the years, I have found that this participation — tinkering with products made by others — puts both makers and fixers at odds with manufacturers. (Apple certainly wouldn't endorse my bionic iPhone.) For the most part, manufacturers would prefer if we all just put down our screwdrivers and got back in line at the store.

By revealing (and reveling in) the secret insides of machines, tinkerers transgress the boundaries of what manufacturers think we should be able to do with our stuff. We alter the code they wrote, we rebuild the hardware they designed, and we find ways of fixing our old stuff instead of buying their new stuff.

For the past 20 years, manufacturers have been waging a quiet war against tinkerers like us. They're using encryption-powered DRM, vague hand-waving claims of proprietary knowledge, DMCA takedown notices, and legal threats to keep people from fixing their tractors, from repairing their Apple products, and even from modifying the software on their calculators. Keurig is even adding a chip to their coffee pods to prevent homebrewers from "reloading" their capsules.

It took two years after 114,000 people signed a We-the-People petition to finally relegalize cell phone unlocking, but we did it! Unfortunately, despite intensive right-to-repair lobbying from the Electronic Frontier Foundation and others, jailbreaking game consoles and bypassing digital locks on tractors is still a crime.

Even the car industry — sacred ground for tinkerers since the rise of the hot rod — has succumbed to the same locked-door policies. These days, cars are made up of as much code as they are nuts and bolts. Tinkering under the hood requires access to service information and schematic systems — information that carmakers don't like to share.

In Massachusetts, voters had to pass a law to force automakers to share internal service manuals, circuit diagrams, and computer codes with independent repair shops and owners.

I think that if you bought it, you own it. I mean *really* own it. You have the right to take it apart, mod it, repair it, tap dance in the code, or hook it up to your personal brand of Arduino kung-fu.

But if you want the right to tinker, you'll have to start fighting for it. Fight for your right to mod and make. Fight for your right to repair. Fight for your right to own your own things.

We live in a brave new digital world, and it's time that we join forces with Cory Doctorow — and other makers leading the movement to free our hardware — when he says, "This has nothing to do with whether information is free or not — it's all about whether people are free." ✇

KYLE WIENS is the cofounder of iFixit (ifixit.org), the free repair manual. iFixit's open-source community has taught millions of people how to fix everything from iPhones to Volvos

Workshops
AND Tools PART 1 >>>>>

You may have first encountered William Gurstelle in the pages of one of his many best-selling books, such as *Backyard Ballistics* or *The Practical Pyromaniac*. If so, you will understand why he calls his workshop "The Barrage Garage."

Bill is a regular contributor to *Make:* magazine, and his workshop articles form the centerpiece of this opening section of *The Best of Make: Volume 2*. Every maker needs a workspace and tools. Whether you decide to convert an existing space or build from scratch like Bill did, you'll find lots of tips for designing, stocking, and organizing you own workshop in his contributions to this collection.

Another long-time contributor to *Make:* magazine, Charles Platt, is the author of the leading book for electronics enthusiasts, *Make: Electronics*. In addition to his well-known passion for electronics, Charles is also accomplished with a wide range of workshop tools. In "Wilderness Workshop," he shares the economical ways he found to furnish his workspace that overlooks the wilds of Northern Arizona.

Master craftsman Len Cullum provides an introduction to basic woodworking tools, as well as the tools he finds most useful for making accurate measurements. In his article, "Japanese Toolbox," he shows you how to build a versatile storage box for your workshop tools. Simplicity and elegance are the hallmarks of Len's work.

Frank Ford also knows something about simplicity and elegance. He is an accomplished maker of stringed musical instruments, and you will find a treasure trove of information about fretted instrument repair, maintenance, and restoration on his website frets.com. Frank turns his ingenuity to a completely different task in his contribution to this volume, as he demonstrates a neat trick for recovering the tiny, hard-to-find things that often get dropped when you're working on a project.

Marc De Vinck, a former director of product development at Maker Media, is now a professor in the technical entrepreneurship program at Lehigh University. While at Maker Media, Marc implemented the popular "Learn to Solder" program that is now a staple at Maker Faire and has taught thousands of people this valuable skill. His "Mini Fume Extractor" project is especially clever because he uses his soldering iron to create a handy little tool to pull fumes away from his soldering projects. Now, that's meta!

Enjoy discovering the tips and tricks of these workshop masters, and get ready to start making your own ideas into reality.

THE ULTIMATE TOOL
BUYING GUIDE

A complete list of tools you need to make almost anything.

If a genie were to grant me my wish for a shed full of tools, this is what I'd ask for. Think of it as an extremely biased guide to outfitting yourself with the ultimate shop for launching your own space program.

Necessity
Priority
Extremely useful
Surprisingly useful
Infrequent but handy
Can do without, better with
You didn't know it was so lovely
Bonus

Tool Name	McMaster #	$ Budget	$ Deluxe
Hand Tools			
Box Knife	3814a11	1	10
Precision Blade	35435a11 35435a71 35515a12	1	10
Claw Hammer	6484a21	10	50
Ball Peen Hammer	6481a31	10	50
Blacksmith's Hammer (Heavy Weight)	6462a24	10	80
Rubber Mallet	5917a8	10	40
Miter Box	4201a11	15	45
Hacksaw	4086a34	5	25
Tight Spot Hacksaw	4060a16	2	5
Hole Punch Tool	3461a22	40	150
Center Punches and Chisel Set	3506a76	25	120
Metric and Imperial Socket Sets	7290a24 5757a35 5582a11	30	1200
Torque Wrench	85555a221	50	300
Hex Key Sets, Imperial and Metric	5541a31 5215a24 7162a13 5215a12	2	80
Torx Key Set	6959a85	2	40
Mini-Hex Drivers	52975a21 7270a59	2	40
Combination Wrenches, Metric and Inch	5314a62 5304a73 5314a25 5772a53	25	800
Vise Grip Long Nose Locking Pliers	7136a19	2	50
Needlenose Pliers, Small and Large	5451a12	2	35
Bull Nose Pliers, Small and Large		2	35
Vise Grip, Large	7136a15	5	60
Vise Grip, Med Curved	5172a17	5	45

Tool Name	McMaster #	$ Budget	$ Deluxe
Adjustable Wrenches	5385a12 5385a15	3	40
Crow Bar / Ripping Bar	5990a2	2	30
Tube Cutter	2706a1	15	80
Glass Cutter	3867a16	2	25
Bolt / Chain Cutter	3771a15	50	150
Sheet Metal Snips	3585a13 3908a11 3902a9	10	40
Finishing Saw	4012a1	10	30
Coping Saw	4099a1 6917a11	4	10
Hole Saw Kit	4008a71	25	120
Pull Saw	4058a52	10	20
Metric / Inch Tap and Die	2726a66	40	1200
Drill Sets	28115a77 31555a55 31555a56 31555a57 8802a11 8802a12 8802a13	5	1200
Deburring Taper	3018a4	5	80
Deburring Tools	4253a16 4289a36	2	25
Drill Stops	8959a16	2	10
Vise	5344a31	10	1500
Clamps	5165a25	2	45
Quick-Grips	51755a7	15	50
Jaw Puller	6293k12	50	180
Files	8176a12 8194a12	2	100
Hydraulic Floor Jack		25	200
Block And Tackle / Lifting Winch		50	500
Screwdrivers, Flat and Phillips	8551a31	1	90
Jeweller's Screwdrivers	52985a21 52985a23	10	40

Tool Name	McMaster #	$ Budget	$ Deluxe
Propane Burner		10	50
Heat Gun		50	250

Power Tools

Tool Name	McMaster #	$ Budget	$ Deluxe
18V Electric Drill	29835a16	25	300
Band Saw	4164a12	250	5000
Reciprocating Saw (Sawzall)	4011a25	120	250
Sliding Compound Miter Saw	3001a21	200	600
Tilting Table Saw	27925a12	300	2000
Drill Press	28865a31	100	2500
Plunge Router	36485a11	100	300
Manual Lathe	8941a12	500	5000
Mig Welder	7899a28	200	1500
Stroboscope	1177t92	25	250
Adjustable Hot Plate	33255k61	50	800
Dremel	4344a42 4370a5	50	150
Angle Grinder	4395a16	50	250
Bench Grinder	20535A654	75	300
Belt Sander	4892a21	100	200
Disc / Belt Sander	46245a49	250	1500
Bridgeport Mill		500	15000
Heisseschneider Hot Knife		50	200
Sewing Machine		25	2500
Air Compressor	4364k3	200	2500
Spot Blaster	31195k11 3210k11	50	500
Vacuum Pump		100	1000+
Oxy / Acetylene Torch	7754a12	250	1500
Plasma Torch		600	3000

Computer Controlled Tools

Tool Name	McMaster #	$ Budget	$ Deluxe
Inkjet Printer		25	250
Large-Format Printer		900	25000
Nc Mill		2500	120000
Nc Lathe		5000	150000
Laser Cutter (Co2)		12000	50000
Plasma Cutter		3000	20000
Wire / Sink EDM		100000	250000
Water Jet		80000	150000
3D Printer (Z Corp, FDM, STL)		25000	250000
Plotter / Cutter (Roland)		1000	25000

Electronics Tools

Tool Name	McMaster #	$ Budget	$ Deluxe
Wire Stripper		2	80
Pliers Set	5323a49	10	120
Work Holder And Magnifier	5007a14	5	100
Multimeter		75	250
Temp-Control Solder Station		150	1000
Hot Air Tool for Point Reflow / Desoldering		30	500
Bench Power Supply, Multi-Output		150	500
Toaster Oven, Adjustable Time / Temp		40	60
Microscope (See Safety / Measurement / Visualization)			
Oscilloscope		500	5000
Micro-tweezer Sets		2	100
Pick-n-Place		3000	25000

Fetish Tools

Tool Name	McMaster #	$ Budget	$ Deluxe
Optics Bench		1000	400000
Mask Writer		50000	1000000
Mini-jector		4000	50000
Thermoformer		1000	20000
ESEM		25000	500000
3D Scanner		5000	100000
Excimer Laser Cutter		100000	1000000
PCR			100000
Micropippettes		20	2000
Spin Coater		500	25000
High Temp / Vacuum Oven		2000	30000
Chemistry Hoods and Glass Equipment		2000	1000000
Ultrasonic Welder		5000	25000
Tube Bender		1000	40000
Tanks for Anodizing, Etching		25	2500
Kiln		500	5000
Anvil		250	1000
Crucible		20	2500
Thin Film Evaporator / Sputterer		5000	100000

Safety, Measurement, and Visualization

Tool Name	McMaster #	$ Budget	$ Deluxe
Safety Goggles	2404t21	1	10
Ear Muffs	9205T6	2	30
Micrometer	2054a75	5	300
Caliper	8647a44	5	500
Head-Mounted Magnifier	1490t3 1509t14	5	120
Feeler Gauges	2070a7	1	25
Spirit Level	2169a4 2169a1	5	50
Tape Measure	19805a74	1	25
Adjustable Stereomicroscope	10705t64	500	25000
Hot Gloves		5	100
Work Gloves		1	40
Welding Mask		15	100
Rules	2042a77 6823a61 20265a36	5	100
Shop Vac	70215t26	60	200

Building the Barrage Garage

The ultimate, multipurpose maker's workshop, built from the ground up.

By William Gurstelle

THE BARRAGE GARAGE

As a city dweller, I've often looked with envy at the spacious outbuildings of my rural friends and relatives. Horse barns, potting sheds, root cellars, equipment garages—plentiful, enclosed, and private space is the one thing that makes me envy those who live beyond the end of the bus line. I think often about what I could make if I had a room of my own: a purpose-built, well-equipped space in which to create.

Apparently I'm not alone in these thoughts. Homebuilders commonly offer two-, three-, and even four-car garages for new homes. But all that space isn't needed simply to shelter the family Chevy. It's needed to keep pace with the explosion in DIY projects and their concomitant material and tool requirements.

Randy Nelson, president of Swisstrax, a manufacturer of workshop and garage floor products, says that garages are quickly evolving into more than simply places where people keep their cars. Installation of the company's special-purpose floor tile in garages and workshops is booming.

"[Spaces for making things] have just about doubled in the last ten years," says Nelson.

"People aren't just stuffing junk in their garages any more. It's become the male domain, the place where they can do their work and have their tools."

There are scores of books providing advice on setting up a wood shop or metal shop, and many others that describe setting up specialty areas such as a paint shop, a photography studio, or a chemistry laboratory.

But what I wanted was not a single-purpose workspace. I was seeking the ultimate, multi-purpose maker's workshop: a versatile, flexible space capable of handling nearly any project I could think of—from building a cedar-strip canoe to compounding fuel and oxidizer for a rocket engine, from soldering a Minty

Boost to developing a model ornithopter.

This series of articles details the creation of a modestly sized yet state-of-the-art maker's workshop, which I named the Barrage Garage.

This installment covers the design and construction of my Barrage Garage, and the considerations behind its doors and windows, floor coverings, and other infrastructure. The parts that follow describe the equipment inside it, such as workbenches, machine tools, hand tools, and my own space-saving tool storage system.

Photography by William Gurstelle

Workshop Design Criteria

» **The first step was to determine which features were the most important and practical.**

Egress A 9-foot-wide, automatic, well-insulated door outfitted with required safety equipment was essential. The huge door makes bringing materials in and out of the workspace a snap.

Fenestration Natural light and a view to the outside were high on my list of priorities. Therefore, the design called for four east-facing sliding windows having a total glass area of 24 square feet.

Organization I devised a plan for a combination of stackable modular cabinets, which, along with a slotted wall storage system, maximize the efficiency and versatility of my space.

CUSTOM CAVE: The obvious advantage of building a workshop from scratch is the luxury of spec'ing it perfectly to your needs.

Surfaces I wanted more functionality and style than a concrete floor could afford. I selected a special-purpose tile floor for workshops and garages that makes walking and standing more comfortable.

Power I needed 240 volts to run the heater and welder, and 120-volt receptacles placed at frequent intervals along all walls on two separate 20-amp, GFI-protected circuits. This ensures a plentiful, safe supply of electrical power to all tools.

Building the Barrage Garage

» My first task was to site the structure. Where should the workshop go?

Initially I considered placing the shop in my basement. Possible, but this would involve far too many compromises. The basement is a low-ceilinged space with marginal access via a narrow stairway. The thought of carrying tools and materials up and down, turning corners, and so forth quickly dissuaded me.

Instead I turned to the nearly forgotten space along the alley in back of my home. Separated from the rest of my yard by a chain-link fence, it was covered with 25-year-old lilac bushes. I loved those fragrant, beautiful spring blossoms, but the space those lilacs grew upon was workshop-perfect: it had room, privacy, and access. So, goodbye lilacs.

City ordinances allowed me a maximum of 240 square feet for the shop. With the city building permit obtained, it was time to push some dirt.

Pushing Dirt

It all starts with a level floor. Every workshop, atelier, pole barn, or garage must have a level floor if great things are to be made in it. It has always been this way.

Four thousand years ago, in the reign of the great Egyptian pyramid builders, construction techniques were rudimentary. Imhotep, legendary architect of the pharaohs, had only knotted measuring ropes stretched taut between stakes, plumb bobs, and sighting sticks.

But Imhotep gave the pharaohs the tools to build monuments capable of withstanding 50 centuries of desert sandstorms. He did that by starting with a perfectly level floor. It's believed that the Egyptians leveled the area under a pyramid by cutting a shallow grid of trenches into the bedrock, then filling them with water. Knowing that the height of water

within connected trenches would be at exactly the same level, the workers hacked out the intervening islands of stone and sand with hoes and stone drills.

The Barrage Garage has a flat floor as well, but my excavators used a 75-horsepower backhoe and modern surveying tools including transits and lasers. My end result is pretty much the same as Imhotep's: a perfectly level slab placed in exactly the right spot.

Concrete Ideas

After excavation, the concrete work began. Concrete is composed of Portland cement, gravel, sand, and water. When freshly poured, concrete is wet and plastic. But within hours it begins to solidify, ultimately becoming as hard as rock.

Most people call that process "drying," but the concrete crew foreman on my job told me that's not really the best choice of words. Concrete does not simply solidify because excess water has evaporated from the slurry. Instead, the water reacts with the cement in a chemical process known as hydration. The cement absorbs the water, causing it to harden and bond the sand and pebbles together, creating the stone-hard material we know as concrete.

Framing the Concept

Prior to the mid-19th century, building was an art that took many years of apprenticeship to learn. There were few, if any, building codes. Quality of work was based largely on the personal integrity and craftsmanship of each builder.

For 2,000 years, the most common technique for building

with wood was the method called *timber framing*. Buildings of that era still exist; typically they are barns and homes with huge wooden beams supporting large open spaces.

In the mid-19th century, building techniques changed. Cheap, factory-produced nails and standardized, "dimensional" lumber from sawmills allowed for a faster, more versatile method of construction called *balloon framing*.

Invented by Augustine Taylor of Chicago, balloon framing revolutionized building construction. It utilized long, vertical framing members called studs that ran from sill to eave, with intermediate floor structures nailed to them. What used to take a crew of experienced timber framers months to join and raise, could be constructed in a fraction of the time by a competent carpenter and a few helpers.

Over time, balloon framing evolved into the current technique known as *platform framing*. The Barrage Garage, like most modern buildings, is built by nailing together standard dimensional lumber—2×4 trusses holding the roof and 2×6 studs forming the walls—at code-defined intervals. Then, plywood sheathing is attached to the lumber frame, and the basic structure is complete.

A Solid Floor

The first order of business after the workshop shell was complete was to install the floor. There are three general options: coatings, mats, and tile. Each has its own advantages and disadvantages.

Most common and least expensive are coatings. There are several types of coating available for concrete floors, including epoxy, polyurethane, and latex.

SCRATCH BUILT: A perfectly level slab is an imperative start. I used platform framing for the structure, and durable, cushioned vinyl tile for the flooring.

Epoxy paint is probably the most widely applied form of floor coating. Epoxy forms a hard, durable surface and bonds solidly to a correctly prepared surface. Because floor coating provides no cushioning, it can be hard on feet and legs. Also, it doesn't last forever: expect to recoat the floor every five years or so.

Polyurethane coatings are also very durable, and they resist chemical spills better than epoxy. But urethanes do not bond directly to concrete, so an epoxy primer coat is required.

Latex garage paint is widely available and inexpensive. It goes on easily and doesn't require the prep work associated with epoxies and urethanes. However, it's less durable.

PVC floor protection mats are another option. They protect the porous concrete floor from staining or corrosive chemicals such as oil, paint, or acid. Mats are typically simple to install, requiring only scissors. Importantly, they add a cushioning layer above the hard concrete.

Special-purpose vinyl tile is the premier flooring option for workshops and garages, and that's what I installed in the Barrage Garage. These floor tiles, from Swisstrax (swisstrax.com), snapped together firmly and were installed without special tools.

Tile handles heavy loads and high traffic. It resists damage caused by chemicals, and it's far more comfortable to stand on than concrete. But best of all is tile's ability to transform a humdrum workshop into a great-looking space.

Choose Your Tools

Outfitting the all-purpose maker's workshop with the tools to tackle most any project.

By William Gurstelle

Photography by William Gurstelle

So far, I've detailed the construction of my all-purpose maker-style work-shop, which I've nicknamed the Barrage Garage. It's turned out beautifully, and as anticipated, it's the envy of my maker friends.

Small? Sure, it's a mere 20 feet by 14 feet, but it has all the space required to do serious creating. It's loaded with features, including a way-cool vinyl tile floor, a high-tech wall storage system, fluorescent lighting, 240-volt power, and lots of electrical outlets.

After the infrastructure was completed, it was time to outfit the Barrage Garage. Choosing tools and supplies is a subjective question to be sure, and one that a dozen people would answer a dozen different ways. My goal was to make the Barrage Garage into the Platte River of workshops: a mile wide and three feet deep. Like the Platte, my workshop covers a lot of different areas but is not particularly deep in any single genre. Flexible as a yoga instructor, it provides an environment in which I can attempt projects in wood, metal, chemistry, home repair, electricity, even the odd bit of pyrotechnics (*see MAKE Volume 13, page 54*).

If you're a maker with dreams of metalworking, woodworking, building electronics projects, customizing your rod, or simply keeping your house up and running, read on. In this installment, we'll examine the must-have tools and equipment that make the Barrage Garage such a maker-enabling space.

The Workbench

Building a workbench was my first consideration, for it's literally the foundation on which all subsequent work will be built. I considered the design carefully, evaluating possibilities ranging from a complex Scandinavian design with a beechwood frame mounted on self-leveling hydraulic cylinders, to an interior door nailed to two sawhorses. I chose something in the middle—a solid, heavy, counterbraced construction made from 2×6 fir lumber.

The work surface is two-thirds wood and one-third granite. From a local countertop maker I was able to inexpensively obtain a beautiful 2'×2' piece of polished granite left over from a bigger job. The ultraflat, smooth granite is perfect for doing fine work or electrical projects. The plywood-covered 2×6s are great for everything else.

I finished the workbench by outfitting it with a wood vise with bench dogs (wooden inserts mounted opposite the vise to hold oversized work pieces), a portable machinist's vise, and a pullout shelf.

The typical advice from experts to novices is to buy the best quality tools you can afford. And I believe it's good advice. Cheap screwdrivers, for example, can be a big mistake; the soft metal

edges of inferior blades can bend or even break under stress, and the plastic handles chip when dropped. For any tool you use frequently, it makes sense to go with quality.

On the other hand, when you've got a one-off job, and you're not sure if you'll ever have another application for piston-ring pliers or a gantry crane, then buying an inexpensive tool may make sense.

Besides raw materials and tools, I stocked up on general supplies: duct tape, electrical tape, transparent tape, powdered graphite, rope or cord, twine, light oil, white glue, super glue, quick-set epoxy, extended-set epoxy, sandpaper, heat-shrink tubing, zip ties, pencils, ink markers, rags, wipes, and towels. Now, on to the tools.

The Tools

BASIC TOOLS

A. Electric drill, cordless or corded A drill with a variety of screwdriver tips and drill bits may well be your most frequently used power tool. In the Barrage Garage, where I have power outlets everywhere, I appreciate the lightness and torque of a corded drill. But many people appreciate the flexibility of a cordless model. The higher the top voltage (e.g., 14.4 or 18 volts) of a cordless drill, the greater its torque and the more it weighs.

B. Files and brushes Flat and round bastard files and a wire brush. (A bastard file refers to one with an intermediate tooth size.)

C. Cutters You'll want diagonal cutters, a utility knife, tinsnips, a wire cutter/crimper/stripper, and a good pair of scissors. You'll find a self-healing cutting mat to be a great help; buy one at any fabric store.

D. Mixing and volume-measuring equipment Sturdy plastic bowls in different sizes, disposable spoons, measuring cups, and measuring spoons.

E. Hacksaw For those occasions that require cutting through something harder than wood.

F. Handsaw Most often, you'll likely be cutting dimensional lumber (2×4s, 2×6s, etc.) to size, so choose a saw with crosscut teeth instead of ripping teeth.

G. Linear measuring gear Tape measure, protractor, and combination square.

H. Socket and wrench set If you work on things mechanical, you'll appreciate the quality of a good socket set. Spend the money and get English and metric sockets, as well as Allen wrenches (hex keys).

I. Pliers come in a variety of shapes. At a minimum, you should have standard, needlenose, and vise-grips.

J. Hammers Start with a claw hammer for nailing and a rubber mallet for knocking things apart.

K. Digital multimeter If you do any electronics work, a volt-ohm meter with several types of probes and clips will be indispensable.

L. Screwdrivers Choose an assortment of high-quality Phillips and flat-headed (and possibly Torx) screwdrivers in a variety of sizes.

M. Scale A triple beam balance or electronic scale is a necessity for chemistry projects and mixing stuff.

POWER TOOLS

N. Belt sander

O. Drill press I simply can't live without my drill press, because it provides far more accuracy than a hand drill ever could.

P. Cut-off saw

Q. Grinder

Beyond these basics, there are hundreds, if not thousands, of tools available, all of which may be useful depending on the project. In regard to stationary power tools, it's a tough call. Because they're expensive and require a lot of shop real estate, it really depends on what you're going to do *most*. I use my table saw all the time. But I know people who consider a band saw an absolute necessity and others who say a scroll saw is their number one power saw priority.

SPECIAL TOOLS

Soldering iron Choose a variable-temperature model with changeable tips.

Magnifying lens You'll find a swing-arm magnifier with a light a very helpful addition to your shop. It mounts directly to your workbench and swings out of the way when not in use. It's great for everything from threading needles to examining surface finishes.

Safety equipment Safety glasses, hearing protection, a fire extinguisher, goggles, a dust mask, and gloves are very important.

All safety glasses, even inexpensive ones, must conform to government regulations, so they all provide adequate protection. However, more expensive ones are more comfortable and look better, making you more inclined to always use them.

Basic Tools

Power Tools

Safety Equipment

General Supplies

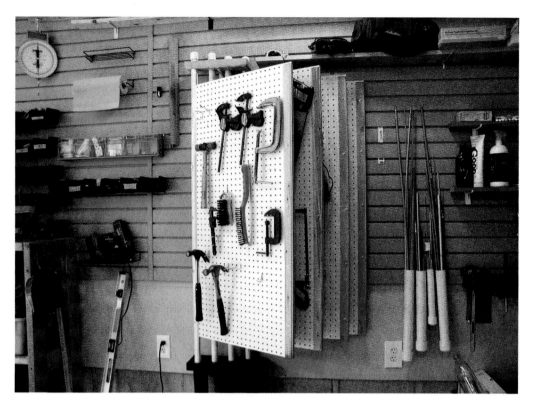

The Tool-Zine

OK, once you've got all this stuff, where are you going to store it? I use a combination of bins hung on StoreWall (storewall.com) panels, and my own contrivance that I call the Tool-Zine. It's easy to build and provides an incredible amount of storage in a small area.

The Tool-Zine is like a magazine for tools; you store your tools on "pages" and simply turn to the correct page when you need a particular tool. You'll be amazed at the convenience and organization it brings to your shop.

A 4-page Tool-Zine provides the equivalent of 64ft^2 of wall space in a space slightly larger than 8ft^2. That's a highly leveraged storage solution!

Conceptually, the Tool-Zine is straightforward. It consists of four 1" PVC pipes slotted lengthwise. A 2'×4' piece of ⅜" pegboard is inserted into each slot and fastened with machine screws. Next, wood lath is bolted to both sides of the pegboard to make it rigid. This entire assembly makes a single page of the Tool-Zine.

Four pages are assembled and then mounted vertically on wooden brackets that are firmly affixed to wall studs, reminiscent of the way the pages in this magazine are bound to the spine.

MATERIALS

» **1" Schedule 40 PVC pipes (4)** 5' long
» **1" PVC pipe end caps (4)**
» **1" wood laths, 4' long (4)**
» **⅜" pegboard in 2'×4' sections (4)** Other pegboard thicknesses might work, but you'll have to adjust the slot width.
» **2×6 lumber** about 8' long
» **¼" machine screws, 1½" long (20)** with nuts and washers
» **#8 machine screws, 1½" long (20)** with nuts and washers
» **Wall anchors** or wood screws

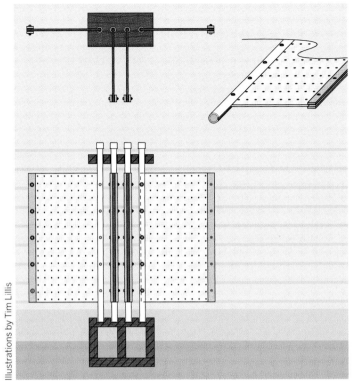

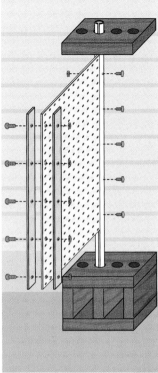

Make Your Tool-Zine

The diagram shown here has all the information you need to build the Tool-Zine. Here are a few pointers to make the task easier.

1. The most difficult part is making the long slot in the PVC pipe. To fit the pegboard, it must be straight, with a constant width. You will likely need a table saw to do this successfully. I bolted the PVC to a 2×2 piece of dimensional lumber so I could use the saw's rip fence to keep the cut as straight as possible.

2. Set the saw blade height on your table saw so that it's just high enough to cut through the bottom of the PVC, but doesn't cut into the bolts used to attach the PVC to the 2×2 guide piece.

3. Depending on the kerf width of your saw's blade, it may be difficult to slide the pegboard into

the slot. If so, use a rubber mallet to pound it in.

4. Be sure to anchor the top bracket firmly into the wall studs.

5. I chose to build a platform to support the lower bracket. The platform rests on the concrete floor and is attached with a concrete anchor. As an alternative, the bottom bracket could be wall-mounted like the top bracket. It you do this, make sure the brackets are securely mounted to structural members that can handle the weight of your tools.

6. The Tool-Zine is customizable. You can easily add additional pages or increase the distance between pages by extending the size of the brackets. However, if you do, be sure the brackets are adequately anchored to the wall studs.

Illustrations by Tim Lillis

Stock Your Shop

Be ready with a basic inventory of the stuff dreams are made of.

By William Gurstelle

In my Barrage Garage workshop, I keep frequently used materials on hand at all times. Stocking a well-considered selection of materials is important when I need to make a simple model or a fast prototype to bring an idea to life, or at least prove to myself that it's worth further exploration.

Choosing materials is an important part of any project. But tradeoffs abound and it can be tricky to decide which raw material is right for the job. One material may be strong, but difficult to machine. Another may be great for use indoors but lose its integrity when placed in the elements.

Over time, I've developed an inventory of basic raw materials that enable me to start, and sometimes even complete, a great variety of projects without the need to visit a lumberyard or wait for the UPS truck.

Dimensional Lumber

Dimensional lumber is the wood commonly sold in lumberyards. It's sized according to standardized widths and depths that are nominally described in whole numbers, but its actual size is ¼" or ½" less than described. For example, a 1×2 board is actually ¾"×1½".

Dimensional lumber comes from softwoods like fir and pine. Cut into 2×4s, it's commonly used for building the frame or supporting structure of a project. Shelves and smaller objects are frequently made from 1"-thick (nominal) boards of various widths and lengths.

It's inexpensive and versatile, so keep ample supplies on hand for spur-of-the-moment projects. I like to stock:

- **1×2×8' (4)**
- **1×4×8' (4)**
- **2×4×8' (4)**

Engineered Wood

Engineered wood products are manufactured from wood components and adhesive. Strong and light, they're just the thing for covering walls and roofs, and they're useful for projects of all types. Engineered woods include the old standby, plywood, as well as particleboard and medium-density fiberboard (MDF).

MDF and particleboard are manufactured from wood pulp and glue pressed into sheets. They're easy to cut and sand, a plus for the inexperienced maker. They're cheaper than plywood, and not as prone to warping.

Plywood, on the other hand, is proportionally lighter and stronger than MDF and particleboard. It holds fasteners more securely, and has far better moisture resistance. Plywood comes in a variety of thicknesses; ⅜" and ⅝" are the most commonly used.

Engineered wood comes in 4'×8' sheets that can take up a lot of storage space. If that's a problem, cut the sheets in half. Stock:

- **⅝" plywood, B or C grade, 4'×4' sheets (2) or a 4'×8' sheet** if you have room

Dowels

When dimensional lumber is too big, dowels (cylindrical rods made from solid wood) are utilitarian wonders. They're great for aligning, fastening, and supporting project parts. Stock:

- **¼" dowel, 3' lengths (4)**
- **⅜" dowel, 3' lengths (4)**
- **½" dowel, 3' lengths (2)**

Hardware Cloth and Expanded Metal

Hardware cloth is wonderfully useful stuff. It's a sheet of stiff, galvanized steel wire welded into a regular square mesh. Flexible and strong, it's perfect for screening off large areas inexpensively.

Expanded metal is a rigid, open mesh made from sheet metal that has been slit and expanded. It is much stronger and more rigid

MATERIAL WORLD: Having a well-stocked selection of basic materials makes it possible to build (and photograph) on the fly.

than hardware cloth.

You can use these materials for animal-proofing, providing support for materials such as plaster or concrete, making guards for dangerous areas, and more. Expanded metal is strong enough to be walked upon. Stock:

Hardware cloth, ¼" mesh, 2'×4' sheet (1)
Expanded metal, 2'×4" sheet (1)

Angle Iron aka Structural Steel

Structural steel or iron, like dimensional lumber, is most often used to build a frame or superstructure. The many types of structural steel are described by the shapes they make when looking at them on-end—angles, channels, bars, I-beams, and more. Angle iron is the king. It's relatively inexpensive, extremely strong, and can be attached to other pieces of iron by fasteners or by welding.

I keep several 8' pieces of galvanized 14-gauge slotted angle iron on hand. Although its galvanized surface makes it nearly unweldable, it's still wonderfully versatile. The slots allow pieces to be cut and joined easily using ⅜" bolts and nuts. Channels and I-beams are useful in situations requiring greater stiffness, although they're

heavier and cost more per linear foot. Stock:

Slotted steel angle, 1¼"×1¼"×8' (2)

Brass Sheet

Brass is an alloy of copper and zinc. Stronger than copper alone, brass sheet is a great prototyping material that can be machined, bent, or cut to close tolerances. It's easily worked with common hand tools and can be joined by soldering or by mechanical fasteners. Stock:

Brass sheet, .032"×6"×12" (1)
Brass sheet, .016"×6"×12" (1)

Plastic Pipe and Pipe Fittings

Because of its low cost, strength, and easy workability, PVC (poly-vinyl chloride) is among the most common plastics in use. PVC pipe comes in many sizes and shapes, with many types of connectors (called fittings). It's useful for making furniture frames, medium-duty support structures, and plumbing for liquids or air. Stock:

PVC pipe, 1" diameter, 8' lengths (2) with 1" elbows (4), tees (4) and caps (4)
PVC pipe, 2" diameter, 8' lengths (2) with 2" elbows (4), tees (4) and caps (4)

Music Wire

Beloved by makers for its ability to be bent and then retain its shape, this tough and springy wire is especially useful in making mechanisms and models. Be forewarned, it's really tough stuff, requiring a special cutter (or a high-speed rotary tool like a Dremel). Stock:

Music wire, .056" OD, 3' lengths (2)

Fasteners

Nearly all projects involve joining parts together, commonly by using mechanical fasteners such as bolts, screws, and nails. Going to the hardware store for every nut and bolt consumes time and fuel, so keep a reasonable selection of fasteners on hand. Stock:

Round-head or pan-head machine screws (12 each):
4-40×¼"
4-40×½"
6-32×⅜"
6-32×¾"
8-32×½"
8-32×1"
10-24×½"
10-24×1"
¼-20×¾"
¼-20×1"
Hex-head cap screws (6 each):
¼-20×1½"
¼-20×2"
⅜-16×1"
⅜-16×1½"
⅜-16×2"
⅜-16×2½"

Nuts (24 each):
4-40
6-32
8-32
10-24
¼-20
⅜-16

Nails (1 box each): 2d and 8d

THE SAFE WORKSHOP

Rules to make by.

By William Gurstelle

Your workshop should be a welcoming and friendly place. The key lies in creating a safe and secure environment. Before embarking on a new project, it's a good idea to take a close look at the working conditions in your shop.

If your project area gives you a vaguely nervous feeling, then now's the time to bring things up to date. Don't delay — inspect, review, and evaluate your space and make whatever changes seem necessary to keep you out of trouble.

Don't know where to start? Here are some ideas from the members of MAKE's Technical Advisory Board to get you started. Have at it!

Wait 12 hours between sketching the plans and starting the construction process. The times people get hurt are usually when they're excited and in a hurry. Slow down, and work deliberately.

The high-decibel noise generated by power tools such as table saws and circular saws can damage your hearing. Protect your ears by using full-sized, earmuff-style protectors.

Wear a particle mask when appropriate to avoid breathing dust and other particulate pollutants that are common in workshops. Sawdust from treated wood and some plastics has known health risks.

Secure your work when using hand or power tools.

Avoid using a table saw when you can. Statistically, it's easily the most dangerous piece of equipment in the shop.

Obtain a pair of well-fitting, cool polycarb goggles, leather work gloves, and a protective lab coat. Make them attractive and stylish so that wearing safety equipment is fun. Pull back long hair.

Aim away from yourself. When cutting with a utility knife, position yourself so that when you slip, the blade doesn't land in your flesh.

Always use clamps, not your hands, to hold a work piece on a drill press table. If the tool binds, the work will spin dangerously.

Don't touch a bare wire, or cut any wire, until you're sure where the other end goes. When in doubt, measure the potential. This will save you from a possible heart-stopping electrical shock.

Always keep a first aid kit in your workshop, and always know where it is. First aid kits can be purchased ready-made, or you can put them together yourself. Essential items include bandages, pads, gauze, scissors, tweezers, and tape.

If you work with heavy things — say, timbers or an angle iron — or are prone to dropping tools, steel-toed safety shoes are a great investment in long-term foot appearance.

Install a smoke detector in your shop and place a fire extinguisher in an easy-to-reach spot. Make sure the extinguisher is rated for all types of fires.

Photography by Jason Madara

DIY

WORKSHOP

HERE'S THE ESCAPEE:
A 0-80 nonmagnetic
stainless steel
socket-head cap screw.

LOST SCREW FINDER

 An easy vacuum attachment that filters small nonmagnetic parts. By Frank Ford

Photography by Frank Ford

MATERIALS
Plastic CD-ROM spindle container
Finally, a use for one of these!
Small wire screen **I cut mine from an old kitchen strainer.**
1" diameter PVC plumbing: two 90° elbows, one 45° elbow, and 2' of pipe
Silicone glue
A vacuum, plus any adapter needed to fit it to the PVC **My shop vacuum's 1½" hose fit the PVC pipe after a bit of filing, no adapter necessary.**

TOOLS
Hacksaw or PVC pipe cutter
Utility knife
Metal file
Scissors or wire cutter
Stapler or staple gun

Ever drop a screw or other teeny part, look around like crazy, and finally have to give up? Wish you had a magic magnet that could attract plastic or brass parts out of the debris on your shop floor?

Well, that's never happened to me ... but if it ever does, I'll be glad I made this little gizmo. It's a vacuum attachment that captures small bits in a little canister so I can sort through them.

1. Using a knife along with a short piece of PVC pipe as a template, mark and cut 2 holes in the base of the CD spindle. Neatness doesn't count for much here, but the holes should just clear the diameter of the pipe and shouldn't be too irregular.

2. Cut ¼" sections off 1 end of each of the 90° elbows. Again, accuracy isn't important. My

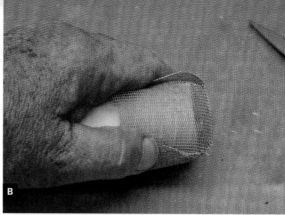
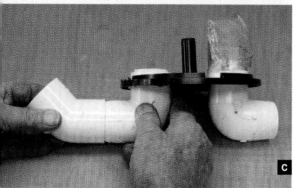

Fig. A: Fit the inlet pipe through one of the holes in the plastic spindle base, and secure it on the inside with more adhesive. Fig. B: Fashion a little screen "boot" that slips over the end of the outlet pipe.

Fig. C: Glue a 2" section of pipe into the inlet elbow, and glue the 45° joint onto the other end, angled back down toward the spindle base. Fig. D: Screw the CD spindle-cover canister into place.

weapon of choice is a hacksaw, but any saw or pipe cutter will do.

3. Using a file, taper the uncut end of one 90° elbow so that it fits the hose of your vacuum or attachment. This is your outlet elbow. My vacuum takes a 1½" input, but your mileage may vary, so here's where you'll improvise.

4. Cut a ¾" section of pipe and squidge on a bunch of silicone glue as you fit it into the cut end of the second 90° elbow. Fit the pipe through one of the holes in the plastic spindle base, and secure it on the inside with more adhesive and the ¼" ring you cut off. This is the inlet pipe (Figure A).

5. Make your outlet pipe the same way, but use a 1½" section of pipe. It should stick in 1" or so into the CD spindle base.

6. Fashion a little screen "boot" that slips over the end of the outlet pipe (Figure B). I hacked away at my screen with scissors, and trial-fit the shape onto a spare piece of pipe. When I got a form that worked, I stapled around the form to hold the shape in place.

7. Glue the screen boot onto the end of the outlet pipe with a lot of that silicone goo, and use more to seal up its ragged edges.

8. Glue a 2" section of pipe into the inlet elbow, and glue the 45° joint onto the other end, angled back down toward the spindle base (Figure C). Fit in a longer section of pipe, to serve as the nozzle. I didn't glue this last section, so that I could switch to different lengths as needed.

9. Screw the spindle-cover canister into place (Figure D), and it's ready for use!

Putting It to Use
Using the Lost Screw Finder is a simple affair. Sweep the area where the errant part was last suspected, round up everything, and check the contents of the dragnet. Most suspects stick to the screen, right where you'd expect them.

Frank Ford is a founder of Gryphon Stringed Instruments in Palo Alto, Calif., where he has been a full-time luthier since 1969. He's a prolific writer, appearing in books, magazines, and his website, frets.com.

MINI FUME EXTRACTOR

Candy tin device helps keep your air clean and your lungs healthy. By Marc de Vinck

A fume extractor uses an activated carbon filter and fan to remove the smoke, and noxious fumes, created from soldering. The average price of a small hobby version is about $100, but this one will run you more like $10. This mini fume extractor won't be as effective as a larger one, but it's definitely better than nothing, and extremely portable. Remember, always work in a well-ventilated area.

1. Build the circuit.

I decided a quick mock-up might be a good idea, and I'm glad I did. At first, I thought that running the case fan off just one 9-volt battery would provide adequate power. In the end I decided that 12 volts "sucked" better, and in this case that's a good thing.

The final circuit (at right) uses a simple switch, two 9-volt batteries, a 40mm case fan, and a 7812

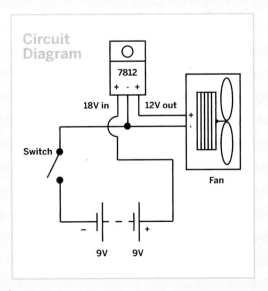

Circuit Diagram

7812
+ - +
18V in   12V out
Switch
Fan
–   –   +
9V   9V

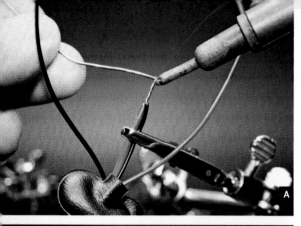

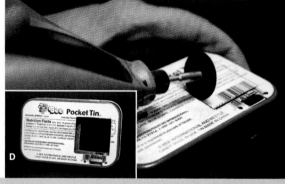

Fig. A: Vinyl 9-volt connectors are low-profile enough to let everything fit. Fig. B: The completed circuit. Make sure to orient the 7812 according to the schematic, and don't forget to slip on heat-shrink tubing prior to soldering, to insulate all connections from the conductive metal tin. Fig. C: It's a snug fit. Fig. D: Use light pressure when cutting the openings; let the tool do the work. The openings don't have to be perfectly aligned.

MATERIALS

7812 voltage regulator IC
Candy tin
Switch, SPST (single pole, single throw)
Case fan, 40mm square
9-volt batteries (2)
9V battery connectors (2) **vinyl, not hard plastic**
Pieces of screen, 50mm square (2)
Piece of carbon filter **cut from a replacement filter**
Heat-shrink tubing
Insulated, threaded hook-up wire
Miscellaneous screws and washers
Paint (optional)

TOOLS

Soldering iron
Rosin-core solder
Dremel with cutoff wheel
Drill and small drill bits
Fine-tip marker
Various screwdrivers
Wire cutters
Safety glasses

⚠ **CAUTION: Wear safety glasses when drilling and cutting metal!**

voltage regulator. The 7812 takes voltage from the 9V batteries wired in series and steps the voltage down from 18V to 12V, which is what the fan requires.

2. Solder the components.

Notice the battery connectors (Figure A); they're the flexible vinyl version, not the hard plastic type. This allows both batteries to fit in the case. The vinyl snaps are only minimally smaller, but it's enough to make the difference.

This is a very simple circuit. Solder it according to the diagram, making sure to attach the component leads to the 7812 properly (Figure B). Don't forget to use heat-shrink tubing on all connections; this is in a metal box ... and metal conducts electricity!

3. Make sure it all fits.

It's a snug fit, but you should be able to stuff everything into the tin, packing the batteries side by side next to the fan (Figure C).

4. Cut and drill the holes.

I used a marker and a paper template for the fan openings, making them 35mm square on each side. After you cut the first fan hole, close the box and

E

F

G

H I

use the template to align the second hole. You can just "eyeball" the placement. There's room for error.

Then I marked the opening for the switch and cut all openings with a Dremel tool and cutoff wheel (Figure D).

Next I marked and drilled 2 mounting holes for the switch screws and one for the regulator.

5. Paint and decorate.

I decided to paint the tin this time, unlike my plain RuntyBoost (makezine.com/2008/03/22/making-the-runtyboost). I chose a nice red Krylon paint. I hot-glued a scrap piece of wood to the inside, so I could hold it while I spray-painted it. Two quick coats and I think it looks good (Figure E). Spray paint can be fairly toxic and flammable, so paint outside and away from everything! I'm happy with how it came out, but it definitely needs some graphics to spruce it up. Any suggestions?

6. Attach the regulator and switch.

First, screw in the 7812 using some washers and a screw to space it slightly away from the side of the tin (Figure F). I used a #6-32 screw and one washer

to keep it from the edge, but you can use anything that fits. The screws and washer will also act as a heat sink.

Finally, screw in the switch.

7. Add the screens and filter.

Here you can see the screen-filter-fan-screen sandwich (Figure G). The screens are 50mm square and the filter is 40mm square. You can buy replacement filters for the commercial extractors at a reasonable price and cut them to size.

Next, just hot-glue or epoxy the corners of the screens to the candy tin, and sandwich the filter and fan in between (Figure H). Compression will ultimately hold it all together. You're done!

8. Test your extractor.

I've run mine continuously for hours and have had no heat buildup from the 7812, and the fan is still running strong (Figure I). It works quite well, and although it's no replacement for a large fume extractor, it will come in handy for small projects. Remember, follow all safety guidelines when soldering, and work in a well-ventilated room, even if you have a fume extractor.

SKILL BUILDER+

EASY

UNDERSTANDING
BASIC WOODWORKING TOOLS

Ten simple hand tools for building almost anything   Written by Len Cullum

LEN CULLUM
is a woodworker living in Seattle, where he specializes in building Japanese-style garden structures and architectural elements. When not woodworking, he teaches at Pratt Fine Arts Center, writes, and dreams of a robot that would sharpen his chisels.

I'm going to focus on what I consider the six basic hand tools for working with wood, plus my four go-to tools for measuring. These are the fundamentals that will allow you to build most anything. Keep in mind that no one tool is right for everyone. The hammer that I love might be the one that makes your wrist sore, or my favorite saw might feel backward. Don't be afraid to try different tools and techniques until you find the ones that feel right and make the most sense to you.

THE HAMMER

Nothing says blunt force like a hunk of metal on the end of a stick.

It's probably the oldest tool in the book. When I first started woodworking, I remember seeing a picture of a guy with his hammer collection, it was a whole room filled with hundreds of different hammers. At the time, I couldn't imagine needing more than one, but I feel much differently now.

Within eyeshot as I type this, I can see nine hammers. Each is different and each sees (fairly) regular use. The one pictured below is easily my favorite. It's a 375g Japanese carpenter's hammer. One face is flat, for driving nails, the other is slightly convex for driving the nail below the surface. I use it for everything from driving chisels and adjusting planes to knocking joints together and closing cans. It's my go-to hammer. The weight is right, and I like its balance.

If your work will require a lot of nailing, a claw hammer might be a better choice. Personally, I would just add a small pry bar to my collection.

CHISELS

Chisels can be used for anything from heavy chopping to light paring or fine carving. While also known to open paint cans, turn screws, and act as a pry bar, these are not recommended uses. Seriously, use a screwdriver. A screwdriver will appreciate the attention. While there are hundreds of chisel sizes and styles, most people can get by with four: ¼", ½", ¾", and 1" standard bench chisels.

There are virtually no chisels that are ready to use right off the shelf; they all need some sharpening to get them to sing. Once you experience a truly sharp chisel, you will understand the difference, not only by what you're able to achieve, but the ease with which you can do it. Below is a heavy patterned chisel called *atsu-nomi* (thick chisel) that's used for cutting joints in large timbers. It's part of a set made for me by master blacksmith Iyoroi, and it's one of my favorites.

HAND PLANES

Historically, hand planes were used mostly (but not exclusively) for smoothing and adjusting the thickness of rough board (called "thicknessing"). These days most stock dimensioning is done by machines, but this doesn't mean the hand plane is obsolete. It remains an incredibly useful tool that no woodworker should be without.

A well-tuned plane can do in minutes what can take a sander an hour, and produce an arguably better surface in the process — allowing you to work while standing in a pile of shavings instead of a cloud of dust. If I had to choose only one, it would be a low-angle block plane, pictured below. It can be used for everything from trimming and shaping stock to finish-planing surfaces. Like chisels, they're rarely ready to use out of the box and need to be sharpened before use.

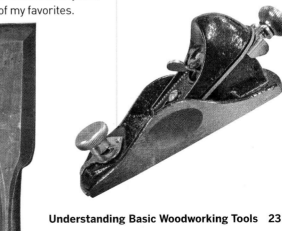

Len Cullum

HANDSAW

As with the hand plane, much of the work a handsaw performs has been picked up by the powered version. Even so, the handsaw remains a useful and necessary part of a woodworker's collection. For cutting wood, there are two basic types: rip saws and crosscut saws.

Rip saws are meant to cut in the direction of the grain and typically have fewer, bigger teeth. Crosscut saws are, as the name implies, for cutting across the grain. They typically have more and finer teeth in order to shear the grain and leave a cleaner cut.

While general-purpose and combination saws exist, they tend to be a little too aggressive for careful work. My choice of handsaw is a Japanese *ryoba nokogiri* (double blade saw), shown above. It has rip teeth on one side, crosscut teeth on the other, and unlike Western saws, it cuts on the pull stroke. While they used to be difficult to find, you can now usually get them at home stores.

CLAMPS

Without clamps, nearly every operation with the preceding tools becomes more difficult. Not only are they good for holding together the final assembly, their ability to keep things here you want them while you work is invaluable. There is little that is more frustrating than trying to work a piece of wood that keeps sliding around. A couple of clamps, are essential and most woodworkers, at least once in their life, have repeated the mantra "you can never have enough clamps." Two 24" bar clamps, like the one shown below, are good. Four are better. Eight are better still ...

WOODWORKING PROJECT LAYOUT TOOLS

Accurate layout work is the critical first step to a successful project. Without precise, repeatable marks, it is very difficult to get everything to come together at the end. So now I'll go over some of the basic tools for measuring, marking, and transferring lines. My big three (actually four) tools for almost all of the work I do are the tape measure, a high-quality 12" combination square, a .005 drafting pen, and a 4" combination square for smaller work.

MEASURING LENGTH

The three most common measuring devices you're likely to find in a wood shop are the tape measure, folding rule, and steel rule. All three have their good and bad points. But as with all tools, find the one(s) that fit your style and make the most sense to you and the way you work.

The familiar tape measure with its spring-steel blade rolled up into a small box is fast and can measure distances that would require a massive folding rule. On the down side, the little hook at the end of the tape can introduce inaccuracy. When new, the hook slides on rivets just enough to adjust for the thickness of the hook's metal. When measuring to the inside of something, the hook is pressed in; when on the outside, the hook is pulled out, keeping the measurements accurate. This works great for a while, but over time, the holes and rivets can wear and get bigger, or the hook can be bent when the tape measure is dropped. To remedy this, most woodworkers "burn an inch." This is where you ignore the hook and start all of your measurements from the 1" mark. This works well and gives accurate results, as long as you remember to subtract 1" from your result. Trust me, no one who uses this method hasn't had a moment of dread after discovering something (or worse, multiple things) didn't fit to the tune of one extra inch. So stay awake out there.

When choosing a tape measure, consider the type of work you are doing. If you primarily work with material shorter than 12 feet, don't buy a 25-foot tape. Those last 13 feet will never see daylight and the extra mass is heavy and cumbersome.

The folding rule (above) overcomes the hook problem by having a fixed metal cap at the end of its wooden rule. This makes for worry-free use, especially when measuring against something. It also has a nifty little sliding rule built into the end to measure depths and interior distances. On the downside, the thickness of the wooden blade means it must be laid on its edge to get accurate results and the way it folds creates a stair step shape that can make it awkward to use over distances.

The steel rule (at left) is a nice balance between the folder's consistency and the tape measure's small size, but its limitations are obvious. It's great for smaller work, but once you get beyond the 6" mark, one of the above will have to take over.

Honorable mention goes to the story pole or story stick. This is usually a long piece of wood that one puts their own marks on for transferring measurements. This can be more reliable because it gets rid of those numbers, and every distance is as marked. Story poles are very useful when you're measuring larger projects with multiple components (like a kitchen or library) or when you need to transfer the same dimension over many parts. It helps eliminate measuring mistakes.

SQUARES

For layout work, a square's primary function is to draw lines 90° perpendicular to a side. As always, there are a few types available but what sets them apart is what else they do. For me, a combination square (at right) is the most useful. Not only does it give me 90° and the occasional 45°, it also transfers measurements from one piece to another, finds the true center of a board, and checks depths and helps set up tools. It's hard to imagine woodworking without it. Definitely spend up when buying one. Get the best one you can afford. A loose, out of square or hard to move blade creates more frustration than it's worth.

The speed square is handy as well, but it is more suited to carpentry. I find the deeply stamped numbers to make for jaggy lines, so I use it mostly for rough layout and marking.

The *sashigane* is the standard square for Japanese joinery. It looks like a Western framing square but has a much thinner, flexible blade. And also like the framing square, it is covered in mysterious, oddly spaced numbers and strange markings that when in the right hands can be used to figure and lay out some pretty complicated joints. Since I have yet to decipher one, those hands are not mine. ◑

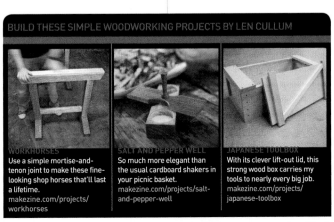

TOOLS & MATERIALS

» **Table saw, chop saw with 12" cut capacity, and handsaw**
» **Drill**
» **Drill bits, 7/64 or 3/32, and countersink, or #6 counterbore** I used a Makita #784815.
» **Acid brush**
» **Tape measure**
» **Adjustable square**
» **Small clamps, 2 or more** I used two 6" and four 12".
» **Low horses** optional but helpful
» **Wood, pine or cedar: 1×12 (4') and 1×8 (8')**
» **Screws or nail equivalent: 1 5/8" trim-head (36), 1¼" (16)** I used drywall screws.
» **Wood glue**

NOTES

As with all woodworking projects, there are many ways to accomplish the same task. The tools and techniques I show here should be viewed as one option, not the only way. Use whatever tools, methods, and materials make sense to you.

You can avoid a lot of frustration by selecting the best boards you can find. Look for pieces that are straight, flat, and free of twist. Sometimes that means going through every board they have; sometimes that means getting 2 boards so you can cut around flaws.

Japanese Toolbox

TIME: 4 HOURS   COST: $$

Build this strong wood box with a clever lift-out lid.

....

Written and photographed by
Len Cullum

The first time I came across a box like this, it was housing a circular saw. It was made from rec room paneling, scrap pine, and roofing nails. While not much to look at, it was strong, and the action of the lift-out lid fascinated me.

A couple of months later I learned that this is also the common design for a Japanese carpenter's toolbox. I knew I had to make one, so with some 1×12s in hand, I had at it.

In the last 15 years, the box has carried my tools to nearly every big job. The trouble is, because I built it to fit in my hatchback instead of to fit my tools, it's kind of big for what I need on most jobs. So instead of continuing to stuff tools into my old messenger bag, I figured it was time to build a new, smaller one.

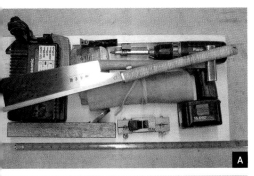

A

B

➕ Sizing

The first step was deciding how big I needed the box to be. Since I like the width of my old toolbox, I used the same 1×12 bottom. Generally, you want the box to accept your longest tool, which in this case is my saw. I found 23" to be a comfortable fit (**Figure A**).

With the length established, it was time to stack some tools. All together, they measured just over 5" high (**Figure B**), and since we lose a little height from the thickness of the lid (and you never know what else you'll want to stuff in there) I went with 1×8 sides. After doing some math and a quick chalk layout, I determined that, with careful cutting, I could get all of my parts out of 2 boards.

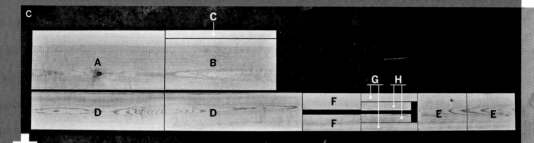

Cutting the Wood

Since I wanted to squeeze everything out of 2 boards, the layout was especially important. **Figure C** shows where each piece came from. Download the cutting diagram from makezine.com/projects/japanese-toolbox.

From the 1×12:
  A: 11¼"×26" (bottom)
  B: 9¹¹⁄₁₆"×21⅛" (lid)
  C: 1⅜"×21⅛" (lid brace)

From the 1×8 (2 each):
  D: 7¼"×26" (sides)
  E: 7¼"× 9¾" (ends)
  F: 3"×11¼" (top ends)
  G: 1½"×11¼" (lid supports)
  H: 1½"×9¾" (grips)

TIP Whenever possible, I try to cut all like-sized pieces at the same time, which ensures that everything will line up properly.

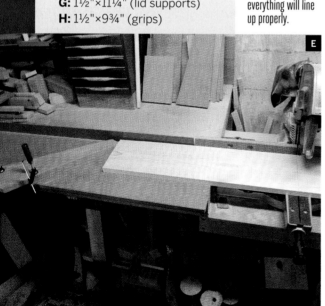

Starting with the 1×12, use a chop saw to trim away about ¼" to get rid of the factory end (**Figure D**); this will give you a cleaner, squarer, more reliable edge. Measure 21⅛" from the fresh end and make the cut.

The bottom and sides are all 26", so to ensure they match, I set up a stop. On the 1×12, mark a line at 26" and align the board on the saw. Place a scrap piece against the cut end and clamp it in place. Check the cut alignment one more time and then make the cut.

Take the 1×8, trim ¼" from the end, slide it carefully up against the stop (you don't want to whack it out of position), and cut the 2 side pieces. Next, lay out 11¼", reset the stop, and cut 2 of those, and then the two 9¾" ends. Leave the stop in this position because you will use it again (**Figure E**).

Next are a couple of rip cuts. On a table saw, rip the lid piece to 9¹¹⁄₁₆" wide, saving the offcut for the lid brace. From one 11¼" piece, rip two 3"-wide pieces. Lastly, from the other 11¼" piece, rip four 1½"-wide sticks. Of these 4 sticks, take 2 back to the

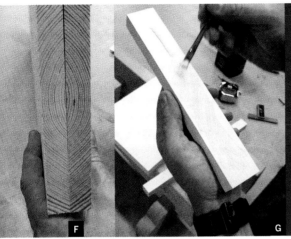

For gluing I like to use an acid brush cut back to about half an inch long.

F

G

H

I

J

K

chop saw, and using the 9¾" stop, cut the 2 grip pieces (keep these offcuts as well). And that's it — you're ready to assemble.

The method of assembly has a lot of options. Over the years I've constructed boxes using screws, nails, glue, joinery, and combinations of the four, all with good results. The point is to approach this however feels right to you. If you prefer nails, great! If you like cutting sliding dovetails, go for it!

The only thing that I would specifically recommend is to try to orient all of the boards with the bark side out (**Figure F**). This side of the board will generally remain smoother and will help keep the outside from developing snaggy bits when you sand or plane it.

Apply a bead of glue and then spread it across the surface (**Figure G**). Don't spread it so thin that it dries before you assemble, but also don't make it so thick that your parts are sliding around and a lot of glue is squeezing out (a little is okay).

1. Attach the grips to the ends.

It's important that the edges are all flush, so check the fit beforehand and take your time adjusting them before you fasten. Apply the glue and then lightly clamp the parts together (**Figure H**). Get them exactly where you want them before you fully tighten the clamps.

Flip the clamped pieces over. Mark the 3 holes ¾" in from the edge: 1" from each side and one centered. Drill and countersink, then fasten using 1¼" screws (**Figures I, J, and K**).

TIPS Slipping a scrap piece under the other end will help keep it stable when adding the screws.

Because I like the look of evenly spaced fasteners, I always make layout lines. Aside from helping make sure I don't miss the board I'm trying to fasten to, it also keeps fasteners from conflicting in the corners.

TIP Putting the other end assembly loosely in place will help keep the sides from shifting.

2. Make the body.

Using an adjustable square set to 1⅛", strike a line parallel to the ends, on the outside of both side pieces (**Figure L**). Then measure in from each edge and mark at 1" and 2¾". Pre-drill and countersink all 16 holes (**Figure M**).

Apply glue to the ends of one end assembly and clamp in place. Again, make sure that everything is properly aligned before clamping tight. To make sure the bottom is set back the proper distance, I use one of those cutoffs from the grips (you saved those, right?) as a guide (**Figure N**).

When everything is in its place, fasten with eight 1⅝" screws. Because my clamp pads are in the way, I fasten the 2 center screws on each side first, then remove the clamps and drive the rest (**Figures O and P**). Repeat on the other end, making sure that the grips are both facing outward in the same direction.

3. Attach the bottom.

Flip the frame so that the grip side is down, and place the bottom. Lay out the fastener locations: for the sides at ⅜" from the edge, 1", 7", and 13" from either end. For the ends, the first one is at 2" (to avoid hitting side screws) and the center one is at 5⅝". Drill.

Glue and clamp the bottom in place, making sure everything aligns, then fasten using 1⅝" screws (**Figure Q**). Don't be surprised if the bottom seems a little wider than the sides. It happens. No one knows why. You can sand or plane it flush later.

4. Attach the top ends.

Following the same procedure, drill, glue, clamp, and fasten the top ends (**Figure R**). Be especially careful when screwing or nailing the outer corners, as the wood is easy to split here. Pre-drilling is highly recommended. My screws are ½" from the end, but ¾" would be safer.

5. Make the lid.

For the lid to work right, make sure everything is positioned accurately. To start, make sure the lid board fits between the sides. It's much easier to make adjustments to it now, before attaching the supports. It should be about ¹⁄₁₆" narrower than the opening and 1⅛" longer.

On top of the lid board, mark a line 1¼" in from one end — this is the long tab. Apply glue to one lid support (stopping about 1" from either end), and lightly clamp it along the line. Using those 2 grip cutoffs again, center the support along the line; it should slightly overhang the scraps (**Figure S**). Tighten the clamps, flip the board, and fasten using three 1¼" screws.

On the other end, mark a line ⅝" from the end (the short tab) and repeat the process. At this point, you can test the lid. Tilt the long tab in first, slide it in until it stops, drop the other end down, then slide it back.

Adding the diagonal is optional. I add it because I like the look, and it makes the lid stiffer and acts as a handle.

To add the lid brace, position it from corner to corner between the lid supports, and clamp it in place (**Figure T**). Using a sharp pencil, scribe a line on the underside (**Figure U**). To keep track of the orientation, after removing the clamps, make a reference mark at one end of both the brace and lid (**Figure V**). This is important because it often only fits one way.

To save the hassle of adjusting the chop saw for these 2 cuts, I prefer to use a handsaw. I place a straight piece of scrap along the line, hold it tight, and run the saw against it (**Figures W and X**). Before cutting the other end, position the brace back on the lid to double check that the cut line is still good. Glue and fasten it in place.

6. Wrap it up.

From here on the rest is up to you. Erase the lines or leave them, flush up the bottom if it needs it (mine did), round edges with sandpaper, and then fill it with tools (**Figure Y**). Or books. Or cupcakes! As for mine, I will stop here, slide it under the bench next to the larger one, and wait for the next job. ◪

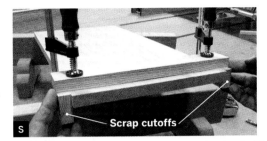

S

Scrap cutoffs

T

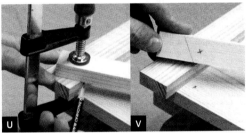

U   V

W   X

Y

Len Cullum (shokunin-do.com) is a Seattle-based woodworker in the Japanese style. He makes shoji doors and windows, Japanese garden structures, and architectural elements.

A NOTE REGARDING WOOD MOVEMENT

Because wood, especially unfinished wood, absorbs and releases moisture from the atmosphere throughout the year, it tends to shrink and swell across the grain. (Movement along the grain is negligible.) This means that a lid that you cut to fit perfectly in a dry week might become too tight during a rainy one. It also means that if you screw and glue a baton across the grain of a wide board (as we do here), when it tries to move the board will cup, making the lid tight during one part of the year and loose during the other.

One way to remedy this is to apply glue to only the center inch or two of the baton; then make slotted pilot holes for the out-ermost screws, to allow the lid board to shrink and swell freely. Another remedy is to use plywood for the lid board. That said, I've never slotted the pilot holes on these box lids. I've just dealt with the lid being tight sometimes. But if you are going to go bigger than this one, movement is some-thing you'll want to plan for.

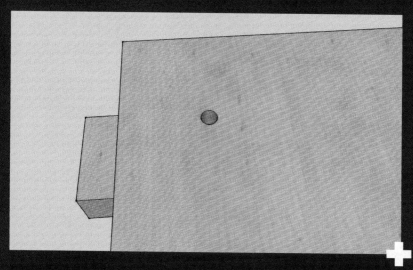

To make a slotted pilot hole, simply drill one hole right next to another, and then file, chisel, or wiggle the drill bit back and forth until the web turns into a slot. Using a pan-head wood screw, attach the baton to the lid snugly, but not so tight that you are crushing the wood beneath the screw head.

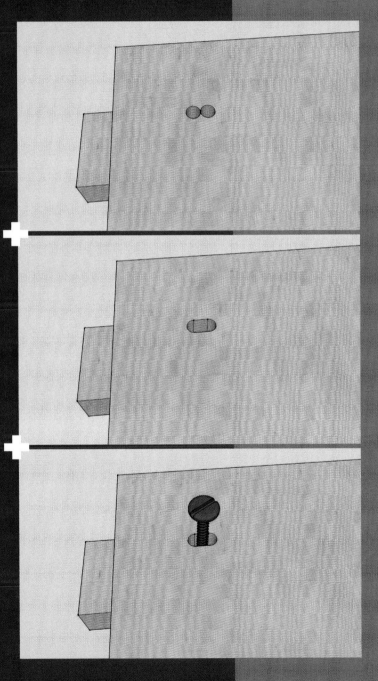

WILDERNESS WORKSHOP

 Build your own inexpensive yet sturdy worktables and shelving. By Charles Platt

Photography and renderings by Charles Platt

Recently I had a problem and an opportunity. The opportunity was to move my little prototype fabrication business from an industrial park in Southern California to a beautiful wilderness retreat in Northern Arizona.

The problem was that I didn't have much time or money. Could I establish an entire workshop within a couple of months, starting from bare earth and finishing with all the tools and benches in place? And after I paid the construction costs, could I install the fixtures for less than $1,000? (I already owned all the tools.)

In some ways this challenge was a blessing in disguise. If you have to be fast, you can't be fancy, and if you have to be cheap, you can't be self-indulgent. This would not be one of those jobs that drag on for months because the details become an obsession in themselves.

I specified a work area of 19'×24' to allow ample space for lifting, rotating, and cutting 4'×8' sheets of plastic and plywood. To minimize heat loss, increase security, and maximize wall space, I decided not to have any windows, but I did include a massive sectional roll-up garage door. The climate where I live is so benign during most of the year, you can work comfortably with a door wide open. And during the winter, a sectioned door on tracks can be quite well-insulated.

After establishing the basics, I stepped back and let the contractors get to work. There was no way I could do the construction myself in the time available.

Free-Standing Benches

In less than a month I had a bare box standing on a concrete slab. It was insulated, drywalled, and painted. Now for the interesting part: I wanted to

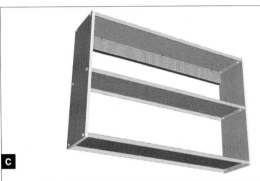

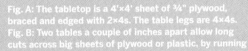

Fig. A: The tabletop is a 4'×4' sheet of ¾" plywood, braced and edged with 2×4s. The table legs are 4×4s.
Fig. B: Two tables a couple of inches apart allow long cuts across big sheets of plywood or plastic, by running the saw through the gap between the tables. The extended table edge allows quick, easy clamping.
Fig. C: The simplest, cheapest, quickest, hang-on-the-wall, non-sagging shelf design.

avoid all the frustrations and errors associated with the workshops I had used previously over the years.

The big central work area allowed me to place 2 free-standing benches of a design that I had always wanted but had never seen. They would be stocky tables, each 4' square. Placing them centrally would allow me to walk around them while building heavy items such as furniture, and a 2" gap between them would facilitate saw cuts.

I stopped using a table saw a few years ago when one of them kicked a piece of plastic at me that almost shattered my arm. Since I don't have enough money or space for the kind of vertical panel saw you see at Home Depot or Lowe's, I like to lay the wood flat and use either a handsaw or a handheld circular saw, which I run along a clamped straightedge. My plan was to align these cuts with the gap between the tables. This would be like using sawhorses, but much more accurate and less aggravating.

With my helper Shawn Hollister, I built the tables lower than a typical workbench, so that we'd be able to reach across them or climb up onto them when making long cuts. We gave them protruding lips so that I could apply clamps easily, and made them heavy to minimize vibration (Figures A and B).

Plastic Bins for Tool Storage

For tool storage, I'm unenthusiastic about the usual options. Tools hanging on pegboard pick up dust and dirt, and when you buy an extra tool, you have to move the others around to make room. As for tool chests, they're expensive, and you have to walk to and fro every time you want something.

My preferred method is so cheap, it's almost embarrassing: plastic tubs from the local big-box store. I group tools in tubs by function, so that when I want, say, a metal file, I pull down the tub containing all the various shaping tools and put it on my worktable. Now I have a full range of options within arm's reach. As for small items such as screwdriver bits and hole saws, I put them in small boxes inside the tubs. At the end of a job, everything is returned to the tubs and stays clean and neat, with the lids snapped on tight.

Shelves That Don't Sag

Where to put the plastic tubs? On shelves, of course, above the side benches where I have a drill press, compound miter saw, band saw, and belt sander, the four tools I consider indispensable for the kind of work I do. But how should the shelves be built? Quickly and cheaply!

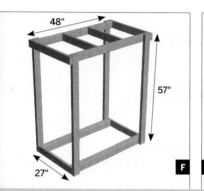
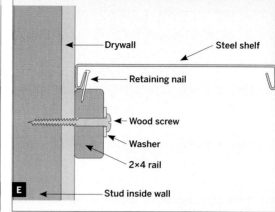

Drywall — Steel shelf

Retaining nail

Wood screw

Washer

2×4 rail

Stud inside wall

D **E**

48"

57"

27"

F **G**

H

Fig. D: Front view of shelf assembly, with the front edge of the shelf cut away to show relevant features. The inset shows consequences if bolts aren't tightened sufficiently: the bolt can chew right through the wooden upright. Fig. E: Cross section showing how the shelf is attached to the wall. Figs. F–H: A wood rack was made from pine 2×4s with 2½" wheels attached. Galvanized wire separators are secured with screws and washers.

Since I don't like the look of sagging wooden shelves, I chose steel shelves of the type sold for warehouses. A standard length is 4', so you don't need many uprights to support them, but they still take heavy loads without bending. You can bolt them to wooden uprights instead of the ugly perforated vertical bars that are normally used.

I chose melamine-coated particleboard for the end pieces, because it's available in exactly the same 11¼" width as the shelves, and it's prefinished, requiring no painting. I cut the melamine board into sections, drilled them to fit the holes in the ends of the shelves, and bolted them on. Then I cut 2×4s into rails 47" long and screwed them into the wooden studs behind the drywall in my conventional framed construction. We hung the shelves on the rails, adding a couple of nails to prevent the shelves from falling off (Figures C–E). That was that.

The horizontal rails must be a full 47" so that the load carried by the shelves is spread across the entire wooden support. Any unsupported metal section will tend to bend.

NOTE: Since melamine board is made from compressed wood chips, it can come apart, so you should use pine boards for uprights if you intend to load your shelves very heavily. Or place an additional 47" rail beneath each shelf.

Tighten the bolts to the max, to take advantage of the friction between the end of the shelf and the upright. Friction is proportional to the force perpendicular to the surface, and it supports a load more effectively than just the shaft of a bolt in a hole drilled through wood.

A Wood Rack on Wheels

Another problem was how to store materials efficiently. I have to stock wood and plastic in bulk, because the nearest retail sources are 50 miles away. I dislike stacking sheets against the wall where I can't pull anything out easily, so my answer was a wood rack on wheels (Figures F and G). I've never seen this elsewhere, but it seems an obvious idea to me. When you don't need it, you roll it out of the way, into a corner. I used heavy galvanized wire to make dividers in the rack, so that I would lose as little horizontal space as possible, and I put a flat top on it, where I could stack small pieces of scrap, with even smaller pieces in some more plastic tubs.

As for seldom-used, bulky tools such as bolt cutters and reciprocating saws, I stashed them all

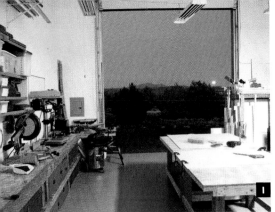

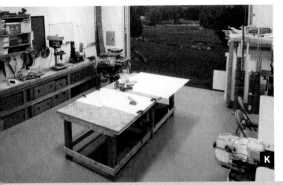

Figs. I–L: With the big door open, it feels as if you're working outside, and eastern exposure gives a nice view of the full moon rising at sunset. Ample floor space will allow the accumulation of miscellaneous junk in the future, or additional work areas. The floor is a concrete slab finished with two coats of epoxy paint. The exterior of the building is covered in Hardie fiber cement siding, which is fire resistant, stable, and durable.

in plastic toolboxes that I placed under the benches against the wall. The boxes aren't strictly necessary; you can just scatter your tools on shelves. But I wanted to keep them clean and categorized. In a shared workshop, when all storage is labeled, you're less likely to misplace things (and less likely to argue with each other when you can't find something).

1.3 Kilowatts of Illumination

The last consideration was in some ways the most fundamental: lighting. If you can't see what you're doing, you can't do good work. I splurged about half of my $1,000 fixtures budget on some GE Ecolux 54-watt high-intensity daylight-spectrum fluorescents, and suspended them from cables stretched from wall to wall below the track of the garage door. When all 24 tubes are glowing they draw almost 1.3kW, and so to reduce energy consumption I installed a separate pull-switch on each fixture, with a chain dangling, so that I can obtain light only where I need it.

During daytime, we don't need the lights at all. We open the huge door and feel as if we're working outside, which is an absolute delight compared with the basement workshops I've used over the years. It's also a lot more pleasant than the industrial

park that was my previous environment. When the breeze wafts in and I can look across 30 miles of national forest to a distant mountain, it definitely alleviates the tedium of cutting and shaping components. Bees from a nearby nest sometimes invade the space, but to discourage them we simply sprinkle some xylene on a rag and leave it lying around. They dislike the smell of this industrial solvent even more than I do.

My workshop isn't going to be featured in Architectural Digest. It was obviously outfitted on a budget. But I couldn't be happier with its functionality. Tools are easily accessible and don't get lost, the space is uncluttered and easy to clean, the lighting reveals every little detail, and as a result, the work flow is fast and accurate. Most important, the pleasure of working in an outdoor ambience is very special indeed.

It certainly justifies the hassle of moving everything from California.

Charles Platt is the author of Make: Electronics, an introductory guide for all ages. He is a contributing editor to MAKE, and he designs and builds medical equipment prototypes.

Electronics

PART 2

Charles Platt, who provided advice on workshop setup earlier in this collection, returns to offer guidance on getting started with your electronics workbench, including what to buy and how to troubleshoot your breadboard and circuits. If you decide to go deeper into electronics, Charles has authored a multivolume *Encyclopedia of Electronic Components* that explains all you need to know about almost every kind of electronic component. Later in this section, he gives you skill-building tips for working with one of those components—the electret microphone. The tiny, inexpensive electret uses very little power and is great for adding an audio component to your electronics projects, such as using sound to trigger an action.

If you want to add motion to your electronics project, Tod Kurt gives you the complete low-down on servomotors and how to put them to work for you. You'll learn how to hack a servo for continuous motion and how to build a simple drawbot. Linear actuators are another type of motor—they push and pull in a straight line. Computer scientist and fan of vintage technology, Andrew Lewis, demonstrates how to make inexpensive linear actuators out of sewing bobbins.

William Gurstelle also appreciates vintage technology, as regular readers of *Make:* magazine will know. Never a fan of programmable microcontroller and development boards, William prefers the simpler method of using prop controllers, or *keybanging*, to bring his projects to life. He'll show you how to set up your own animatronic show and coordinate it to music.

Diana Eng is not just a brilliant fashion designer who appeared on season two of Bravo's *Project Runway*, but she's also an innovative electronics enthusiast and cofounder of Brooklyn's NYC Resistor maker space. Here, Diana shows you how to fabricate a homemade Yagi beam antenna to listen to satellites and even the International Space Station.

John Iovine's "Desktop Digital Geiger Counter" is a challenging electronics project with many subassemblies that come together to form a whole. If you're up for the challenge, you'll find clear steps that walk you through every procedure, and John even offers boards and other components on his website to help smooth the way. Along the way, you'll also learn a lot about the physics of radioactivity and how scientists measure it.

Taking your skills to yet another level is Scott Driscoll's guide to surface-mount soldering. You may be comfortable soldering components to a breadboard, but soldering a surface-mount device to a printed circuit board can be a bit tricky, especially when it's an integrated circuit with no leads. In his overview, Scott provides thorough guidance to three approaches to surface-mount soldering.

Your Electronics Workbench

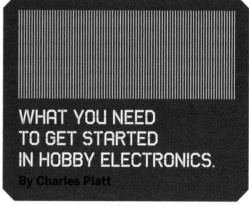

WHAT YOU NEED TO GET STARTED IN HOBBY ELECTRONICS.

By Charles Platt

Illustration by Damien Scogin

THE BASICS

First, you will need a breadboard. You can, of course, call it a "prototyping board," but this is like calling a battery a "power cell." Search RadioShack online for "breadboard" and you will find more than a dozen products, all of them for electronics hobbyists, and none of them useful for doing anything with bread.

A breadboard is a plastic strip perforated with holes $1/10$" apart, which happens to be the same spacing as the legs on old-style silicon chips — the kind that were endemic in computers before the era of surface-mounted chips with legs so close together only a robot could love them. Fortunately for hobbyists, old-style chips are still in plentiful supply and are simple to play with.

Your breadboard makes this very easy. Behind its holes are copper conductors, arrayed in hidden rows and columns. When you push the wires of components into the holes, the wires engage with the conductors, and the conductors link the components together, with no solder required.

Figure 1 (on page 42) shows a basic breadboard. You insert chips so that their legs straddle the central groove, and you add other components on either side. Figure 1 also shows the bottom of a printed circuit (PC) board that has the same pattern

of copper connectors as the breadboard. First you use the breadboard to make sure everything works, then you transpose the parts to the PC board, pushing their wires through from the top. You immortalize your circuit by soldering the wires to the copper strips.

Soldering, of course, is the tricky part. As always, it pays to get the right tool for the job. I never used to believe this, because I grew up in England, where "making do with less" is somehow seen as a virtue.

When I finally bought a 15-watt pencil-sized soldering iron with a very fine tip (Figure 2), I realized I had spent years punishing myself. You need that very fine-tipped soldering iron, and thin solder to go with it. You also need a loupe — the little magnifier included in Figure 2. A cheap plastic one is quite sufficient. You'll use it to make sure that the solder you apply to the PC board has not run across any of the narrow spaces separating adjacent copper strips, thus creating short circuits.

Short circuits are the #2 cause of frustration when a project that worked perfectly on a breadboard becomes totally uncommunicative on a PC board. The #1 cause of frustration (in my experience, anyway) would be dry joints.

Any soldering guide will tell you to hold two metal parts together while simultaneously applying solder and the tip of the soldering iron. If you can manage this far-fetched anatomical feat, you must

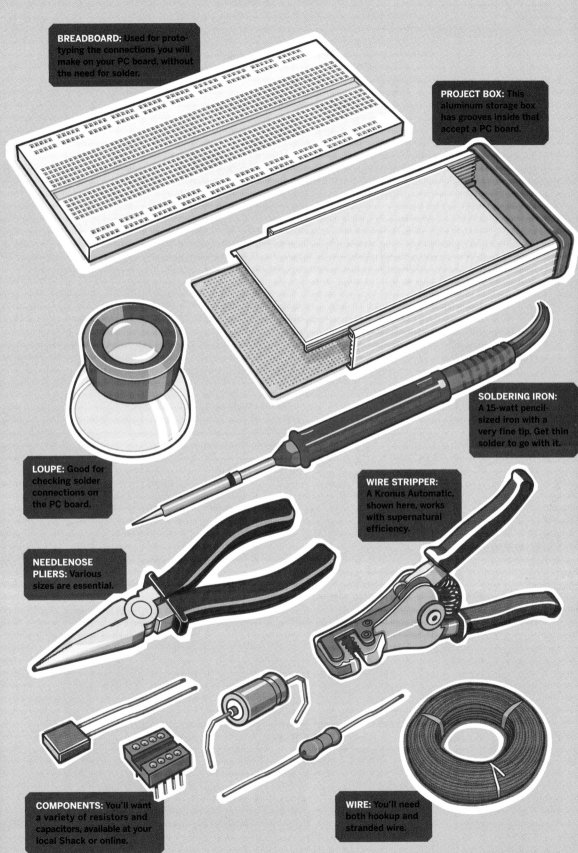

BREADBOARD: Used for prototyping the connections you will make on your PC board, without the need for solder.

PROJECT BOX: This aluminum storage box has grooves inside that accept a PC board.

SOLDERING IRON: A 15-watt pencil-sized iron with a very fine tip. Get thin solder to go with it.

LOUPE: Good for checking solder connections on the PC board.

WIRE STRIPPER: A Kronus Automatic, shown here, works with supernatural efficiency.

NEEDLENOSE PLIERS: Various sizes are essential.

COMPONENTS: You'll want a variety of resistors and capacitors, available at your local Shack or online.

WIRE: You'll need both hookup and stranded wire.

also watch the solder with supernatural close-up vision. You want the solder to run like a tiny stream that clings to the metal, instead of forming beads that sit on top of the metal. At the precise moment when the solder does this, you remove the soldering iron. The solder solidifies, and the joint is complete.

You get a dry joint if the solder isn't quite hot enough. Its crystalline structure lacks integrity and crumbles under stress. If you have joined two wires, it's easy to test for a dry joint: you can pull them apart quite easily. On a PC board, it's another matter. You can't test a chip by trying to pull it off the board, because the good joints on most of its legs will compensate for any bad joints.

You must use your loupe to check for the bad joints. You may see, for instance, a wire-end perfectly centered in a hole in the PC board, with solder on the wire, solder around the hole, but no solder actually connecting the two. This gap of maybe $1/100$" is quite enough to stop everything from working, but you'll need a good desk lamp and high magnification to see it.

A FEW COMPONENTS AND TOOLS

Just as a kitchen should contain eggs and orange juice, you'll want a variety of resistors and capacitors (Figure 3). Your neighborhood Shack can sell you prepackaged assortments, or you can shop online at mouser.com or eBay.

After you buy the components, you'll need to sort and label them. Some may be marked only with colored bands to indicate their values. With a multimeter (a good one costs maybe $50) you can test the values instead of trying to remember the color-coding system. For storage I like the kind of little plastic boxes that craft stores sell to store beads.

For your breadboard you will need hookup wire. This is available in precut lengths, with insulation already stripped to expose the ends. You'll also need stranded wire to make flexible connections from the PC board to panel-mounted components such as LEDs or switches. To strip the ends of the wire, nothing

Fig. 1: Breadboard (left); upturned PC board.

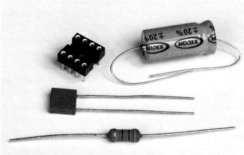
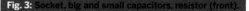
Fig. 3: Socket, big and small capacitors, resistor (front).

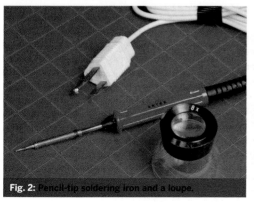
Fig. 2: Pencil-tip soldering iron and a loupe.

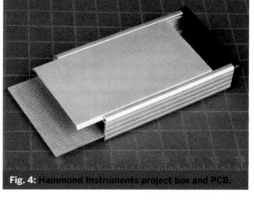
Fig. 4: Hammond Instruments project box and PCB.

Photography by Charles Platt

> ## " ELECTRONICS IS A MUCH CHEAPER HOBBY THAN MORE VENERABLE CRAFTS SUCH AS WOODWORKING, AND IT CONSUMES VERY LITTLE SPACE. "

beats the Kronus Automatic Wire Stripper, which looks like a monster but works with supernatural efficiency, letting you do the job with just one hand.

Needlenose pliers and side cutters of various sizes are essential, with perhaps tweezers, a miniature vise to hold your work, alligator clips, and that wonderfully mysterious stuff, heat-shrink tubing (you will never use electrical tape again). To shrink the heat-shrink tube, you'll apply a Black and Decker heat gun.

If this sounds like a substantial investment, it isn't. A basic workbench should entail no more than a $250 expenditure for tools and parts. Electronics is a much cheaper hobby than more venerable crafts such as woodworking, and since all the components are small, it consumes very little space.

For completed projects you need, naturally enough, project boxes. You can settle for simple plastic containers with screw-on lids, but I prefer something a little fancier. Hammond Instruments makes a lovely brushed aluminum box with a lid that slides out to allow access. Grooves inside the box accept a PC board. My preferred box has a pattern of conductors emulating three breadboards put together (Figure 4). This is big enough for ambitious projects involving multiple chips.

T he final and perhaps most important thing you will need is a basic understanding of what you are doing, so that you will not be a mere slave to instructions, unable to fix anything if the project doesn't work. Read a basic electronics guide to learn the relationships between ohms, amperes, volts, and watts, so that you can do the numbers and avoid burning out a resistor with excessive current or an LED with too much voltage. And follow the rules of troubleshooting:

LEARN THE RULES

» LOOK FOR DEAD ZONES. This is easy on a breadboard, where you can include extra LEDs to give a visual indication of whether each section is dead or alive. You can use piezo beepers for this purpose, too. And, of course, you can clip the black wire of your meter to the negative source in your circuit, then touch the red probe (carefully, without shorting anything out!) to points of interest. If you get an intermittent reading when you flex the PC board gently, almost certainly you have a dry joint somewhere, making and breaking contact. More than once I have found that a circuit that works fine on a naked PC board stops working when I mount it in a plastic box, because the process of screwing the board into place flexes it just enough to break a connection.

» CHECK FOR SHORT CIRCUITS. If there's a short, current will prefer to flow through it, and other parts of the circuit will be deprived. They will show much less voltage than they should.

Alternatively you can set your meter to measure amperes and then connect the meter between one side of your power source and the input point on your circuit. A zero reading on the meter may mean that you just blew its internal fuse because a short circuit tried to draw too much current.

» CHECK FOR HEAT-DAMAGED COMPONENTS. This is harder, and it's better to avoid damaging the components in the first place. If you use sockets for your chips, solder the empty socket to the PC board, then plug the chip in after everything cools. When soldering delicate diodes (including LEDs), apply an alligator clip between the soldering iron and the component. The clip absorbs the heat.

Tracing faults in circuits is truly an annoying process. On the upside, when you do manage to put together an array of components that works properly, it usually keeps on working cooperatively, without change or complaint, for many decades — unlike automobiles, lawn mowers, power tools, or, for that matter, people.

To me this is the irresistible aspect of hobby electronics. You end up with something that is more than the sum of its parts — and the magic endures.

Charles Platt is a frequent contributor to MAKE, has been a senior writer for *Wired*, and has written science fiction novels, including *The Silicon Man*.

I-SOBOT

T00392

HEXTRONIK
HXT500
5GR SERVO

HEXTRC

Photography by Tod Kurt

SERVO-MOTORS

Beef up on robotic musculature.

By Tod E. Kurt

You've seen them in robots and toys, or at least heard the distinctive *zzt-zzt-zzt* sound that accompanies their movement. R/C servomotors, designed for use in radio-controlled hobby cars and planes, are a common tool for robotics, movie effects, and puppeteering.

Servos don't spin like normal motors, but instead rotate and stop at a commanded position between 0 and 180 degrees. They're one of the easiest ways to add motion to a project, and there are many different kinds of servos to choose from.

Servos can also be hacked to create high-quality, digitally controlled, variable-speed gearmotors, with a few simple modifications. In this article I'll explain the basics of how to use servos, and how to hack them to make continuous rotation servos as well.

UNDERSTANDING SERVOS

R/C servomotors (or just "servos") are packed with technology, including a DC motor, gear train, sensor, and control electronics. They are a kind of *servomechanism*. Servomechanisms use a *feedback control loop* to adjust how a mechanism functions. One example of a servomechanism is a thermostat-controlled heating system. The temperature-sensing thermostat is the feedback-providing sensor, and the heating element is the output. The heater gets turned on and off based on the temperature sensing.

For R/C servos, the sensor input is a potentiometer (or "pot" for short) that's used to measure the amount of rotation from the output motor. Control electronics read the electrical resistance of the pot and adjust the speed and direction of the motor to spin it toward the commanded position. Figure A shows an exploded view of a standard servo and the workings of the servo's closed feedback control loop.

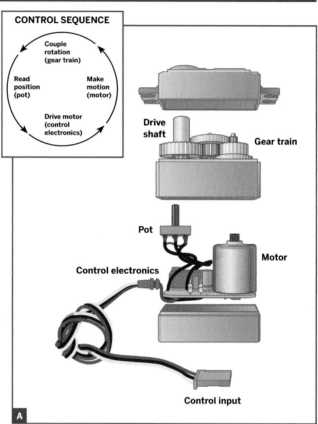

CONTROL SEQUENCE

Couple rotation (gear train) → Make motion (motor) → Drive motor (control electronics) → Read position (pot) → Couple rotation (gear train)

Drive shaft

Gear train

Pot

Motor

Control electronics

Control input

A

SELECTING SERVOS

Servos come in a variety of shapes and sizes (Figure B). The most common class of servo is the *standard* servo (1). The smallest servos are the *micro* class (2, 3), and the largest are the *high-torque/winch* class (4). All of these have the same 3-wire control, so it's easy to move to a bigger or smaller servo as your needs change.

Besides their dimensions and weight, servos are characterized by two main attributes: *torque* and *speed*, determined by the gearing and the motor in the servo.

The torque is essentially the strength of the servo. A standard servo torque value is 5.5kg/cm (75oz/in) when

B

Illustrations by Damien Scogin

operating at 5V. The speed is how fast the servo can move from one position to another. A standard speed is 0.20 seconds to move 60° at 5V. In general, the larger servos are slower but more powerful.

Until you know exactly what you need, it's easier to pick a size class (standard/micro/winch) and then choose the cheapest one in the list. For these projects, I use the micro HexTronik HXT500 servo, rated at 0.8kg and 0.10sec. I got them for under $4 each at hobbyking.com. They're cheap and they work well.

SERVO MOUNTING AND LINKAGES

To move something with a servo, you need two things: a mount for the base and a linkage to join the drive shaft to what needs to be moved. Servos come with mounting holes in their bases for screw mounting. For experimenting, it's easier to use hot glue or double-sided foam tape to hold a servo in place.

To connect the drive shaft, servos come with a collection of adapters called *servo horns*. These fit over the shaft and have an arm with mounting holes. By connecting a rod or wire to a mounting hole, you can turn the servo's rotary motion into linear motion. Choosing different servo horns or mounting holes will provide larger or smaller amounts of motion.

Figure C shows a collection of different types of servo horns. The 4 white ones in front of the servo are stock horns, and the 4 on the right are DIY horns cut from plastic on a laser cutter. The 2 at far right are a combination of horn and mount, allowing the chaining together of servos.

Creating custom horns is easy. Use a vector art program to make a star shape with the diameter and number of points matching the drive shaft of your servo. That star becomes the shaft connector, and any other kind of custom linkage can be added around that (Figure D).

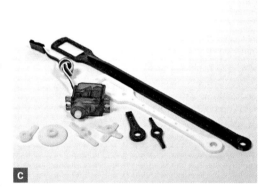

C

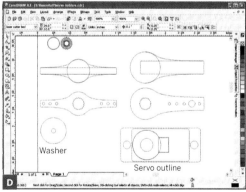

Washer

Servo outline

D

HOW TO CONTROL A SERVO

Servos have a 3-wire interface, as shown in Figure E. The black (or brown) wire connects to ground, the red wire to +5V, and the yellow (or white or orange) wire connects to the control signal.

The control signal (shown in action in Figure F, following page) is a type of *pulse-width modulation* (PWM) signal, easily produced by all microcontrollers. For this article I used the common

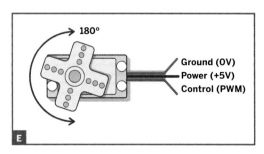

180°

Ground (0V)
Power (+5V)
Control (PWM)

E

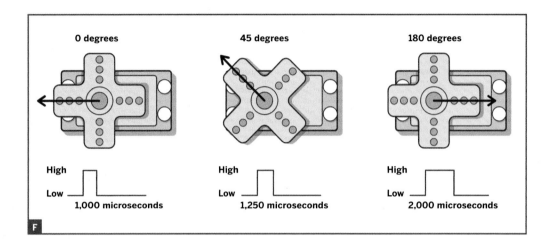

0 degrees   45 degrees   180 degrees

High   High   High

Low   Low   Low

1,000 microseconds   1,250 microseconds   2,000 microseconds

F

Arduino microcontroller.

The HIGH part of the pulse lasts between 1 and 2 milliseconds (ms), equal to 1,000 to 2,000 microseconds (µs). At 1,000µs, the servo will rotate to its full anti-clockwise position. At 2,000µs, it will rotate to its full clockwise position. Some servos accept shorter or longer pulses and have a correspondingly larger rotational range.

The LOW part of the control pulse lasts for 20 milliseconds. Every 20ms (50 times per second), the HIGH pulse must be received again or the servo will de-energize and not hold its position. This is useful if you want your project to "go limp."

Below is a complete Arduino sketch that will continually position the servo at its midpoint. Controlling a servo is pretty easy.

```
int servoPin = 9;
int servoPosition = 1500; // position in microseconds
void setup() {
pinMode(servoPin, OUTPUT);
}
void loop() {
digitalWrite(servoPin, HIGH);
delayMicroseconds(servoPosition);
digitalWrite(servoPin, LOW);
delay(20); // wait 20 milliseconds
}
```

Wire up the servo as in Figure G. Red and black wires go to the Arduino's 5V power and Gnd pins. The control wire goes to digital input/output pin 9.

G

The problem with this Arduino sketch is that it spends most of its time stuck in the delay commands. Fortunately, the Arduino's built-in Servo library lets you control 2 servos (on pins 9 and 10) using its built-in timers, and frees up your code to do other things. The same sketch using the library would be:

```
#include <Servo.h>
Servo myservo;
void setup() {
 myservo.attach(9); // servo is on pin 9 like before
 myservo.write(90); // set servo to 90 degree position
}
void loop() {
 // free to do anything, our servo is still being driven for us
}
```

USING SERVOS: PAN-TILT HEAD FOR WEBCAMS

Here's a quick project using 2 servos and an Arduino: a pan-tilt head for a webcam. Hot-glue a servo horn to the bottom of the webcam. Hot-glue another horn to the side of one servo. Hot glue the other servo to a base. Then plug all the servo horns into the servos and you've got a pan-tilt webcam.

Figure H shows an example: a homebrewed network pan-tilt webcam built from an Asus wi-fi router running OpenWrt Linux. Both the webcam and the Arduino controlling the servos are connected via a small USB hub to the router's USB port.

The code to control 2 servos from the Arduino's USB/serial port is shown below. The sketch waits for 2 bytes to arrive over the serial port, then treats the first byte as the 0–180 value for the pan servo and the second as the 0–180 value for the tilt servo.

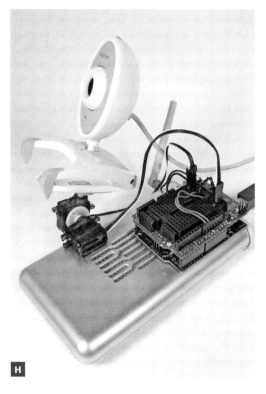

H

```
#include <Servo.h>
Servo servoPan;
Servo servoTilt;
void setup() {
 servoPan.attach(9); // pan servo is on pin 9
 servoTilt.attach(10); // tilt servo is on pin 10
 servoPan.write(90); // home both servos to center
 servoTilt.write(90); // home both servos to center
}
void loop() {
 if( Serial.available() >= 2 ) { // two bytes waiting for us
 int pan = Serial.read(); // 1st byte is Pan position
 int tilt = Serial.read(); // 2nd byte is Tilt position
 servoPan.write(pan);  // move pan servo
 servoTilt.write(tilt);  // move tilt servo
 }
}
```

MODIFYING SERVOS FOR CONTINUOUS ROTATION

Any servo can be turned into a bidirectional, variable-speed gearmotor. Normally, controlling a motor's speed and direction requires a motor driver chip and other parts. A servo already contains all these parts. Hacking it is one of the best-known and cheapest ways to get a digitally controlled gearmotor for use in robotics — a *continuous rotation servo*.

The modification is part mechanical and part electrical. The electrical mod replaces the pot with 2 fixed resistors of equal value. The mechanical mod removes the stops that prevent the motor from rotating fully.

I

NOTE: If you don't want to open up a servo, Parallax (maker of the BASIC Stamp) has a ready-to-use, standard-sized continuous rotation servo.

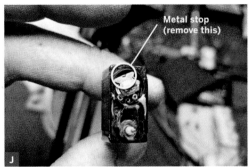

Metal stop (remove this)

J

Plastic stop (remove this)

K

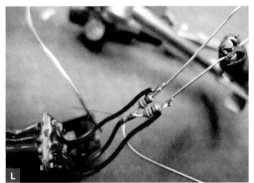

L

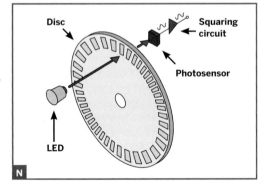

M

First, open up the servo. The HTX500 servo case is made of 3 plastic pieces press-fit together. You can use a small blade screwdriver or similar thin wedge to separate them. From the top, pull off the gears (note which ones go where). From the bottom, carefully pull out the servo's circuit board (Figure I).

There are 2 mechanical stops. Remove the metal stop at the shaft base by bending it with needlenose pliers (Figure J). Remove the plastic stop on the top case with diagonal cutters (Figure K).

Replace the 5kΩ pot with 2 fixed resistors that add up to near 5kΩ. A pair of 2.2kΩ resistors works well. Unsolder the 3 wires from the servo's pot and solder them to the 2 resistors as pictured in Figure L. Wrap this new assembly in electrical tape or heat-shrink tubing (Figure M). Tuck it and the rest of the electronics back into the servo case and snap it all back together.

That's it for the hack. Calibrate your finished continuous servo by finding where the *zero point* is. If the 2 fixed resistors were exactly identical, a 90° angle sent to the servo would make the motor stop. Your servo's value will likely be a bit off. You can use

Disc

Squaring circuit

Photosensor

LED

N

the previous sketch to experiment in finding what angle does stop the motors. Remember that value, because it'll be different for each servo.

Hobby servos use potentiometers to measure shaft rotation. Servos used in larger systems like industrial robots and CNC machines use rotary encoders. Optical rotary encoders attach a disc with slits and count the number of slits that go by with an LED and photosensor (Figure N). This is also how mechanical ball computer mice measure movement.

5-MINUTE DRAWBOT PROJECT

With two continuous rotation servos, you can start making robots. Figure O shows a drawbot made from two servos, a 9V battery, a small breadboard, an Adafruit Boarduino (Arduino clone), a Sharpie marker, and a couple of plastic discs.

The circuit is the same as for the pan-tilt head, and all the parts can be held together with hot glue. For wheels, any disc that's 1"–3" in diameter should work, such as plastic screw-top lids. To increase traction, wrap the wheel edges with duct tape.

The servos are set up as before. The sketch uses variables that contain the experimentally determined zero values to stop the motors. (Your zero values will be different.) The logic of this sketch runs one motor in one direction for a period of time, then switches to the other motor. The result is a Spirograph-like shape (Figure P).

O

P

```
#include <Servo.h>
Servo servoL;
Servo servoR;
int servoLZero = 83;  // experimentally found to stop
L motor
int servoRZero = 91;  // experimentally found to stop
R motor
boolean turnleft = false;
void setup() {
  servoL.attach(9);
  servoR.attach(10);
  servoL.write(servoLZero);  // start out not moving
  servoR.write(servoRZero);  // start out not moving
}
void loop() {
  turnleft = !turnleft;
  if( turnleft ) {
    servoL.write( servoLZero - 10 );
    servoR.write( servoRZero );
    delay(1000);
  } else {
    servoL.write( servoLZero );
    servoR.write( servoRZero + 10 );
    delay(4000);  // turn more one way than the other
  }
}
```

NOTE: The Sharpie is a permanent marker, so be sure to run the drawbot on top of cardboard or layers of butcher paper, or substitute water-soluble markers.

RESOURCES

» **Hobby People** (hobbypeople.net) **is a good U.S.-based vendor of servos.**

» **HobbyKing** (hobbyking.com)**, formerly HobbyCity, is a China-based servo vendor with a huge selection.**

» **Adafruit** (adafruit.com) **carries Arduino and Boarduino microcontrollers, and Parallax continuous servos.**

» **Trossen Robotics** (trossenrobotics.com) **carries Parallax continuous servos and many other neat robot things.**

» **Oomlout SERB robot** (oomlout.com/a/products/serb) **is an open source robot base that uses continuous servos.**

» **Servo hacks on Instructables** (instructables.com)**: Do a search for "servo" to see many more servo projects.**

Tod E. Kurt (todbot.com/blog) is co-founder of ThingM, makers of the BlinkM Smart LED (blinkm.thingm.com), and author of *Hacking Roomba* (hackingroomba.com), an introductory robotics course disguised as a set of vacuum cleaner hacks.

Listening to Satellites

Tune in to space with a homemade yagi antenna.

BY DIANA ENG

Photography by Diana Eng

One of my favorite things to do is talk with other ham radio operators through satellites or the International Space Station (ISS). To do this, I stand on a rooftop and tune a handheld multiband radio while tracing the orbit of a satellite or the ISS with my homemade yagi antenna.

Orbiting satellites such as SO-50 act as repeaters, relaying signals from low-power transceivers like mine back to hams elsewhere on the planet. So if you know where to aim the antenna, you can communicate around the world via space.

The ISS also has a repeater, and occasionally, when we're lucky, the astronauts themselves exchange transmissions to communicate with hams on the ground.

To listen to these signals from space, you don't

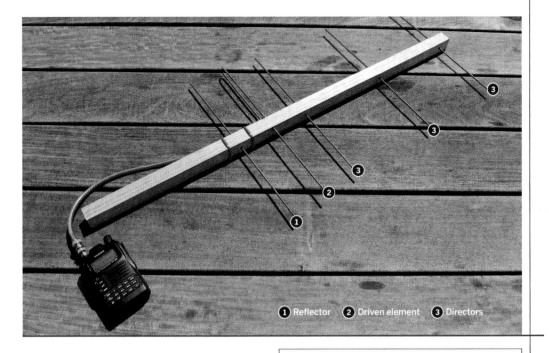

① Reflector   ② Driven element   ③ Directors

have to be a licensed ham radio operator, or even stand on the roof. You can do it in your own backyard with an off-the-shelf UHF FM radio. The whip antenna on the radio might let you hear satellites and the ISS, but you'll get far better reception by making your own yagi antenna, which takes about an hour and costs less than $25 (not including the cost of your radio) using materials from your local hardware store.

If you do have a ham radio license and a UHF/VHF transceiver, you can upgrade this antenna with VHF elements so that it can both send and receive transmissions.

A yagi antenna has three types of elements, consisting of metal rods of varying lengths and quantities. The *driven element* is a dipole antenna that's connected to the radio and receives the signal, just like a whip antenna. The *reflector* is positioned behind the driven element, where it acts as a mirror by bouncing signals from the satellite forward to the driven element. *Directors* are one or more rods that act like a lens, focusing the incoming signal onto the driven element. Both the reflector and the directors improve reception from whatever direction the antenna points.

The antenna design I use comes from Kent Britain's (WA5VJB) "Cheap Antennas for Low Earth Orbit" (available at wa5vjb.com/references.html), which is a great reference for building many different types of yagi antennas.

MATERIALS

UHF FM radio like a police scanner, such as the Uniden BC72XLT handheld scanner, amazon.com, $85. Or, if you have a ham radio license, a UHF transceiver such as the Yaesu VX-7R, universal-radio.com/catalog/ht/0777.html.

Square wooden dowel, 1"×1"×30" or longer if you want a handle longer than 10" for attaching to a tripod or mounting the radio. Approximately $2–$3 at hardware or craft stores.

Brass rods, ⅛" diameter, 36" long (3) Uncoated brazing rods work, but almost any brass rod or tube will do as long as it's approximately ⅛" diameter; $3–$4 each at a hardware or craft store.

Coaxial cable, RG-58, with BNC connectors, 3' universal-radio.com/catalog/cable/cable.html, order #4616, $4

Nylon cable ties (2) aka zip ties

TOOLS

Hacksaw
Soldering iron and solder
Glue gun and glue
Wire strippers and cutters
Ruler
Marker
Vise
Drill, or drill press, and ⅛" drill bit
File, smooth cut, flat, approx. 10"
Wooden dowel or broomstick, approx. ¾" diameter
Computer with an internet connection
Pencil

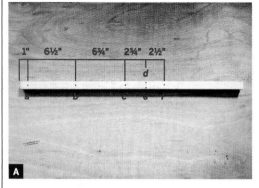

A

B

C

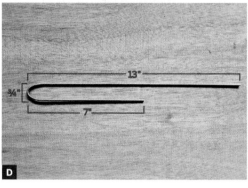

D

» Build the Yagi Antenna

Time: 1–2 Hours
Complexity: Building Antenna = Easy;
Receiving Signals = Medium

1. Measure and cut.

Use a pencil and ruler to draw a centerline down
one long side of the wooden beam. Then measure
and mark hole locations on the centerline (except
holes *d* and *e*) at the following intervals: hole *a*
1" from one end; hole *b* 6½" from *a*; hole *c* 6¾"
from *b*; holes *d* and *e* 2¾" from *c*, ⅝" apart and
equidistant from the centerline; and hole *f* 2½"
from *d* and *e*.

Drill ⅛" holes completely through the beam at
each point (Figure A). Be careful when drilling
d and *e* not to blow out the sides of the beam.

Use a marker and ruler to mark 5 pieces of brass
rod at the following lengths: 21", 13½", 12½", 12¼",
and 11¾". Secure the rod in a vise, cut to the measured lengths using a hacksaw, and file the ends so
they're no longer sharp and dangerous (Figure B).

To make the driven element, place the 21" rod in the
vise, mark it 13" from one end, center the mark on the
broomstick, then bend it 180° around so it's J-shaped
(Figure C). Trim the rod so it measures 13" from one

end to the center of the ¾" curve, and 7" from the
other end to the center of the curve (Figure D).

2. Assemble the parts.

Insert the 11¾" element into hole *a*, the 12¼"
element into *b*, the 12½" element into *c*, the
J-shaped (driven) element into *d* and *e*, and the
13½" element into *f*. Center all the elements, and
secure them in place with hot glue (Figure E).

To prepare the coaxial cable, cut off one of its
connectors and strip 3" of outer insulation off that
end, being careful not to cut the wires. Separate the
outer wires, twist them to one side, and strip 2" of
insulation off the inner wire (Figure F).

Connect the coax cable to the 2 parts of the
driven element near where they enter the wooden
beam. Wrap the cable's inner wire around the short
leg of the J, and the twisted outer wires around the
long leg. Solder the wires in place (Figure G).

Secure the coax cable with a couple of zip ties
(Figure H). Your antenna is done (Figure I)!

3. Receive signals from space.

To use your antenna, you need to find out where to
point it and what frequency to tune in to. To find
a good satellite target, visit heavens-above.com.

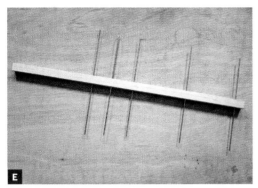

To specify your location, select a Configuration option (map, database, or manual), plug in the necessary info, then click Submit. From your new location-specific homepage, select "All passes of the ISS" to track the International Space Station or "Radio amateur satellites" to track a ham radio repeater satellite (Figure J, following page).

On the Radio Amateur Satellites page, click on one of the radio satellites you want to track from the Satellite column (such as AO-Echo, aka AO-51; SaudiSat 1C, aka SO-50; or AO-27), then show its pass chart by selecting "Passes (all)" above the globe (Figure K).

The pass chart lists all the satellite passes for the next few days. Each pass is listed by its times and locations in polar coordinates, for its start, maximum altitude, and end, with each pass typically taking about 10 minutes. The start and end points are defined as when the "bird" appears 10° above the horizon, and the maximum altitude (in degrees above the horizon) will vary. The azimuth for each location is listed in compass points (Figure L).

Make sure your location is listed correctly on the chart, and pick a pass during which the satellite will come close to directly overhead. Look for max altitudes that are 45° or higher — the higher, the

better. In the example here, the second pass, on July 17 at 3:50, looks good since its altitude reaches 75°, but the first pass, on July 16 at 16:55, only comes up to 18°, which is very close to the horizon and difficult to pick up.

Next, find the frequency to tune in to. Satellite repeaters work with 2 different frequencies — an uplink and a downlink. You listen to signals received via the downlink. (If you wish to transmit, you'll need to program in the uplink frequency as well.)

To find a radio satellite's current frequencies, you have to refer to the authoritative web page for each individual satellite. Some references online, including AMSAT (amsat.org), aggregate frequency information for multiple satellites, but these can be incorrect and you often need to dig deeper.

What you want is a current update or schedule with uplink and downlink frequencies, and this data

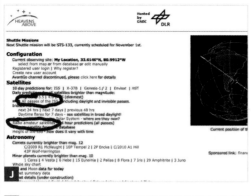

HEAVENS ABOVE

Hosted by GSOC DLR

Shuttle Missions
Next Shuttle mission will be STS-133, currently scheduled for November 1st.

Configuration
Current observing site: **My Location, 33.6146°N, 80.9912°W**
select from map or from database or edit manually
Registered user login | Why register?
Create new user account
AvantGo channel discontinued, please click here for details

Satellites
10-day predictions for: ISS | X-37B | Genesis-1/2 | Envisat | HST
Daily predictions of all satellites brighter than magnitude:
3.5 | 4.5 | 5.5 (datmost)
All passes of the ISS including daylight and invisible passes.
Iridium flares
next 24 hrs | next 7 days | previous 48 hrs
Daytime flares for 7 days – see satellites in broad daylight!
Radio amateur satellites · Solar System - where are they now?
Whole sky chart database
Height of the ISS – how does it vary with time

Astronomy
Comets currently brighter than mag. 12
C/2009 R1 McNaught | 10P Tempel 2 | 2P Encke | C/2010 A1 Hill
43P Wolf-Harrington
Minor planets currently brighter than mag. 10
1 Ceres | 4 Vesta | 6 Hebe | 15 Eunomia | 2 Pallas | 8 Flora | 7 Iris | 29 Amphitrite | 3 Juno
Whole sky chart
and Moon data for today
et summary data
et details (under construction)

Current position of th

Sponsored link: finan

J

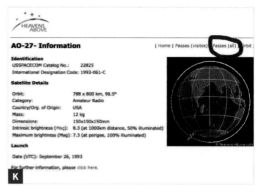

HEAVENS ABOVE

AO-27- Information   [Home | Passes (visible) | Passes (all) | Orbit

Identification
USSPACECOM Catalog No.:   22825
International Designation Code: 1993-061-C

Satellite Details

Orbit: 788 x 800 km, 98.5°
Category: Amateur Radio
Country/Org. of Origin: USA
Mass: 12 kg
Dimensions: 150x150x150mm
Intrinsic brightness (Mag):   8.3 (at 1000km distance, 50% illuminated)
Maximum brightness (Mag): 7.3 (at perigee, 100% illuminated)

Launch

Date (UTC): September 26, 1993

For further information, please click here.

K

L

HEAVENS ABOVE

AO-27 - All Passes   [Home | Info. | Orbit | Prev. | Next |

Search Period Start: 00:00 Friday, 16 July, 2010
Search Period End: 00:00 Monday, 26 July, 2010
Observer's Location: My Location (33.6146°N, 80.9912°W)
Local Time: Central European Summer Time (GMT + 2:00)
Orbit: 788 x 800 km, 98.5° (Epoch 15 Jul)

Click on the date to get a ground track plot.

Date	Starts			Max. Altitude			Ends		
	Time	Alt.	Az.	Time	Alt.	Az.	Time	Alt.	Az.
16 Jul	16:55:31	10	SW	16:59:03	18	WSW	17:02:35	10	WNW
17 Jul	03:50:03	10	NNE	03:55:10	75	E	04:00:24	10	S
17 Jul	14:49:24	10	E	14:51:30	12	ENE	14:53:37	10	NE
17 Jul	16:25:54	10	S	16:30:47	40	WSW	16:35:40	10	NW
18 Jul	03:22:18	10	NE	03:26:55	33	ESE	03:31:34	10	SSE
18 Jul	05:03:26	10	NW	05:06:13	15	WNW	05:09:01	10	WSW
18 Jul	15:57:14	10	SSE	16:02:29	88	SW	16:07:45	10	NNW

is unfortunately not published in a standardized manner. I like to Google the satellite's name to find its authoritative source.

At the time of this writing, SO-50 is the only active FM satellite. But EO-80 and FOX-1A will be active soon.Once you've determined your target sat's current downlink frequency (example: 436.7950MHz FM), tune your radio to that frequency, and you're ready to go out and listen. Aim your yagi antenna directly at the satellite, with the shortest rods (directors) closest to the satellite and the longest rod (the reflector) farthest away. When the pass starts, aim the yagi toward the satellite (Figure M), then sweep it right and left slightly until you hear something. You can also move the antenna up and down slightly as you sweep right and left. Also try rotating the antenna by twisting your wrist, adjusting its polarity to receive a stronger signal.

If you're using a whip antenna, hold it perpendicular to the satellite, and keep it perpendicular while you rotate it to get a clearer signal (Figure N).

Trace the path of the satellite's orbit according to the pass chart, so that at its maximum altitude and its end time, the antenna is pointed in the corresponding locations. In our example, the antenna should be pointed east at 75° above the horizon at 3:55, and south at 10° above the horizon at 4:00. It can be difficult trying to catch the satellites, and you may spend a lot of time not hearing anything. The best method is to move the antenna around in small side-to-side and up-and-down motions until you hear a bit of audio.

The Doppler effect makes the frequency vary by 0.010MHz, so as you trace the satellite's path you'll also need to twiddle the tuning a bit. Add 0.010MHz to your target frequency early in the pass, then gradually dial it down until it's approximately 0.010MHz less than the listed downlink frequency by the end time.

The FM satellites repeat whatever they receive, so you'll hear whoever's signal is strongest. (Another type of satellite, linear transponders, can handle multiple conversations at once, but these are harder to use and require a more expensive single sideband, not FM, radio.)

Hamspeak

When you eavesdrop on ham radio satellites and ISS, you'll probably hear a lot of letters, numbers, and strange words, like "KC2UHB Foxtrot November three one … roger roger." One reason is that ham operators use a phonetic alphabet to make themselves clear through the static and interference, so that "P" sounds nothing like "T," for example.

The ham ABCs are: Alpha, Bravo, Charlie, Delta,

Echo, Foxtrot, Golf, Hotel, India, Juliet, Kilo, Lima, Mike, November, Oscar, Papa, Quebec, Romeo, Sierra, Tango, Uniform, Victor, Whiskey, X-ray, Yankee, and Zulu.

Also, orbits don't last very long, so radio operators extending their reach via satellites tend to communicate quickly, following the same general dialogue. Here's an example:

"Kilo Charlie two Uniform Hotel Bravo." (Hi, my call sign is KC2UHB, does anyone want to talk to me?) A call sign is like a screen name assigned to ham radio operators when they receive their license. Some operators have vanity call signs like NE1RD.

"KC2UHB from Whiskey two Victor Victor please copy Foxtrot November three one." (I hear you KC2UHB, my call sign is W2VV and I am in Maidenhead location FN31.) The Maidenhead system divides the Earth into grid squares as shorthand to describe locations, and FN31 covers most of Connecticut and some of New York State. You can look up grid square locations online at levinecentral. com/ham/grid_square.php and elsewhere.

"W2VV, QSL this is KC2UHB, Echo Mike eight nine." (W2VV, I received your transmission, my location is EM89.) KC2UHB is in central Ohio.

"QSL. Thank you for the contact. 73." (I received your transmission. Thank you for the contact. Goodbye.)

"73." (Goodbye.)

Just as we text each other abbreviations like OMG, BRB, TTYL, LOL, BF, GF, and <3, ham operators have their own, much older shorthand that was originally based on Morse code but became spoken with the advent of voice transmissions — much like when people say "Oh em gee" or "Be eff eff" today. Here are some ham abbreviations you may hear:

73 = goodbye, best wishes
88 = xoxo
OM (old man) = a friendly term for a male ham, a boyfriend/husband if described as "my OM"
YL (young lady) = a female ham, a girlfriend if described as "my YL"
XYL = wife
QSL = confirmation of message received
QRP = operating with low power
HT (handy talky) = a walkie-talkie

▣ To learn more about how a yagi antenna works, watch Diana Eng's MAKE video on directional antennas, aka "Seeing Radio Waves With a Light Bulb," at makezine.com/go/yagi.

Diana Eng (dianaeng.com) is a fashion designer who works with technology, math, and science. She is author of *Fashion Geek: Clothes, Accessories, Tech* (North Light Books, 2009) and is the ham radio correspondent for Make: Online (makezine.com).

EASY

THE ECLECTIC
Electret
MICROPHONE

Written and photographed by Charles Platt

PCB TRACES
RADIATING "FINGERS" CONNECT CASE TO GROUND

PCB SUBSTRATE
WITH HOLES FOR TRANSISTOR LEADS

PLASTIC CASE
HOUSES INTERNAL AMPLIFER

PLASTIC SPACER
MAINTAINS INSULATING GAP

ELECTRET MEMBRANE
BONDED TO METAL

AMPLIFYING TRANSISTOR
SILICON JFET N-CHANNEL TYPE, WITH GROUND ON SOURCE, PICKUP PLATE ON GATE, AND SIGNALOUT ON DRAIN

CASE
CRIMPED AT BACK TO RETAIN CONTENTS

PICK-UP PLATE
FORMS CAPACITOR WITH MEMBRANE

DUST COVER
CLOTH OR PAPER

Time Required:
1 Hour
Cost:
$10-$20

Materials

» **Electret microphone** Most 8mm or 10mm electrets will work in this circuit.
» **Ceramic capacitors: 0.1µF (2), 0.68µF, 10µF (2)**
» **Electrolytic capacitor, 330µF**
» **Resistors: 22Ω, 1K (2), 1.5K, 3.3K, 10K, 100K (2)**
» **Trimmer potentiometers: 10K, 100K**
» **Integrated circuits: LM741, LM386** Manufacturers may precede these generic part numbers with additional letters or numbers.
» **Loudspeaker: 2-inch or 3-inch, 50Ω–100Ω**
» **9V batteries (2) and connection clips**

THE MICROPHONES IN OLD-FASHIONED WIRED TELE-PHONES were relatively heavy, large, and expensive, and their sound quality was terrible. Thanks to materials science developments in the 1990s, electrets are now tiny, high-quality, and available from some sources for less than $1 each.

The electret microphone performs its magic with a pair of thin electrostatically charged membranes. When sound waves force one closer to the other, a tiny transistor in the microphone amplifies the fluctuations in electrical potential. We can amplify them further, in our circuits, and use them for many purposes.

Testing, Testing ...

Most electrets have two terminals. They may have leads attached or just solder pads for surface-mount applications. Since the pads are reasonably large, you can easily add your own leads if necessary.

Your first step is to distinguish the positive and negative terminals. They are not usually marked in any way, and the datasheets can be surprisingly uninformative. However, if you look at the back of the microphone, you should see metal "fingers" radiating outward to the shell from one of the terminals (**Figure A**). These "fingers" — embedded in a translucent sealing compound — identify the negative side of the electret, which should be connected to ground.

Connect the other terminal through a 3.3K series resistor to the positive side of a 9VDC power supply, and you should see the electret responding to sound when you apply a meter (**Figure B**). Don't forget to set your meter to measure AC, not DC. A range of 1mV to 40mV is typical.

Amplification

We can use an op-amp to turn millivolts from the electret into volts. **Figure C** shows a circuit using the LM741. While many simpler circuits exist, this one minimizes oscillations and distortion. The LM741 outputs to an LM386, a basic power amplifier chip that can drive a small loudspeaker.

Notice we use a "split power supply" consisting of +9V DC, –9V DC, and 0V (represented by a ground symbol). To set this up, you can use a pair of 9V batteries in series, as shown in **Figure D**. But why is it necessary?

Consider how sound waves are created. All around us is static air pressure, which can be imagined as an absence of sound. When you speak, you create waves that rise above the ambient level, separated by troughs that drop below it. An amplifier must reproduce these fluctuations accurately, and relatively positive and relatively negative voltages are the most obvious way.

In **Figure C**, a 0.68µF capacitor couples the microphone through a 1K resistor to the op-amp. The capacitor blocks DC voltage, to stop the op-amp from trying to amplify it. But the capacitor is transparent to the alternating audio signal, which we *do* want to amplify. The mic signal induces fluctuations in a neutral voltage provided by a voltage divider, and the op-amp amplifies the difference between these fluctuations and a second input, which has a stable reference voltage.

This reference is created with negative feedback from the op-amp output, adjusted with the 100K trimmer. Negative feedback keeps the op-amp under control, so that it creates an accurate copy of the input signal. To learn more signal processing with op-amps, look out for *Make: More Electronics*, the sequel to my book *Make: Electronics*.

Making It Work

The two-battery split supply has some limitations. The output won't be loud, and it may be scratchy, especially if your two 9V batteries deliver unequal voltage. Use the 100K trimmer to minimize the distortion and the 10K trimmer to maximize the volume. You'll get better results if you have a proper split power supply, or two 12V AC adapters connected through separate 9VDC voltage regulators.

A 50Ω-100Ω loudspeaker is preferred. I got really good results when I used alligator clips to connect the output from the circuit to the mini-jack plug on my computer speakers, but if you make a wiring error, you may damage your speakers.

You may be interested in other ideas, such as using sound to switch on a light or start a motor. For this purpose, instead of an audio amplifier such as the LM386, the output from the op-amp can trigger a solid-state relay. Add a 100µF capacitor between the op-amp output and ground to smooth the signal (so that the relay doesn't "chatter") and adjust the 100K trimmer until the sensitivity of the circuit is appropriate.

An op-amp can feed its output (through a coupling capacitor) to a microcontroller input pin. You'll have to discover the digital value that the analog-digital converter inside the microcontroller assigns to various levels of sound, but after that you can program different outputs for different sound levels.

Analog audio circuits can be trickier than digital circuits. Learning how to use an electret is a great introduction! ◗

CHARLES PLATT is the author of *Make: Electronics*, an introductory guide for all ages. He has completed a sequel, *Make: More Electronics*, and is also the author of the *Encyclopedia of Electronic Components* in three volumes.

To avoid oscillations and other noise, keep all wires as short as possible, and pack the components tightly together. The pairs of red and black wires connect with 9V batteries, while the yellow wires go to a loudspeaker.

Three electret microphones viewed from below. Upper right: from RadioShack. Upper left: from mouser.com. Bottom: from allelectronics.com. The ground terminal is on the right-hand side in each case.

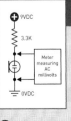

Detecting the output from an electret microphone.

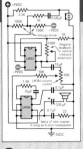

An audio amplifier circuit.

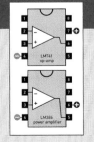

Basic pinouts of the LM741 op-amp and LM386 power amp. Unlabeled pins have additional functions; see datasheets for details.

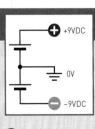

The split power supply required by the circuit can be provided by two 9V batteries wired in series.

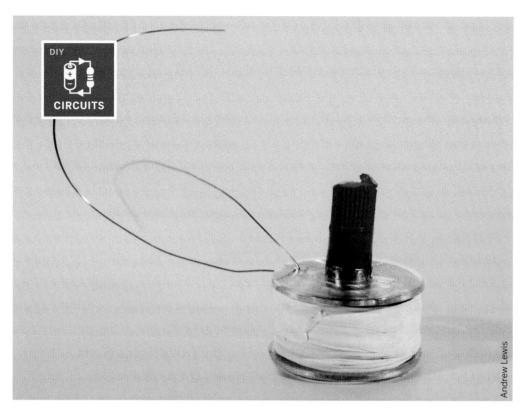

The Bobbinator

Make inexpensive linear actuators out of sewing machine bobbins.

By Andrew Lewis

LINEAR ACTUATORS ARE MOTORS THAT work in a straight line. They're very useful for controlling valves and levers, building robots, and retrofitting old mechanisms for digital control. The simplest form of linear actuator has only 2 positions, while more complicated actuators can be positioned just like stepper motors and servos.

I needed 32 linear actuators for my latest project, and I was shocked to discover how much they would cost. After a little thought, I decided that I only needed simple on/off actuators, and I would try to make them myself using off-the-shelf parts.

Wind and Wrap the Bobbin

1. Place an empty bobbin onto the bobbin filler on the sewing machine. Feed the end of the copper wire into the hole at the top of the bobbin. I left about 1" of wire sticking out of the top, so I could solder the coil in place later.

2. Feed the wire through the thread guide nearest to the bobbin filler. Do not use the tensioning wheel or the other guides. Place the spool of wire somewhere that it won't snag, and run the sewing machine slowly. Watch the spool fill with copper wire (Figure A), and apply gentle tension to the wire with your hand. Don't grab the wire tightly, or you'll cut yourself.

3. When the bobbin is full, stop the machine. Hold the wire in place on the bobbin with your thumb. Cut the wire about 1" from the side of the bobbin, and then remove the bobbin

MATERIALS AND TOOLS

Bobbins, plastic, for sewing machine
Magnets, rare earth, 6mm diameter × 3mm
Rod, plastic, 6mm
Wire, copper, enamelled, 38 SWG (35.5 AWG)
PTFE tape aka Teflon plumber's tape
Thin plastic or laminated card
Epoxy glue
Cable cutters
Sewing machine

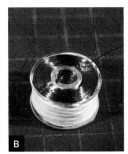

from the bobbin filler. Remember to keep your thumb on the side of the bobbin, or the wire will unravel.

4. Wrap a couple of layers of PTFE tape around the side of the bobbin. Make sure that the wire at the side of the bobbin is roughly aligned to the wire poking out of the top (Figure B).

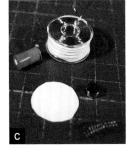

Add Plastic, Spring, and Magnet

5. Cut a circle of thin plastic and glue it to the bottom of the bobbin (Figure C). Make sure the hole through the bobbin is blocked off on the side opposite to the wires.

6. Take a short length of the copper wire, and wind it around a thin screwdriver shaft to make a small spring (Figure D). The spring should be about 4mm diameter, and about ¾ the height of the bobbin.

7. Glue a 6mm magnet onto the end of a length of metal or plastic rod. I needed about ½" of plastic rod, so I used the shaft from an old plastic potentiometer. Don't get too much glue around the edge of the shaft, or it may get wedged inside the bobbin when you fit it. Finally, add another dab of glue to the magnet, and glue it to the spring (Figure E). Fit the spring inside the bobbin center, and fix with a bit more glue.

You should now have a simple linear actuator that pulls the shaft inside the bobbin when you apply a voltage to the coil.

You can reverse the action of the actuator by reversing the voltage, and you can reverse

the action of the spring by using a thinner shaft with a second piece of thin plastic as a washer on the end of the bobbin.

You can also use multiple coils to create a longer stroke for the actuator, and position the shaft by varying the current between each of the coils. I find that each coil has a resistance of between 8Ω and 9Ω, and works well between 6V and 12V. Increasing the voltage will up the power, but will also increase the heat generated by the coil. Experiment to find the maximum voltage for a particular application, and with long duty cycles, stick a thermal fuse next to the coil for safety's sake. ◢

Andrew Lewis is a keen artificer and computer scientist with interests in 3D scanning, computational theory, algorithmics, and electronics.

Keybanging

Enjoy code-free home automation using prop controllers.

BY WILLIAM GURSTELLE

Despite the unfortunate experiences of Dr. Frankenstein, I think adding life to inanimate things is one of most interesting types of automation. Thanks to the current generation of easy-to-use programmable animation controllers, making an articulated figure move like a person, giving jets of water "personality" as they stream out of a fountain, or sequencing holiday lights to musical beats isn't just for theme park set designers anymore.

Of course, making all that happen requires computer programming. I may be limiting my opportunities to tackle the most involved projects, but I don't want to be a computer programmer. If a project includes stuff like:

```
void setup(){ pinMode(ledPin, OUTPUT);
```

then count me out.

Sure, microcontrollers like Arduino and BASIC Stamp can perform a million automation tasks a million different ways, but usually that's overkill for what I need. Instead, I use a far simpler method of adding life to props, displays, and room environments that doesn't require coding or digital programming.

It's called *keybanging*. A keybanger is a standalone, controllable programming device that allows the automation of complex tasks through a simple pushbutton programming interface. Born in the animatronics industry, these devices are also known as *prop controllers*, *programmable switches*, or *effects programmers*. The simplest devices have a single trigger circuit, control a single device, and

cost about $50. More elaborate models use a microcontroller chip for a brain and include onboard audio amplifiers, output relays for 8 or more devices, interfaces for servomotors, and (shudder) a programmable digital interface to supplement the buttons.

Besides animatronics, keybangers are used to automate water fountains, fireworks, lighting displays, sound systems, slide and movie projectors, window displays, signs, or just about anything else that can be controlled by an electrical on-off signal.

Say, for example, you had a 6-pump water fountain and you wanted to coordinate the fountain sprays to the rhythm of a particular piece of music. If you used a typical microprocessor, this would require some elaborate I/O programming. Further, turning the water jets on and off to coordinate with the music equates to some fancy sequencing and timing.

But programming a keybanger for this job is quite simple. Just play the music and then push the buttons when you want the fountains to

L-Dopa

spray; release them when you want them off. All keybangers allow you to record at least a 2-minute-long sequence, and some allow for far longer programs.

Input/output schemes on keybangers are straightforward. There are one or more trigger inputs on the front of the box. When the circuit is closed (or opened, depending on the type of trigger) the device begins running its program. The circuit closure could be a simple pushbutton, a switch hidden under a floor mat, or a passive infrared sensor, for example. When the circuit closure is made, the keybanger detects it and starts running its program.

Once the program starts, electrical contacts on the device open and close according to the instructions you programmed into it. When the keybanger is triggered, its little computer brain turns the attached gadgets on and off in accordance with the program. Since the unit is fairly intelligent (at least in a specialized way), it can easily control complicated or lengthy sequences.

A number of small electronics designers produce keybanger logic systems. An internet search will turn up dozens of websites with products made by companies ranging from part-time garage operations to large companies with customer support staff and full product lines. Most, but not all, started in the commercial haunted house business. In fact, animating Halloween props is by far the most common application.

The simplest and cheapest devices sequence a single relay. A well-known device of this sort is the Animation Maestro from Haunted Enterprises. AMs are great for simple Halloween props such as the beloved Trash Can Trauma (see MAKE's Halloween

MATERIALS

Prop controller, Monster Guts Nerve Center Other controllers would work equally well; however, you may need a separate amplifier to drive the speaker.

Speaker, 8Ω Use one with a ⅛" phono plug to connect to the Monster Guts Nerve Center.

Switch, momentary pushbutton, normally open (NO)

Fountain pump, submersible

Water container

Light source such as an incandescent bulb, LEDs, etc.

Extension cords, 6' (2)

Wire, miscellaneous

Digital music player (e.g., iPod)

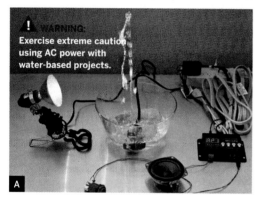

⚠ WARNING: Exercise extreme caution using AC power with water-based projects.

A

B

Gregory Hayes

Special Edition from 2007, makershed.com item #9780596514297.PDF). An Animation Maestro costs about $60. The simplest model has a single trigger input and a single output. Programming the device takes about a minute to learn.

Gilderfluke & Co. designs and manufactures a wide variety of animatronic devices. Their MiniBrick controller models can control 4, 8, or 32 relays, and can interface with standard DC servomotors. As keybangers go, MiniBricks are the most complicated to program, but once you figure out the nuances, their capabilities are nearly endless. An 8-output MiniBrick with 2 servo outputs runs $210.

My personal favorite is the Monster Guts Nerve Center for about $70. The Nerve Center controls 2 on-off relays and has a small LED display that makes it a snap to program. It runs multiple programs, so the animation can behave differently on succes-

sive triggers. Best of all, it has an onboard audio amplifier and enough memory to record 8 seperate sound tracks, so if your project includes sound, just add a speaker (no amp needed) and you're good to go. Here's how I used it to make an animated water fountain.

START
1. BUILD YOUR FOUNTAIN TABLEAU
Place a submersible fountain pump inside a water-filled container and plumb the device so that the water jets upward when the pump is energized. Attach a light source to the container so you can light up the water spray (Figure A).

2. SET UP THE AUDIO
Connect a music source (e.g., iPod) to the audio input on the Nerve Center, and connect a speaker to the audio out jack (Figure B).

Select the Enter Sound command and record your musical selection into memory.

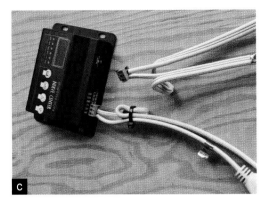

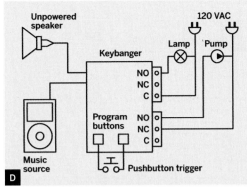

Unpowered speaker

120 VAC

Keybanger

Lamp   Pump

NO
NC
C

Program buttons

NO
NC
C

Music source

Pushbutton trigger

MONSTER GUTS
NERVE CENTER

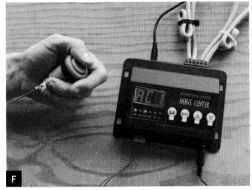

3. CONNECT THE PUMP AND LIGHT

The Nerve Center connections are simple relays that make or break according to the programmed instructions. The Nerve Center doesn't provide AC power; it merely switches it on or off. That means that to control power to the pump and light, you'll need to break into one conductor on each extension cord so they connect to the Nerve Center relays (Figure C) as shown in the schematic (Figure D).

4. RECORD YOUR SEQUENCE

This is the fun part. Select the Action Control command on the LED display. The music recorded to the device in Step 2 begins to play.

Use the 2 programming buttons on the front of the controller to bring your fountain to life. When you depress the button corre-spond-ing to the pump, the pump will run. Releasing the button will turn it off. The light control button works similarly. You can sequence your entire show just by pressing and releasing the 2 buttons on the front panel (Figure E). The fountain spray

and light work independently, so you can devise all sorts of routines.

5. ADD A PUSHBUTTON

Wire the pushbutton switch between the key-banger's NO (normally open) and Common trigger contacts (Figure F).

6. ANIMATE!

Select the Scene Control command and press the pushbutton to trigger your animated sequence. Want to make changes? Just tweak the program for best effect, then sit back and enjoy the show.

The applications for keybangers are limited only by your imagination. By replacing the pushbutton with a sensor (motion, heat, humidity, temperature, etc.) you can use the keybanger to automate any number of household systems. ◪

William Gurstelle is a contributing editor of MAKE. Visit williamgurstelle.com for more information on this and other maker-friendly projects.

PRIMER

Surface Mount Soldering

Techniques for making modern circuits.

By Scott Driscoll

When cellphones were housed in briefcases, manufactured electronics had easy-to-solder leads. Now phones fit in pockets, and the smaller surface-mount devices (SMDs) inside are driving through-hole components into extinction.

SMDs can cost less than their old-school equivalents, and many newer devices, including most accelerometers, are only available in SMD format.

If you design printed circuit boards, using SMT (surface-mount technology) and putting components on both sides makes them cheaper and smaller. This may not matter on a robot, but it helps a project fit into a mint tin or hang off a kite.

SMDs are designed for precise machinery to mass-assemble onto densely packed PCBs. Their tiny leads may look impossible for human hands to work with, but there are several good, relatively inexpensive methods that don't require a $1,000-and-up professional SMT soldering station.

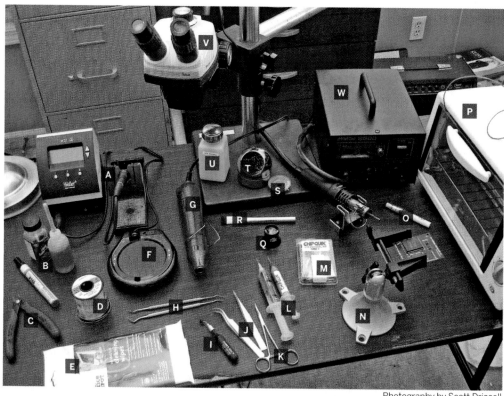

Photography by Scott Driscoll

TOOLS

What you need depends on what you're doing and how much of it (see story).

[A] Soldering station

[B] Flux felt pen, brush bottle, or needle bottle

[C] Flush wire cutters

[D] Solder

[E] Lint-free wipes

[F] Hot plate or coffee pot warmer or skillet

[G] Embossing heat tool from an art store

[H] Dental picks

[I] Vacuum pickup

[J] Tweezers

[K] Hemostat

[L] Solder paste

[M] Chip Quik SMD removal kit

[N] PanaVise

[O] Temperature-indicating marker or thermocouple

[P] Toaster oven

[Q] Loupe or lighted magnifying glass

[R] Acid brush

[S] Desoldering braid

[T] Dry tip cleaner or sponge

[U] Isopropyl alcohol

[V] Stereo zoom microscope, 30x

[W] Hot air station

[NOT SHOWN]

Soldering tip

X-Acto knife

Mylar stencil

Small squeegee

TYPICAL SMD PACKAGES

[a] QPF208
[b] QPF44
[c] PLCC
[d] SOIC
[e] Electrolytic capacitor
[f] SOT23
[g] QFN
[h] Tantalum capacitor
[i] 805 resistor
[j] 603 resistor
[k] 402 resistor

IRONS, HOT AIR, AND TOASTER OVENS

We'll look at 3 methods of SMD soldering. The easiest components have feet or other accessible contacts that lay flat on the board's pads. These you can connect with a soldering iron. A quick touch of the tip, and a bit of solder will naturally flow under the foot and make the connection. This is the magic of SMD soldering — capillary action does most of the work for you.

Other SMD packages have their contacts on the underside, out of reach. You can solder these in 2 ways: individually, using solder or solder paste and a jet of hot air, or en masse by positioning all components on the board with solder paste between each contact and its pad, and then heating the board on a skillet or in a toaster oven to "reflow" the board (melt the paste) and make all the connections.

BASIC SMD SOLDERING

Each method has its own tools and supplies. Here are the ones you'll need for iron-soldering the simplest SMDs: resistors, capacitors, and IC (integrated circuit) packages with leads.

» Fine-tipped industrial **tweezers** let you pick up and align small components. Also helpful are **hemostats**, **dental picks** (for fixing bent leads), and an **X-Acto knife**.

» **Flux** is the secret sauce in surface-mount soldering. It removes oxides from the connections so that solder can bond to them, and also helps to distribute heat. During normal through-hole soldering, you heat the joint with an iron and then melt solder wire against it, which lets the flux in the solder's core melt out and clean the joint. With surface-mount soldering, solder is often melted *on the iron and then transferred to the joint* — a mortal sin in regular soldering. The flux tends to boil off during this transfer, so you need to add more to the connection directly. Flux comes in 3 types of container: felt pen, brush bottle, and needle bottle.

» You can solder all but the most finely pitched components using a **lighted magnifying glass**, and you can use a $10 **loupe** with 10x magnification for the finest. If you expect to do a lot of SMD work, get a **stereo zoom microscope** with up to 30x magnification (try eBay).

» I recommend getting a temperature-controlled **soldering station**, at least 50 watts, which will probably cost $50–$120. A cheap 15W iron will work on some things, but will be slower and more frustrating. A good soldering iron is especially important if you're using lead-free solder, which requires higher heat.

» Soldering stations include a sponge, but a **dry tip cleaner** lets you clean a soldering tip without lowering its temperature.

» **Soldering tip** selection is a matter of personal preference. I prefer a small ⅟₃₂" (0.8mm) chisel or screwdriver tip because it can hold a bit of solder at its end. I don't recommend tips smaller than 0.6mm, as solder tends to draw away from the point. Bevel/spade/hoof tips are designed to hold a small ball of solder at the end, which is useful for the drag-soldering technique explained later.

» Use 0.02" or 0.015" diameter, flux-cored **solder**. To get the hang of SMT, I'd recommend starting off with lead-based solder, which is slightly easier to work with.

» **Desoldering braid** or wick is a fine mesh of copper strands that you can use to remove excess solder.

» For removing SMDs without a hot air station and myriad special nozzles, use the Chip Quik **SMD removal kit** (item #SMD1, $16 from chipquik.com). The kit contains a low-melting-point metal that when mixed with existing solder causes it to remain molten for a couple of seconds — long enough to flick off the component.

» A small vise such as a **PanaVise**.

Install a 1206 Resistor

Now we're ready to install a surface-mount resistor. Note that the resistive element in an SMT resistor is exposed and colored, and it should face upward to dissipate heat. The number 1206 means that the package measures 0.12"×0.06". A 603 package is 0.06"×0.03", and so on. Let's get started.

1. Add flux to the pads (Figure A). This may not be necessary for 1206s, but is helpful for 603s and 402s, where melting solder wire directly on the connection will likely deposit too much. A lightly tinned tip may provide all the solder necessary. As a rule, if you're melting solder wire directly onto a connection, you don't need additional flux, but if you're carrying solder to the joint with an iron, you do.

2. Add a small amount of solder to 1 of the 2 pads (Figure B).

3. Use tweezers to hold the 1206 in place while touching the junction between chip and pad with the iron. You should feel the chip drop into place as the solder liquefies underneath (Figure C).

4. Solder the other side by holding the iron so it touches the chip and board and adding a small amount of solder (Figure D).

Install a QFP (Quad Flat Package)
QFPs are square IC packages with leads all around. The distance between the leads, called the *pitch*, is typically 0.5mm or 0.8mm, but some are 0.4mm.

1. Flux the pads (Figure E).

2. Align the QFP over its pads with tweezers or dental picks (Figure F).

3. Add a small drop of solder to the tip of the iron. This part is key: you want a small drop to hang off the end (Figure G).

4. Tack 1 corner by sliding the tinned tip up against the toe of the lead (Figure H). The solder should quickly wick under the lead. Check alignment and tack an opposite corner. Sometimes I add more flux on top of the leads after tacking.

5. Continue touching the toes of the leads with the iron to complete the chip. You should be able to solder several leads with 1 load of solder on the tip. With practice, you can slowly drag the tip over the feet and "drag-solder" an entire row with 1 pass (Figure I).

6. Use the loupe to check for bridges and sufficient solder (Figures J and K).

7. Remove any shorted or bridged connections by touching the leads with a clean iron tip or applying solder wick (Figure L).

Alternately, there's the "flood and wick" method, which involves flooding all the leads with solder and then removing the bridges with wick. Surface tension holds some solder under the leads even after wicking. I hate to argue against something that works, but folks in the industry don't recommend this technique because it can overheat the board or component, and the wick might detach pads.

Install a PLCC (Plastic Leaded Chip Carrier)
PLCCs have legs that fold back under the package rather than sticking outward. The steps are similar to soldering a QFP: flux the pads (Figure M), align the part, tack some corners, flux some more, and solder. Keep the iron in contact long enough for

the solder to wick around the back of each pin. I like to lay a length of 0.02" solder along the pins and then press it into each pin with the iron (Figure N).

SOLDERING NO-LEAD SMDs

The following tools let you handle IC packages without leads, like QFNs (quad flat no-lead) and BGAs (ball grid array), that defy soldering with an iron.

» You can buy a **hot air station** with temperature- and flow-controlled air for under $300 from Madell (Figure O, background; madelltech.com). Instructables.com also has a wonderful array of DIY hot-air machines. If you're feeling less adventurous, a $25 arts and crafts **embossing heat tool** (Figure O, foreground) also gets the job done. Avoid ordinary heat guns; their nozzles are too big and they're too hot for SMD work.

» **Solder paste** consists of tiny solder balls floating in flux gel. It comes in 2 forms: in syringes, for applying to contacts individually, or in jars, for applying en masse with a **mylar stencil** and **squeegee** (see sidebar). Some distributors require fast shipping on solder paste, since its lifespan decreases without refrigeration.

» A **hot plate** can preheat the board to 212°F–250°F in order to limit the time and energy required when applying solder or hot air. This is optional, but it mimics the large-scale manufacturing process and reduces the risk of damaging boards or components. Preheating is especially helpful if you're using lead-free solder or if the board contains large,

SOLDER PASTE TYPES

Solder paste comes in either syringes or jars. With a syringe, you should apply small, Hershey's Kiss-shaped drops to individual pads on the PCB, and thin lines on packages with rows of pins. I like a 22-gauge needle. In the oven, the paste will wick to the connections and avoid bridging (for the most part). Don't bother trying to put paste on every little contact individually, because it will slump (spread out) anyway when it heats. You can buy solder paste syringes from Chip Quik, Zephyrtronics, smtsolderpaste.com, and others.

Paste in jars retains its form, and you can quickly apply it to all the pads on a board using a squeegee and a laser-cut mylar stencil. Getting the right amount of paste — between having too little solder and bridging leads — takes some trial and error. For stencil material, try stencilsunlimited.com.

Both types of paste come in either "no-clean" or water-soluble formulas. With water-soluble paste, the flux residues are corrosive and must be removed.

heat-absorbing ground planes. Preheaters are also available from Madell or Zephyrtronics (zeph.com), but a $7 Mr. Coffee hot plate works for small, single-sided boards.

» You can reflow a board in a **toaster oven**. Look for one that can heat up to 480°F (250°C) in less than 5 minutes, which will let it reflow all the solder without baking the board. Since toaster ovens don't have their 0-to-480°F speed marked on the outside of the box, I'd advise using a small one, or a large one that's more than 1,400 watts.

As an alternative, sparkfun.com has tutorials and blog entries that recommend using a skillet instead of a toaster oven for boards that carry both plastic and large metal connectors. The downside of a skillet is that it only works with 1-sided boards.

» Stencilsunlimited.com sells **temperature-indicating markers** that change color when a particular temperature is reached, to let you know when to stop applying heat. You can also monitor temperature with a **thermocouple**.

» An **acid brush**, isopropyl alcohol, and **lint-free wipes** clean up flux residues. I keep the alcohol in a pump bottle that dispenses as needed and prevents the rest from evaporating.

» A **vacuum pickup tool** can help place larger components that tweezers can't hold, although fingers do a decent job, too.

Install a QFN (Quad Flat No-Lead)

The recommended method with these chips is to use a stencil with solder paste, but you can also get by with regular solder and hot air.

You needn't apply solder to a chip's bottom-side heat sink, which is present on many motor amps and voltage regulators, but if you do, it shouldn't exceed 0.01" in thickness.

Also, you'll probably need to reflow it individually with a direct shot of hot air or solder it through a hole drilled underneath.

1. Flux and tin the bottom connections on the QFN (Figure P).

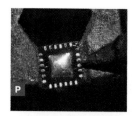

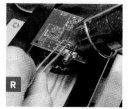

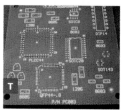

SMD PROTOTYPING

Prototyping with SMDs is more difficult than quickly plugging through-hole components into a solderless breadboard, but SchmartBoard (schmartboard.com) carries breakout boards that port any SMD to standard 0.1"-spaced through-hole pins.

For prototyping, you still have to solder the chip onto the breakout board and then remove it later to install on the final board (unless you just solder in the breakout board, which takes up space). But the breakout boards are perfect if you have a limited number of SMDs that you need to interface with through-hole components, and you aren't making your own PCB.

In my experience, it's faster to skip the breadboarding stage and go straight to a PCB prototype of the whole circuit. You can fix mistakes by scraping traces and jumpering with small, 30-gauge "green" wires. I've found that drawing schematics on a computer is more reliable than dealing with a million breadboard wires, although it's less immediate.

2. Flux and tin just the outer pads (Figure Q).

3. I recommend preheating, especially if you're soldering the heat sink.

4. Apply hot air about ¾" away in a circling motion until you feel the chip drop. Surface tension from the molten solder should pull the chip into alignment.

You can also nudge the chip with tweezers to make sure it's correctly seated; it should spring back into position (Figure R).

5. Check the sides with a loupe to make sure the markers line up with the pads (Figure S).

SOLDER A DOUBLE-SIDED BOARD

If the board has components on both sides, you need to use a toaster oven rather than a skillet.

1. Apply solder paste, using either a syringe or a stencil and squeegee (see "Solder Paste Types"), to whichever side of the board has lighter components (Figure T; PLCCs are the heaviest).

2. Place the components using tweezers, fingers, or a vacuum tool. It's alright if the smaller components aren't perfectly aligned; they'll snap into place during reflow (Figure U).

3. Reflow the board in the toaster oven. I use binder clips to suspend it above the rack (Figure V). Paste and component manufacturers recommended a precise 3-phase sequence:

3a. Preheat and evaporate solvents in the paste at 300°F (150°C).

3b. "Soak" between 300°F and 350°F (150°C–180°C) for 1–2 minutes to let the flux remove the oxides.

3c. Run up to about 425°F (220°C) for 1–1½ minutes to melt the solder.

What I do is simply turn the oven on max, wait for all the solder to melt, then count to 15 and open the door.

More complex boards and BGAs might require greater precision. A thermocouple or temperature-indicating marker lets you see when you've reached your target temperature.

For more control, sites like articulationllc.com sell controllers that plug into toaster ovens and let you program and run time-temperature sequences, although most toasters don't heat up fast enough to give a controller much to work with.

4. After the first side is cooled, apply solder paste, place components, and cook the other side. Surface tension will hold the lighter bottom components in place.

My results with the project photographed here were about 25 bridged connections on the 208-pin QFP with 0.5mm pitch, and a couple here and there on the other packages, but the majority turned out OK.

Scott Driscoll (scott@curiousinventor.com) is an IPC-certified soldering specialist, has master's degrees in mechanical engineering and music technology from Georgia Tech. He researches and writes how-to guides at curiousinventor.com.

DESKTOP DIGITAL GEIGER COUNTER

By John Iovine

A Geiger counter is a radiation detection instrument. Since the nuclear disaster in Fukushima Japan in 2011, radiation hazards and radiation detection instruments like Geiger counters were brought to the public consciousness. The news media coverage of the disaster and its subsequent safety hazards created an increased demand for Geiger counters that is still felt today.

This article will show you how to build a laboratory-quality Desktop Digital Geiger Counter.

John Iovine is a scientist and electronics tinkerer and author who owns and operates Images SI Inc. (imagesco.com), a small science company.

LOOKING FOR THE TRAIL

Radiation is emitted by the decay of unstable atoms. Geiger-Müller (GM) tubes detect 3 forms of this energy:

α **ALPHA RADIATION** Free-traveling helium nuclei, can penetrate a few sheets of paper.

β **BETA RADIATION** Free-traveling electrons or positrons, can penetrate 3mm of aluminum.

γ **GAMMA RADIATION** High-energy photons, can penetrate several centimeters of lead.

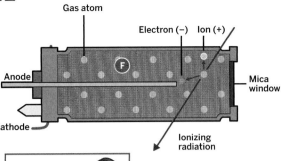

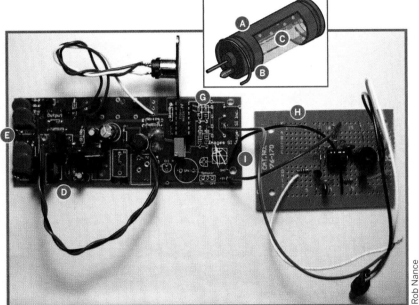

Rob Nance

(A) The **wall** of the GM tube is a cylinder of thin metal or conductor-coated glass. With metal cylinders, one end is mica, which (unlike metal) lets some alpha radiation through.

(B) The tube is sealed and filled with **halogen and noble gases**.

(C) An **electrode** runs through the middle of the tube.

(D) The **power supply** takes power from a 9V battery or 12V DC power plug.

(E) A **high-voltage power supply subcircuit** uses transistorized feedback with a step-up transformer to create a "ringing choke converter" that converts a 6V, 14.1kHz oscillation induced in the primary coil of the transformer into 325V across the secondary winding. A voltage multipli-er boosts this voltage in steps up to about 600V DC. Three Zener diodes let you configure the output voltage to between 300V and 600V DC, to match the GM tube used.

(F) The high voltage is applied across the tube's wall and electrode, turning them into a cathode and anode, respectively. In the tube's normal resting state, the resistance between its anode and cathode is very high.

When a **radioactive particle** passes through the tube, it ionizes gas molecules in its path, which momentarily creates a **conductive trail** through the gas (like a vapor trail in a cloud chamber). This lowers the tube's resistance, in a momentary pulse.

(G) In the **detector subcircuit**, the tube's cathode directs the momentary pulse to the input of a comparator, alongside a reference voltage tunable by a potentiometer. The comparator cleans up each pulse by outputting a digital 1 while the tube output exceeds the reference voltage and a 0 when it's lower. This first comparator down-streams its output in parallel to the 3 other comparators on the same quad chip, which simply act as buffers.

(H) One downstream comparator feeds to the base of a transistor, which powers the **light and sound output** — an LED flash and speaker click — when each particle is detected.

(I) The other 2 downstream compar-ators connect to an **output connector** for output to other devices.

After each detection, the high-voltage subcircuit takes some time to recharge the tube. During this "dead time" (which varies with differ-ent tubes; 90µs for an LND712), the instrument cannot make detections.

SET UP.

This Desktop Digital Geiger Counter contains three separate PCBs. Each board has its own schematic. Care must be taken when building the individual boards, as each board may share common component names, such as R7. So be sure, when you choose a component, which PC circuit board you are building.

The main circuit is the Geiger counter shown on the left in the "Looking for the Trail" image on the previous page. The HV section of the circuit is built around a single transistorized (TIP3055) ringing choke converter that has been around for years. This particular derivative of the circuit was designed by Sam Evans to minimize current draw. The important element in this circuit is the transformer.

In this configuration, the primary winding of the

MATERIALS

All the PC boards, wands, PDF files, and other miscellaneous components used to build the Desktop Digital Geiger Counter are available on the Images SI, Inc., website. Part numbers are included where appropriate. For more information, visit imagesco.com/geiger/desktop-geiger-counter-kit.html.

Geiger Counter Circuit (#GCK-02)

» PCB printed circuit board (1): (#PCB-52)	
» C5 C6 C7 C8	0.01µF, 1KV capacitors (4)
» C1	0.1µF, 100V capacitor (1)
» C2	330µF, 16V capacitor (1)
» C3 C4	10µF, 16V capacitors (2)
» D1–D6	1N4007 diodes, 1000V (6)
» D7	5.1V Zener diode (1): (#1N751)
» D8	100V Zener diode (1): (#1N5271B)
» D9 D10	200V Zener diodes (2): (#1N5281B)
» Q1	TIP 3055 regulator (1): (#ALT.H1061)
» R2	33Ω resistor (1)
» R3 R7	1 MΩ resistors (2)
» R4	470KΩ resistor (1)
» R6	1 MΩ, multi-turn potentiometer (1)
» R8 R9 R10 R11 R14	10KΩ resistors (5)
» R12	100Ω resistor (1)
» R13	100KΩ resistor (1)
» T1	Mini step-up transformer (1): (#HVT-06)
» U1	7805 voltage regulator (1)
» U2	LE33 (1)
» U3	Bridge rectifier (1): (#W01M)
» U4	LM339N (1)
»	14-pin IC socket (1)
» J1	2-pin header (1)

Digital Meter Adapter (#DMAD-04)

»	Printed circuit board (1): (#PCB-74)
» C1	0.1µF, 50V capacitor (1)
» C2 C3	10µF, 16V capacitors (2)
» D1	SMH-16 (1)
» D2	1N4007 diode, 1000V (1)
» P1 P2	SMH-02 (1)
» R1	33Ω ¼W resistor (1)
» R4 R5 R6	10K ¼W resistor (3)
»	LCD display (1): (#LCD-01-16×2)
» U1	Preprogrammed PIC16F88 (1)
»	18-pin IC socket (1)
» U2	Voltage regulator (1): (#LDO-5V)

Soundboard

- » Prototyping board (1)
- » 10K ¼W resistor (1)
- » 220K ¼W resistor (1)
- » 330Ω ¼W resistor (1)
- » 1K ¼W resistor (1)
- » 220Ω ¼W resistor (1)
- » 0.1µF, 50V capacitors (2)
- » 2N3904 transistors (2)
- » Piezoelectric speaker (1): (#SPK-05)
- » LM555 timer (1)
- » 8-pin IC socket (1)

Other Components

- » GCW-01 Geiger counter wand (1)
- » SPST 2-position toggle switches (4): (#SW-10)
- » DPDT 3-position, center off toggle switch (1): (#SW-18)
- » Red panel mount 5mm T1 LED (1)
- » Green panel mount 5mm T1 LED (1)
- » 20K panel mount potentiometer and knob (1)
- » Panel mount TTL-stereo jack (1)
- » Panel mount mono jack (1)
- » Panel mount power jack (1): (#Jack-18)
- » Panel mount 3-pin mini DIN socket (1)
- » 22 gauge wire, various colors

Hardware

- » Bud IP-6130 enclosure with stand (1)
- » 4-40 x 1" nylon screws (8) (for mounting PCBs)
- » 4-40 nylon hex nuts (16)
- » #4 x 9/16" nylon spacers (8)
- » 2-56 x 1" nylon screws (4) (for mounting LCD)
- » 2-56 nylon hex nuts (4)
- » 2-56 × ½" black nylon screws (2) (for mounting mini DIN connector)
- » 2-56 black nylon hex nuts (2)
- » 9VDC wall transformer (1)

Tools

- » Soldering equipment
- » Multimeter

transformer and the feedback winding of the transformer are arranged so that the circuit begins a sustaining oscillation when power is applied. If you checked the output of the oscillator you would find the waveform's duty cycle is symmetrical.

The high voltage output from this stage is regulated by Zener diodes. In the latest revision of the PCB, there is a jumper J1 that shorts out the D8 Zener. You would only use this jumper to power a GM tube that requires 400 VDC.

We are using a 500 VDC output because our wand contains a LND712 GM tube. The cathode of the tube is connected to a 100K resistor. Each time a particle is detected, a voltage pulse is generated across this resistance. The reason I didn't use the speaker on this board is I didn't like the volume of the sound. It was too faint. So I created a soundboard. The difference is that the soundboard stretches the pulse, creating a better click. The soundboard follows the construction of the GCK-02 board.

MAKE IT.

BUILD YOUR GEIGER COUNTER

Time: A Weekend

Complexity: Moderate

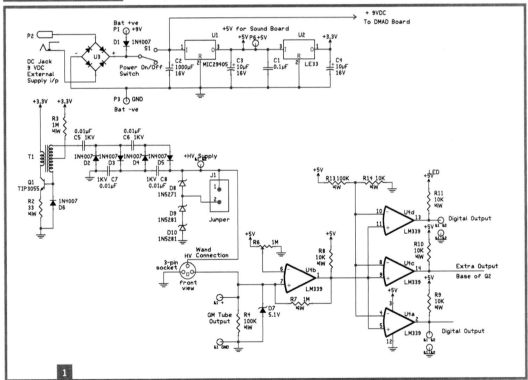

1. Assemble the PCB

You may hardwire this circuit to a breadboard or use the available PCB. Although you do not need the PCB, it will make construction easier (see Figures **1** and **2**).

When using the PCB, it's merely a matter of mounting and soldering the components in their proper position. All the parts are outlined on the top

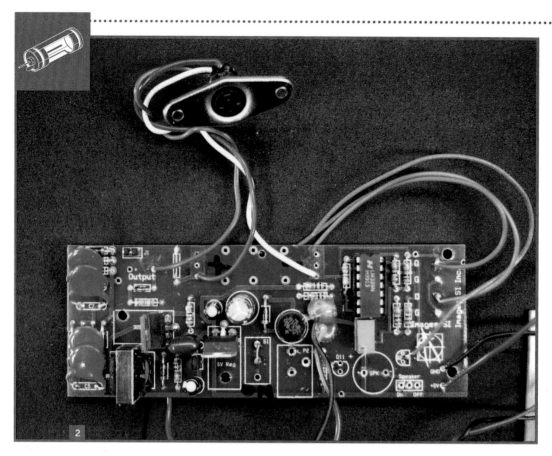

2

silk screen (Figure **2**). For our application, however, we can leave quite a few components off the board.

Begin by mounting and soldering all the resistors:

» R2 — 33 ohm (color code — orange, orange, black, gold)

» R3 and R7 — 1 mega-ohm (color code — brown, black, green, gold)

» R4 — 470 K-ohm (color code — yellow, purple, yellow, gold)

» R5 — Jump with a small length of wire (in this unit a 10 Mega ohm resistor is housed inside the wand)

» R8, R9, R10, R11 and R14 — 10K-ohm (color code — brown, black, orange, gold)

» R12 — 100 ohms (color code — brown, black, brown, gold)

» R13 — 100 K-ohm (color code — brown, black, yellow, gold)

Next, mount and solder diodes D1 through to D10. Make sure to orient the line on the diodes with the line on the silkscreen diode drawing on the printed circuit board.

» D1, D2, D3, D4, D5, and D6 — 1N4007

» D7 — 5.1V Zener

» D8 — 100V Zener

» D9 and D10 — 200V Zener

Install and solder the regulator 7805 and the 3055 (or H1061) transistor. Mount and solder the capacitors:

» C1 — 0.1µF-50V

» C2 — 330µF-16V

» C3 and C4 — 10µF-16V

» C5, C6, C7 and C8 — 0.01µF — 1KV

Solder a small piece of wire connecting the top two holes of S1 as shown in Figure **2**. Mount and solder the mini-step-up transformer T1. Then mount and solder the 14-pin IC socket for the LM339, lining up the notch on the IC socket with the silkscreen outline on the PCB.

Mount and solder U2, making sure the profile matches the silk-screened outline of the part

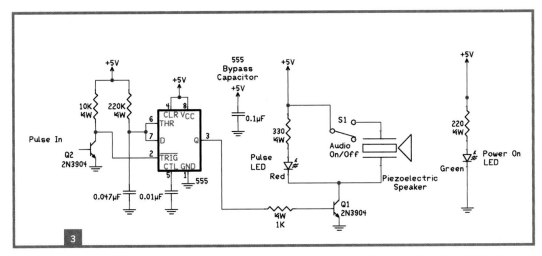

before soldering. Next, mount and solder the bridge rectifier U3. Align the positive terminal of the bridge rectifier with the + pad on the PCB. Mount and solder the 1-Mega-ohm potentiometer in the R6 position on the PCB.

Finish the board by mounting and soldering the two-pin header in the J1 position. J1 is used to vary the voltage to GM tube. By shorting J1, diode D8 is shorted out and the voltage to the GM tube drops from 500VDC to 400VDC.

2. Connect the Boards

Next, we must add wires to connect this PCB to the other boards used in the assembly, as well as to panel-mounted components. To simplify the construction, we have color-coordinated the majority of the wires. Where power is involved, a red wire is used for + and black for ground, in most instances. As a standard, eight-inch pieces of wire are used and may be trimmed down as needed.

Solder a red wire to P1, and black wire to P3 for power. Also, solder a red wire on the lower right-hand side of the PCB, in the pad marked +5V,

and a black wire in the pad marked GND. These wires will connect power to the soundboard.

Solder a purple wire to the third hole down on the right side of the PCB marked with the letter D. This will connect to the digital output that will go to the input of the soundboard. Solder another wire to the other "D" output that will connect to the Pulse Input of the DMAD.

Next solder three wires to the 3-pin mini DIN socket: in the photograph a dark green wire to the — output, a red wire to the + pad, and a white wire to the — pad on the top part of the PCB. Connect these wires to the appropriate connections of the panel-mount 3-pin mini DIN socket as shown in the schematic.

Solder a wire into the right-most hole of Q2 (this is the base of Q2, not shown in picture). This is a digital output that will connect to the 1/8" jack for the digital pulse output on the front panel.

3. Make the Soundboard

Figure 3 shows the schematic, while Figure 4 shows the soundboard constructed using point-to-point wiring on a prototyping bread-

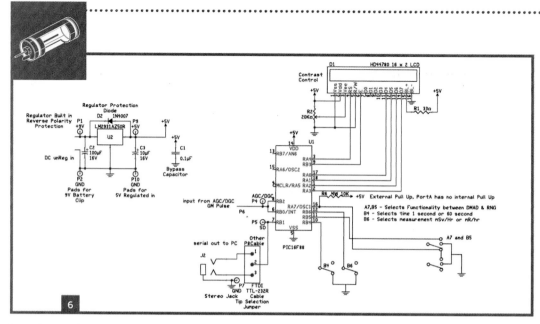

board as outlined in the schematic.

The Pulse In connection is fed from the purple wire of the Geiger counter circuit to the soundboard. The +5V power and ground connection for the soundboard come from the wired connections on the lower right-hand corner of the Geiger counter circuit. Figure **5** shows the GCK and soundboard connected.

Brown wires to the soundboard are for the Audio on/off switch, S1. Connect these wires to an SPST toggle switch (not shown in the photograph).

4. Make the Digital Display Board

The schematic for the Digital Meter Adapter (DMAD) is shown in Figure **6**.

Begin by mounting and soldering the 16-pin header on the bottom of the board (the side with *no* silk screen). Now, on the top silk screen side, mount and solder the resistors:

» R1 — 33 ohm (color code — orange, orange, black)

» R4, R5, and R6 — 10K ohm (color code — brown, black, orange)

Solder a small piece of wire to connect the center and right holes of S1. Repeat for S2. This jumps these switches in the on position.

Next, mount the 1N4007 diode, D2, making sure to align the stripe on the diode with the line on the silk screen. U2 is the regulator. When mounting the regulator, be sure the flat side is oriented with the flat side of the silk screen.

Mount and solder a 2-pin header at P1 and P2. A wired cable will attach here to provide

power to the entire unit; P1 is +V and P2 is ground. Mount the 3-pin header marked P8.

Now, mount C1, the 0.1uf capacitor, and the two 10uf 16V capacitors marked C2 and C3. When mounting these capacitors be sure the longer lead is oriented to the hole marked positive.

Next, mount and solder 18-pin socket U1 to the PCB making sure that it is oriented according to the outline on the silk screen. Insert the prepro- grammed 16F88 microcontroller matching the notch on the chip to the notch on the socket.

Connect the following wires to the PCB before attaching the LCD module to the PCB. Connect 3 wires to the holes at R2, shown as 3 yellow wires in the photograph. These will connect to a pan- el-mounted potentiometer to allow you to adjust the contrast of the LCD screen.

Connect three wires, shown as dark green wires in the photograph (Figure **7**) to the larg- est of the pads at J2. The center pad is ground. These wires connect to the stereo Serial Digital TTL output socket on the front panel.

Connect a wire, shown as a gray wire, in Figure **7** to the right-most hole at J1. (You may choose to connect this wire at P4 instead.) This wire will connect to the "D" digital output of the main Geiger counter circuit. Connect a black power ground wire to the center hole at P3, and a red wire for the +9VDC to the right-most hole. These will connect to the power supply.

Connect eight wires, shown as white wires to each of the holes at S5. These will connect to the three switches labeled B4, B6, and A7, B5 switches in the schematic and provide several different functions of the unit.

Finally, mount and solder the LCD module to the 16-pin header.

5. Attach Switches to the DMAD Board

Begin by wiring switches to the white wires connected to S5 of the digital display board. (Refer to Figure **7**.) All of the bottom wires are connected to ground.

Connect the top and bottom wires of B6 to an SPST toggle switch. This switch is used to select between mR/hr and mSv/hr.

Connect the top and bottom wires of B4 to an SPST toggle switch. This switch is used to change between one-second and sixty-second time selections (CPS and CPM).

The wires from A7 and B5 are connected to a DPDT toggle switch with center off for Random Number Selection and range. Connect the top wire at A7 to the left-side top and bottom prongs of the DPDT switch. The top wire of B5

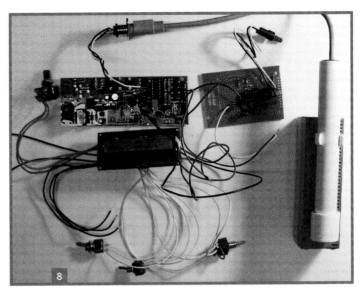

is connected to the top right prong of the DPDT switch. Both bottom wires for A7 and B5 are connected to the appropriate center prong of the DPDT switch, see Figure **7** for switch details.

Next, connect the yellow wires attached to R2 of the digital display to the appropriate connections of a 10K panel-mount potentiometer.

Power Connections

Pulse In

Yellow Wires to external potentiometer

To Stereo Jack for Serial Out to USB Cable
Set output with jumper on P8
Alternative:
Use pad P7 for Ground
Use pad P5 for Serial Our
(Wires Not Shown)

White Wires to Switched
B6 Wires to mR/hr mSv/hr selection
B4 Wires to 1Sec or 60Sec time selection
A7 and B5 to DPDT switch with center off for Random Number Selection & Range

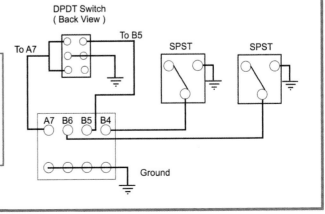

DPDT Switch
(Back View)

To A7

To B5

SPST

SPST

A7 B6 B5 B4

Ground

⚠ The author and publisher do not make any warranties (express or implied) about the radiation information provided here for your use. All information provided should be considered experimental. Safety and health issues and concerns involving radioactive contamination should be addressed, confirmed, and verified with local and national government organizations or recognized experts in this field.

7

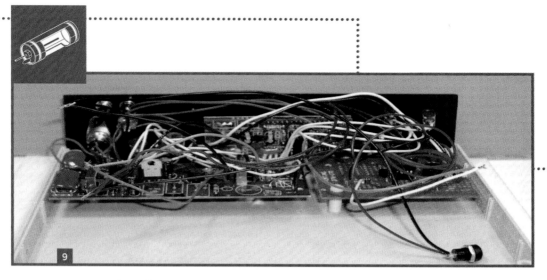

9

NOTE: You have two sets of power leads into the DMAD board. The two leads you just soldered at P3, and the header you soldered on P1 and P2. You can feed the main power into either one of these power leads. The power leads you do not use are then used to feed power to the GCK-02 Geiger counter board.

6. Calibrate Your Device

After you have everything wired and before mounting anything into the case, power up the unit and make sure everything is functional. See Figure **8** on the previous page.

Now is the proper time to calibrate your Geiger counter!

To calibrate your Geiger counter, you need a radioactive source. See "Radioactive Sources" on page 84.

Set the digital meter to CPS. You will need to adjust potentiometer R6 on the Geiger counter board. Turn on the Geiger counter. If you have a radiation source, bring it very close to the front of the wand — but BE CAREFUL! The front of the GM tube inside the wand has a thin mica window that is easily damaged. Make sure you do not touch the mica window. Adjust R6 to obtain the highest CPS reading on the digital display. If the clicking sound is annoying, use the audio switch to mute the sound.

You can use a simple procedure to get a ballpark calibration for the Geiger counter. The difficulty in calibrating the digital meter has much to do with the variables in play. The tube's response can vary +/- 20 %. The strength of the radioactive source can also vary in addition to variances in the electronic components. All these factors affect accuracy. With this being said, you can still proceed to get that ballpark approximation for the digital display meter.

This calibration procedure uses a 10 uCi Cs-137 source. The source is held at specific distances from the front of the GM tube in the wand without any shield.

Distance from front of GM tube:

> Approximately 1 mR/hr = 6"
> Approximately 10 mR/hr = 2.5"
> Approximately 100 mR/hr = 0.5"

7. Mount the Unit Inside the Case

Once the circuit has been tested, you can begin mounting the unit inside the case. Begin by lining up the boards inside your case the way you want them orientated, and then mark and drill the mounting holes. The holes I used for attaching the PCBs to the bottom of the case are $1/8$."

Insert (8) 4-40 x 1" Phillips-head nylon screws into the mounting holes on the bottom of the case and secure with nylon hex nuts. Four $9/16$" nylon spacers are used to elevate the Geiger circuit and soundboard inside the case.

Secure the PCBs to the case with hex nuts. Figure **9** shows the end result.

The front panel hole guide, Figure **10**, is provided as a drilling template. This assumes you are using the BUD enclosure listed in the parts list.

The drill sizes for the holes are detailed in the diagram. The decal used on the front Panel is

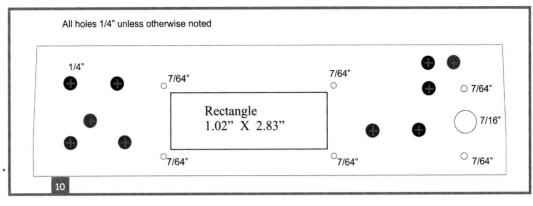

All holes 1/4" unless otherwise noted

1/4"

7/64"

Rectangle
1.02" X 2.83"

7/64"

7/64"

7/64"

7/16"

7/64"

7/64"

10

shown in the Figure **11** on the following page.

Full-size PDF files for both the holes and the decal are available for free download at the Images SI Inc. website (imagesco.com).

When assembling this unit into the case, some of the components will need to be mounted on the face panel before being wired; others, like the switches, can be wired and then mounted to the face panel.

Mount the TTL serial and digital output sockets into the appropriate holes in the front panel. Connect the three wires, shown as dark green wires in Figure **7** from the digital display to the appropriate connections on the TTL serial socket. The center wire is ground.

Connect the wire from the base of Q2 on the Geiger counter circuit to the + lead of the digital output socket. Use a small piece of wire to connect the ground of the two sockets.

Insert the green power LED and red pulse LED into the appropriate holes on the front panel. Connect the red and green LEDs, using

wires, to the appropriate connections as diagrammed in the soundboard schematic.

Power is fed to the completed unit from a 9VDC wall adapter. It is fed through the panel-mounted SPST toggle power switch. After the switch, the +9V is fed to the DMAD power input leads. The secondary power leads connect to and power the GCK-02 Geiger Counter board. The power supply is in parallel powering both the Geiger counter circuit and digital display. The soundboard receives +5V from the Geiger counter circuit.

The power jack must be attached to the back panel before wiring. Attach a red wire to the positive lead of the power jack. Connect the opposite end of the wire to the center prong of an SPST toggle switch. The other pole of the switch is then connected to P1 of the DMAD. The positive lead of P3 on the DMAD connects to the positive power input on the Geiger counter circuit labeled P1. A black wire is connected to the ground of the power jack is connected to

How Much Radiation Is Safe?

In the United States the U.S. Nuclear Regulatory Commission (NRC) determines what radiation exposure level is considered safe. Occupational exposure for workers is limited to 5000 mrem per year. For the general population, the exposure is 500 mrem above background radiation in any one year. For long term, multi-year exposure, 100 mrem above background radiation is the limit set per year.

Let's extrapolate the 100 mrem number to an hourly radiation exposure rate: 365 days/yr x 24 hr/day equals 8760 hours. Divide 100 mrem by 8760 hours, and it equals .0114 mrem/hr or 11.4/hr microrem. This is an extremely low radiation level. The background radiation in my lab hovers around 32 uR/hr. Am I in trouble? No. Typically background radiation in the United States averages 300 mrem/yr, or 34 microrem/hr. The NRC specification is for radiation above this 34 urem/hr background radiation.

Notice that my lab readings are in microrads (uR/hr) and the exposure limit is given in microrems (urem/hr). I do not know what type of radiation (a , b or y) the Geiger counter is reading in my lab at any particular instant, so I do not know the Q factor of the radiation and therefore can not calculate the mrem. For general purposes, I consider them the one and the same. The digital Geiger counters I use are calibrated using a Cs-137 radioactive source. Therefore, the highest accuracy in reading radiation levels will be from Cs-137 sources. See makezine.com/go/bom2 for more information.

Exposure	Source Dose (conventional)	Dose (SI)
Flight from LA to NY	1.5 mrem	.015 mSv
Dental X-ray	9 mrem	.09 mSv
Chest X-ray	10 mrem	0.1 mSv
Mammogram	70 mrem	0.7 mSv
Background radiation	620 mrem/year	6.2 mSv/year

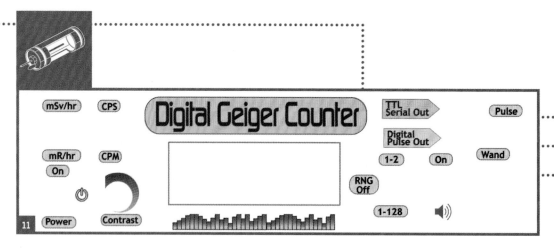

the P2 ground on the DMAD. The ground connection of P3 on the DMAD board is connected to the ground of Geiger counter circuit board (P3). (See diagram.)

Secure the LCD display to the front panel of the desktop unit with (4) 2-56 x 1" nylon screws and hex nuts. Attach the mini DIN connector to the front panel with (2) 2-56 x ½" nylon screws and hex nuts. For the face of the unit, black nylon screws are preferred but not required.

Now insert your switches into the appropriate holes and check that they are orientated correctly before securing in place.

The front and back panels of the unit will slide into the groves on the case. Slide on the top of the case, and insert screws into holes in the bottom of the case to secure.

You can now plug in your Geiger counter wand and connect the Desktop Digital Geiger Counter to a power source.

FINISH ☒

NOW GO USE IT »

Radioactive Sources

Uranium ore is available from a number of sources including eBay. You can find a source list on the imagesco.com website. Small amounts of radioactive materials are available for sale encased in 1-inch diameter by ¼" thick plastic disks. The disks are available to the general public, license-exempt. This material outputs radiation in the micro-curie range and has been deemed by the U.S. Federal government to be license-free and safe. The cesium-137 is a good gamma ray source and has a half-life of 30 years.

USE IT. » THE HIDDEN LANDSCAPE OF RADIOACTIVITY

Test for Radioactive Contamination

If you're testing for contamination, your GM tube should be sensitive to alpha radiation as well as beta and gamma. Geiger counters can only test for gross levels of contamination that show up clearly above background radiation; they are not the proper instruments for detecting low-level contamination. That said, here is how to test for radioactive levels above background:

1. Establish the background radiation level by measuring CPM for at least 20 minutes. Longer is better. Note the lowest and highest levels and then average them all to establish the baseline minimum, maximum, and average.

2. Position the GM tube very close to the top surface of the material you're testing, and run the counter, recording the CPM output. The longer the run, the more accurate the results.

3. Compare the radiation output of your sample against your baseline.

Microcontrollers
AND Microcomputers

PART 3 >>>>

Before the Arduino came along, programming microcontrollers and integrating them into electronics projects was a job most hobbyists left to professional engineers. The genius of the Arduino project was to mount a microprocessor onto a development board with built-in inputs and outputs and develop an easy programming environment for it. This development put programming microprocessors within the reach of hobbyists, artists, and young students. Within a few years, the Raspberry Pi put the full power of a computer with a Linux operating system on one of those little boards. The revolution continues, and *Make:* magazine has been at the forefront from the beginning.

Dan Rasmussen is among many other *Make:* authors who are fascinated with vintage electronics—he even has a website called Retro-Tronics—and also with the latest cutting-edge technology. Dan's "Million Color HSL Flashlight" project integrates the old and the new. It was inspired by his childhood love of flashlights that had colored lenses. But he updates that childhood toy by hacking a big 6-volt lantern with an Arduino and NeoPixel RBG LEDs to create a lightshow you can hold in your hand.

Jim Newell shows you how to use your Arduino to hack an X10 home automation module and expand its capabilities. Would you like your mailbox to alert you when the mail arrives? Why not have you outdoor lights turn on automatically when your car pulls into the driveway? The coding to get you there is easier than you might think, and—once you're in—the possibilities are, as they say, limited only by your imagination.

Getting an Arduino to produce audio that goes beyond beeps and buzzes is a bit of a mystery for many. So it's ironic that international man of mystery

Jon Thompson steps in to *de*mystify how it's done.

Musician and artist Peter Edwards introduces you to the fun of circuit bending and electronic music in "Hack Electronic Pushbuttons."

With Sam Freeman and Wynter Woods' "Raspberry Pirate Radio," we move from the Arduino to the Raspberry Pi single-board computer. It's a clever title, but the cleverness doesn't stop there. Sam and Wynter show you how to stick a wire into your Raspi to turn it into a transmitter, then modify the PiFM code to play back your music files continuously. All you have to do is tune your radio to the right channel, and you're receiving! They even drew up a 3D-printable case for your transmitter to keep it stylish and safe.

You may have seen Limor Fried's picture on the cover of *Wired* magazine. If you are into DIY electronics you're undoubtedly familiar with her company, Adafruit Industries, which is a critical go-to source of products and information for makers. Limor and Adafruit creative director Phillip Torrone team up to show you how to turn your Raspberry into an anonymous Tor proxy.

Million Color HSL

Written by Dan Rasmussen

Flashlight

Bring fun back to the flashlight with Arduino and full-color NeoPixel LEDs

DAN RASMUSSEN
is an avid collector, fixer, and hacker of vintage technology. He's a software engineer who lives in Groton, Massachusetts with his wife and three kids.

WHEN I WAS A KID THERE WERE NO SMART-PHONES, INTERNET, HOME COMPUTERS, OR ARDUINOS. Sure, we had TV and radio, but—believe it or not—we had fun with flashlights too. They often came with colored lenses, or we made our own with paper or plastic.

These days flashlights are mostly for practical purposes and they usually only make white light. Even high-end flashlights are pretty much the same thing,

just super bright and rugged for inspecting the levee during the hurricane. Bright and rugged is great, but a one-button, small, efficient, vanilla light isn't much fun, is it? Let's bring the fun back to flashlights.

THE HSL FLASHLIGHT

In this project you'll hack an old-style 6-volt lantern to become a Million Color HSL Flashlight. It's big and bulky

Materials

- » A big 6V lantern flashlight to hack old or new
- » Arduino Pro Mini 328 microcontroller board, 5V, 16MHz Maker Shed #MKSF8, makershed.com, or Adafruit #2378, adafruit.com
- » Right angle male headers, 6 pin cut them from Adafruit #1540 or similar.
- » NeoPixel RGB LED ring, Adafruit #1643
- » Potentiometers, 10kΩ (3)
- » Hookup wire, 22 gauge solid core Maker Shed #MKEE3 or Adafruit #1311. It's nice to have lots of colors for this project.
- » Batteries, NiMH recharge-able, AA size (4) and charger. Do not use alkaline batteries.
- » Battery holder, 4xAA
- » Resistor, 300Ω, ¼W
- » Capacitor, 1,000μF, 6.3V or higher
- » Rotary switch, 10-position such as SparkFun #13253
- » SparkFun Rotary Switch Breakout board (optional) SparkFun #13098
- » Potentiometer knobs (4) such as Adafruit #2046
- » Heat-shrink tubing various diameters
- » Tic Tacs, any flavor you like
- » Solder

Tools

- » Soldering iron, temperature controlled such as Maker Shed #MKME01, set to 700ºF
- » Wire strippers such as Adafruit #527 or, for fun, Maker Shed #MKLTM2-ES4
- » FTDI Serial TTL-232 USB cable such as Adafruit #70
- » Dental pick (optional) handy for attaching wires to posts
- » Hot glue gun
- » Computer with Arduino IDE software free download from arduino.cc/downloads
- » Project code Download the Arduino sketch *HSLFlashlight.ino* from the project page, makezine .com/projects/million-color-hsl-flashlight.
- » Hobby knife
- » Adhesive tape
- » Heat gun or butane lighter for heat-shrink tubing
- » Drill and drill bits
- » Digital voltmeter or multimeter
- » Sharpie marker

Time Required: 4-6 Hours

Cost: $40-$60

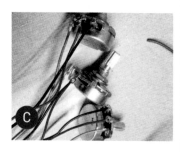

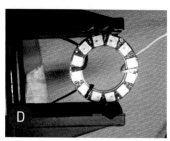

Dan Rasmussen

and different. Not quite as bright as some, but it has crazy color that's so easy to set—just dial your way around the rainbow. Lots of knobs and modes, and you can write code for it too.

The great thing about HSL is that it's an intuitive way to select a color. HSL stands for *hue*, *saturation*, and *lightness*. It's like the 360-degree color pickers available in many computer applications. Our HSL flashlight has a knob for each component: hue selects the color by allowing you to dial your way around the rainbow, saturation selects how deep and rich the colors are (fully desaturated is white, fully saturated is pure color), and lightness behaves like a dimmer.

The flashlight also has a 10-position mode switch, just for fun. The following are the modes I programmed—watch the video on the project page at makezine.com/projects/million-color-hsl-flashlight for a demonstration—but you can always program your own!

1. White
2. Manual HSL selection
3. Auto hue: rotates through the color spectrum (hue knob controls speed)
4. Multicolor all-pixel auto-rotate
5. Multicolor tri-pixel auto-rotate
6. Cylon (aka Larson Scanner)
7. Full-color strobe
8. Alternate pixel multicolor 180° color alternation
9. Half-moon multicolor 180° color alternation
10. Alternate pixel-pair multicolor 180° color alternation

HARDWARE

Arduino is a great platform for prototyping your projects. Once you're ready to build, though, the standard Arduino board is kind of big and its jumper wires provide only fragile connections. So I used the Arduino Pro Mini: inexpensive, very small, and reliable.

I chose Adafruit NeoPixel RGB LEDs because their 12-element ring fits nicely in the reflector bowl in most 6V lanterns, and they come with an Arduino library that's easy to use.

Old-school "6-volt lantern" flashlights are perfect for this project because the reflector bowl and the battery compartment are both huge. Cool old lanterns are easy to find at flea markets or eBay, but you can also buy new ones.

1. SOLDER HEADERS ON THE ARDUINO

Tack a single pin of the 6-pin header, then check to make sure it's flush. If not, reheat and adjust.

Now solder the other pins. Rework the tacked pin and add solder if necessary.

2. PROGRAM THE ARDUINO

Download the Arduino sketch HSLFlashlight.ino from the project page, makezine.com/projects/million-color-hsl-flashlight. Connect the Arduino Pro Mini to your computer using your FTDI cable or other compatible device. Open the Arduino IDE software and select the correct board from the Tools → Boards menu. Then open the sketch and verify/compile/upload it to the Arduino.

This tiny board has an LED attached to pin 13. The HSL program will flash it 5 times when it starts. This is a good way to verify that the program has been properly uploaded. (Note that the upload process itself will flash this LED a couple of times.)

Disconnect the Arduino and set it aside.

3. WIRE THE 10-POSITION ROTARY SWITCH

The optional breakout board makes it easier to attach and manage the switch's 11 wires. Tack one pin, make sure it's flush, then solder all the connections (Figure **A**).

Cut eleven 8" lengths of 22-gauge wire. Use different colors of wire, or use a light color and mark the wires with a Sharpie to indicate the pin number. Strip about ⅛" of insulation from each end.

I always use black for common/ground and red for raw power. Here I attached black to the common pin, then used different colors for the first few positions (Figure **B**). After that I used all gray, and marked the other end of the gray wires with dots (4 dots for position 4, 5 for position 5, and so on).

Tame the mess of wires with some heat-shrink tubing. I covered about 60% of the wires.

If you omit the breakout board, solder the wires directly to the 10-position pot.

4. WIRE THE POTENTIOMETERS

Cut three 8" lengths of wire for ground (black), 3 more for Vcc (I used orange), and 3 for analog input (blue). Connect one set of wires to each potentiometer as shown (Figure **C**). First wrap the wire around the post, then solder it.

Mark the potentiometers as 1, 2, and 3, then mark the other end of each analog input wire with 1, 2, or 3 dots.

5. PREPARE THE NEOPIXEL RING

Cut about 12" each of black, red, and white wire and solder these

---

> **CAUTION:** Some flashlights (especially old ones) use metal reflector bowls, so it's important to insulate these connections. I simply covered them with small blobs of hot glue.

to the NeoPixel's power, ground, and signal in pads, routing the wires from the top as shown (Figure **D**).

Hot-glue the NeoPixel to the bowl of your flashlight and route the wires through the old bulb hole.

Solder the 300Ω resistor to the other end of your signal wire (white), and insulate with heat-shrink, leaving about ⅛" exposed.

> **IMPORTANT:** Before connecting a NeoPixel strip to power, connect a large capacitor (1,000µF, 6.3V or higher) across the + and – terminals. This prevents the initial onrush of current from damaging the pixels.

6. WIRE THE ARDUINO

Solder a 12" red wire to the Arduino's raw power input pin; a 12" black wire to the adjacent ground (GND) pin; and a 12" orange wire to the regulated power (VCC) pin.

Solder the 3 blue wires from the potentiometers to the Arduino's analog inputs A1, A2, and A3 (indicated by the red box in Figure **E**), matching the dots you marked. A1 is for hue, A2 for lightness, and A3 for saturation.

Solder the 10 signal wires from the switch to the Arduino's digital inputs 2 through 11, matching position 2 on the switch to Arduino pin 2, position 3 to pin 3, etc. Match position 1 to pin 11.

Solder the 300Ω resistor on the NeoPixel signal wire to digital input 12 (Figure **F**).

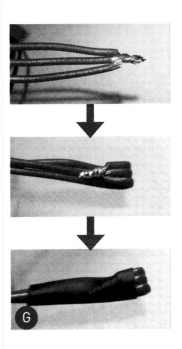

> **IMPORTANT:** The HSL code doesn't use pin A0, so be sure to connect to the correct 3 pins.

Dan Rasmussen

7. MANAGE POWER, GROUND, AND VCC WIRES

Tie all 3 raw power wires (red) together as shown—from the Arduino, the NeoPixel, and the battery pack—then solder and finish with heat-shrink (Figure **G**).

Connect all the VCC power wires (orange) together the same way. Then all the ground wires (black).

8. INTEGRATE FLASHLIGHT'S ON/OFF SWITCH

Find the wires that go to each side of the switch in your flashlight. Cut them and strip ½" from the ends.

> **IMPORTANT:** Leave the switch in the OFF position for now.

Cut the red wire from your battery pack and strip ½" from each end. Solder each end to one side of the flashlight switch. Patch in more red wire as necessary to give you some flexibility (Figure **H**).

9. MOUNT THE ARDUINO IN THE TIC TAC BOX

Empty the Tic Tac box, remove the white plastic dispenser, and cut it down a bit to accommodate the bundle of wires. Then insert the Arduino into the box and seal it with strong tape (Figure **I**). This will insulate the Arduino from any metal inside the flashlight.

10. TEST

No doubt you're anxious to see if this thing is going to work! First, center all 3 pots (to be sure the lightness is not all the way down.) Load 4 fully charged NiMH batteries, and power up the electronics by turning on the flashlight switch. Watch for the 5 flashes on the Arduino's LED. Shortly after that you should see some activity from the NeoPixel ring—no matter where the mode switch is set.

If you see no activity on the Arduino, then there's probably a short somewhere—turn off the power right away and find the short. In my case it was the NeoPixel ring shorting against the metal bowl of my flashlight.

11. MOUNT THE ELECTRONICS IN THE FLASHLIGHT

Drill 3 holes in the flashlight case for the pots and one for the rotary switch, then mount them and attach the knobs. I put the HSL controls on one side and the rotary mode switch on the other (Figure **J**).

> **TIP:** If you're drilling a metal case, first use a center punch (Figure **K**) to keep your drill from wandering. If you've got a plastic case, you can do the same thing with your soldering iron tip.

Now stuff the Arduino, batteries, and wires into the flashlight. These 6V lanterns have plenty of room. You might find that all your heat-shrink is making the wires hard to bend. Just go slowly and it will all work (believe me, it's better than a rat's nest of wires).

CONGRATULATIONS

You've built a fully programmable Million Color HSL Flashlight that will keep you (and maybe even your kids) entertained for years. We'd love to see how you modify it, and what new modes you come up with! ⊘

Get more photos, tips, and video, follow this project, and share your build—all on the project page at makezine.com/projects/million-color-hsl-flashlight.

Hack Electronic Pushbuttons
Tap into your electronic devices and take control.
By Peter Edwards

IN THIS TUTORIAL I'LL EXPLAIN HOW you can easily hack the controls of almost any electronic device. Why would you want to? Maybe you want to rewire a Nintendo joystick to your computer so you can control Mario via Max/MSP. Maybe you want to set up magnetic sensors to steer a remote control car while you're tap-dancing. Or wire the buttons in your TV remote to big, arcade-style buttons mounted in your coffee table. Or modify a musical instrument (as I'll show here). There are countless possibilities.

This guide applies to most button-hacking projects but there are always exceptions to the norm and baffling anomalies. These techniques will work for many circuits but not all.

What's a Button?
A pushbutton is a simple electromechanical device that makes or breaks an electrical connection when activated. Hold 2 pieces of wire in your hands. Connect the ends together, now disconnect them. You just performed the functions of a pushbutton.

There are many varieties of pushbuttons but the most common (and simplest) is the *momentary SPST (single-pole, single-throw) switch*. This will often be listed as "(NO)" which stands for "normally open." The parentheses around NO tell you it's a momentary switch — the circuit stays closed only while you keep your finger on the button.

This button has 2 connection terminals or nodes. Activating (or pushing) the button connects these terminals together. That's it! This sends a signal to the circuit telling it to do something specific. It doesn't matter how the nodes are connected; all the circuit knows

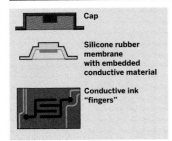

Cap

Silicone rubber membrane with embedded conductive material

Conductive ink "fingers"

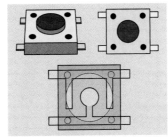

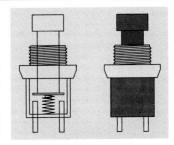

CONDUCTIVE RUBBER BUTTON

The cheapest and most common style of button used in consumer electronics. It's easy to install and can be made in a wide range of shapes. When you press the button, it pushes the conductive rubber membrane against 2 interlinked sets of "fingers," which are printed on the board in conductive ink.

Pros: Cheap and reliable.

Cons: This style of button requires special fabrication techniques that are unavailable to hobbyists. It's also the most difficult style to hack, and because it's made of conductive rubber it can handle only very low current.

TACTILE BUTTON

Used where conductive rubber isn't feasible, or in higher-current applications. This button is cheap, easy to use, and easy to hack. A tactile switch has 4 legs; the legs across from each other are typically connected internally.

The only drawback for hobbyists is that it must be board-mounted, which presents challenges over panel-mount pushbuttons.

Pros: Cheap and reliable, easy to hack, low resistance, can pass higher current than rubber switches.

Cons: Small, must be mounted on a circuit board.

PANEL-MOUNT BUTTON

Rarely found in consumer electronics but a favorite with hobbyists. These switches are the most versatile but also the most expensive and, relative to tactile and rubber switches, the most delicate.

Pros: Easy to use, available in many different configurations, can handle high current, very easy to hack.

Cons: Much more expensive than other kinds of buttons. More moving parts, therefore more delicate.

Illustration by Peter Edwards

⚠ **CAUTION:** As always, only work with battery-powered electronics or circuits that are powered with a wall wart adapter. Don't tinker with circuits that plug directly into the wall unless you know what you're doing and are qualified to work with deadly voltage levels.

is that they are. That means anywhere there's a button, you can replace it with another button, a sensor, a relay, or any other means of passing and breaking current flow. As long as the nodes connect and disconnect, it'll work.

Just make sure the button you use can handle the electrical current that will pass through it. This article only covers low-current circuits and buttons.

Let's Hack Some Buttons

I like making music, so I decided to hack the buttons in a bunch of musical toys so that I could control them all with a sequencer. I selected 3 Casio keyboards — one for the drumbeat and 2 for the melody — and 3 voice memo recorders to sample and play back the sound from the keyboards, introducing all kinds of interesting variables into the music.

MATERIALS AND TOOLS

Soldering iron, solder, and insulated wire
NPN transistors general purpose
Resistors, 10kΩ (optional) and 100kΩ
Arduino microcontroller Maker Shed, makershed.com
Breadboard
Multimeter with continuity tester
Mini screwdrivers for opening electronic devices

1. Identify the buttons you want to hack.

Let's look at the Casio SA-38. This keyboard is useful because it has 5 big drum buttons (Figure A, following page). I can hack these and sequence them to make drumbeats.

2. Open it up.

Once you open up the keyboard, you'll see 3 circuit boards (Figure B). The green board handles power and audio. Underneath it is a brown board with a big chip on it. This chip is the main brain of the keyboard, so I'll call this the brain board. This board and the third board alongside it hold all the conductive rubber pushbuttons. The brain board has the function buttons, including the drum buttons I want to get at (shown by the red box). The third board has the keyboard buttons.

3. Find your buttons.
Unscrew the brain board and turn it over. Now you can see all the button contact points. The 5 drum button contact points each have 2 nodes (Figure C). These nodes look like 2 hands with their fingers interlaced.

4. Connect your wires.
The contact nodes are printed on the board in conductive ink, which is impossible to solder to. Follow the leads trailing away from each node and you can see that one node of each button is connected to a solder point nearby.

You can also see that the bottom nodes of the 3 leftmost buttons are connected together, and so are the 2 right buttons. This is a common trick used in digital circuits to trigger several functions with just a few pins. There's no obvious solder point near these nodes, so use your continuity tester to find their 2 solder points elsewhere on the board.

Solder a wire to each node's solder point, then test it by touching the ends of your wires together to trigger each sound. Finally, write down the button configuration. I use colored ribbon cable to make this easy (Figure D).

5. Hack more buttons!
The Casio SK1 keyboard lets you record sequences then play them back one note at a time by pressing One Key Play. I hacked this button so I can control playback (Figure E).

On the voice recorders, I hacked the Record and Play Back buttons (Figure F). Any simple voice recorder like the Velleman-mk195 can be used (see Maplin, KitStop, etc.).

6. Choose switches to interface with.
To replace a conductive rubber switch, you can use any of the following:
High-Voltage/Low-Resistance Switches
» Panel-mount and tactile pushbuttons
» Magnetic reed switches
» Relays

A

B

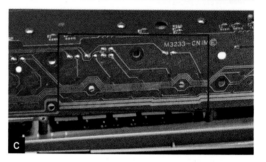

C

D

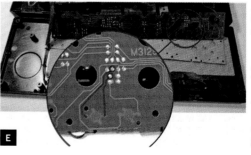

E

Photography and diagrams by Peter Edwards

Low-Voltage/High-Resistance Switches

» Transistors (My favorite! See Step 7.)

» Switching ICs (such as the CD4016)

Panel-mount pushbuttons and relays are necessary for low-resistance, mid-voltage (more than 1V) applications such as passing audio signals or powering circuitry.

For the sequencer shown in Figure G, I used relays as my switching mechanism to allow for the greatest variety of application. Plans for building your own analog step sequencer are available at makezine.com/go/sequencer.

7. Connect hacked buttons to the Arduino.

I use an Arduino microcontroller and some additional transistors to trigger the buttons I hacked. This is a very flexible method that I also use to interface devices with my modular synthesizer.

Transistors are small, cheap, and amazingly powerful. I use general purpose NPN style, 2N3904 or 2N2222 (Figures H and I). I'm sure lots of other transistors will work.

Wire up the switch as shown in Figure I, and test it on a breadboard before soldering. This configuration will work in most cases, but experimentation may be necessary. If your hacked device is triggering erratically, install a 10kΩ resistor to ground (as shown in Figure I). If it's still acting up, try increasing the value of the 100kΩ resistor. If it's not triggering, reduce the value of the 100kΩ. If it still doesn't work, try different transistors. There's a combination out there that will work!

Next connect one of the digital outputs of your Arduino to the transistor through the 100K resistor shown in Figure I. When the digital output pin goes HIGH it will trigger the hacked button. Start with the Arduino "blink" code to get started. Connect pin 13 of the Arduino to the transistor. Now the hacked button will be triggered every time the light on the Arduino blinks. ▣

Peter Edwards is a circuit-bending and creative-electronics pioneer in Troy, N.Y. He builds electronic musical instruments for a living at Casper Electronics (casperelectronics.com).

NOTE: You must connect the ground point on the circuit you're hacking to ground of the Arduino. If you're using battery-powered electronics, just connect the negative battery terminal to the GND pin on the Arduino. If you don't do this, it won't work.

F

G

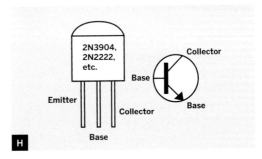

H

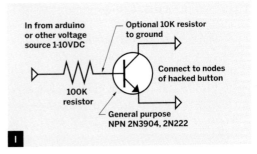

I

X10 Arduino Macro Module

Maximize your X10 control by tapping into a vast open source code library.

BY JIM NEWELL

With a little creativity, you can do some amazing things around the house.

» Place a motion detector in your mailbox, and have it turn on a light and sound a chime whenever the box is opened.

» Add a contact sensor and an X10 Powerflash module to your garage door, to sound a

chime if the door is left open for 5 minutes or more.

» Put a proximity tag in your teenager's car, so that when they drive up, the porch light comes on and interior lights illuminate a path to their room — or to the refrigerator.

L-Dopa

As a home automation buff, I've used X10 Powerline control products for over 30 years. And although the technology has some shortcomings, I still believe X10 is the cheapest and fastest way to automate a home.

One of the most powerful additions to any X10 system is macro capability, in which a controller monitors the power line for trigger signals and responds by executing timed sequences of additional X10 commands. Unfortunately, commercial solutions for running X10 macros are lacking. One choice is Windows software that must run on a dedicated, always-on PC, and is incapable of defining macros with nesting or conditional logic. Another choice is standalone X10 controller modules such as the CM11a and CM15a, but I've found these unreliable and prone to frequent lock-ups.

I have long wanted a small, reliable X10 macro module that I could program in C++, to maximize my options for coding and algorithm development. By standing on the shoulders of giants, I have created one — and the controller looks so good housed in a small jewelry box (Figure A) that my wife allows it to reside on our bedroom dresser.

My X10 macro module has 2 main components: an Arduino microcontroller and a PSC05 (or equivalent TW523) X10 Powerline Interface Module (Figure B). The Arduino runs macros that you write in C++, calling in to a great open source code library of X10 transmit and receive commands. The PSC05

is a module designed to plug into a wall socket and translate in both directions, like a router, between X10 signals carried over a 120V AC power line and wires carrying standard 5V DC encoding, as used in digital electronics.

On the low-voltage side, the module has a 4-conductor RJ11 jack for connecting it via telephone cable to X-10 control devices like home automation control consoles or wireless remote control receivers. But for this project, we'll connect these 4 wires to the Arduino.

X10 PROTOCOL

To understand how this project works, you need to know the basics of X10. The X10 protocol was first developed in the late 1970s to support home automation commands carried over house wiring, so that appliances could be turned on and off and lamps dimmed or brightened remotely, with no need for dedicated control wires. X10 commands are injected onto the power line in a binary format, where a binary 1 is represented by a 1ms burst of a 120kHz signal, and a binary 0 is the lack thereof.

Because home power wiring is noisy, these signal bursts are injected only at the moments when the 60Hz (in the U.S.) AC

A

B

Jim Newell

power wave crosses 0, to maximize signal integrity. This happens 120 times per second (Figure C), and to accommodate artifacts from the 3-phase long-distance power grid transmission before it's split into single-phase house current, each bit is transmitted redundantly 3 times, once for each phase.

To allow remote operation, each light and appliance unit in an X10 network has its own unique 8-bit address, and X10 command sequences consist of one or more unit addresses followed by a 4-bit code for the command that is requested. The command codes include on, off, dimmer, and brighter; query and response codes for checking device status; and special commands for switching on all lights or turning off all units that share the same house code.

Device addresses break down into a 4-bit house code, usually designated by a letter A–P, and a 4-bit unit code designated by a number 1–16. You assign a unique address to a plug module by turning 2 dials on the front (Figure D), and then plug in a light or other device.

The module will listen on the power line for its address and respond to any commands sent to it. On the control side, Figure E shows a typical X10 wireless remote system. You plug the wireless receiver into the power line and turn the dial at the bottom of the remote to your house code. Then the remote's 2 columns of buttons turn on and off your 16 devices, 1–8 with the slider switch to the left and 9–16 with the slider to the right.

PSC05 TRANSCEIVER WIRING

A PSC05 transceiver plugged into a wall socket detects every zero-crossing on our power line, and can read or transmit any X10 data associated with the crossing. Incremental work by Tom Igoe, BroHogan, and Creatrope has produced a free, open source Arduino library for interfacing with this device, letting us communicate via power line X10 as we wish. Using this library, we don't need to know any X10 binary to program our macro module; we just need device addresses

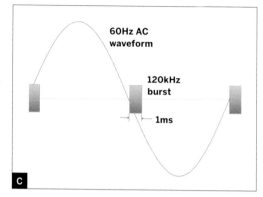

C

60Hz AC waveform

120kHz burst

1ms

Lamp Module

D

Transceiver Module

E

and desired functions.

The 4 connections to the PSC05 transceiver's RJ11 phone jack are, in pin 1–4 order: zero-crossing detect, which reads high (5V) for 1 millisecond after each zero-crossing of the AC wave; ground reference (GND, 0V); data receive (RX) from the power line; and data transmit (TX) to the power line. The Arduino X10 library expects these to be connected to the Arduino's digital I/O pin 2, ground, digital pin 4, and digital pin 5, respectively (Figure F).

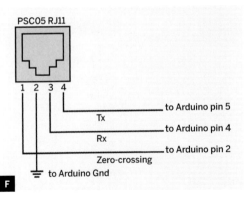

PSC05 RJ11

1 2 3 4

— to Arduino pin 5

Tx

— to Arduino pin 4

Rx

— to Arduino pin 2

Zero-crossing

⏚ to Arduino Gnd

F

G

H

I

START
ASSEMBLE THE HARDWARE

To break out the four RJ11 jack contacts on the PSC05 for connection to the Arduino, I plugged in a short phone cord, plugged the other end to a wall-mount phone jack receptacle, removed the receptacle cover, and wired the receptacle contacts to the Arduino headers using 3" lengths of 20-gauge wire (Figure G). Assuming standard phone wire color-coding, connect yellow to Arduino pin 2, green to GND, red to pin 4, and black to pin 5.

Mount the RJ11 receptacle onto the back of the Arduino using the double-sided adhesive patches that came with the receptacle (or double-stick foam tape), and snap or screw its cover back on, making sure the wire connections route nicely (Figure H). Your macro module is complete; now we'll package it.

Place the module in your box and mark good locations for the phone cable and power cord to run through one side of the box, to plug into the phone receptacle and Arduino.

Drill ⅜" holes at the marked locations (after removing the module). Thread grommets over each cable, run the cables into the box, and push the grommets into the holes. Replace the module and plug in the cords (Figure I).

For a finishing touch, add rubber bumpers to the bottom of the box (Figure J, following page).

It looks so good housed in a small jewelry box that my wife allows it to reside on our bedroom dresser.

INSTALL THE SOFTWARE

Download and install the Arduino IDE (integrated development environment) software from arduino.cc/en/Main/Software, and if it doesn't include a USB serial driver for the onboard FTDI chip, download and install the latest driver from ftdichip.com/drivers/vcp.htm. Launch the IDE and select your Arduino board type under the Tools → Board menu.

Download Creatrope's X10 send/receive library from makezine.com/projects/x10-macro-module and copy the X10 folder to your Arduino libraries folder, the *Contents/Resources/Java/libraries* subfolder under your Arduino application folder (to navigate to the libraries folder, right-click on your Arduino application and select Show Package Contents). The X10 folder should now sit alongside *EEPROM*, *Ethernet*, and other included library folders.

TEST AND PROGRAM

Now comes the acid test: checking the Arduino's X10 send and receive functions. Disconnect and remove the macro module

from its box, and connect a USB cable between it and your computer. If you use a Diecimila, ensure that the plastic jumper near the square USB port is positioned on the 2 pins closest to the port, to supply power to the Arduino (this is not necessary with a Duemilanove). Connect a phone cable between the module's RJ11 jack and the PSC05, and plug the latter into a wall outlet.

Restart the Arduino IDE and load the *X10_receive* example code by selecting it under the File → Examples → X10 menu (Figure K). Click the Verify button at the far left of the Arduino IDE menu bar, wait for the *X10_receive* code to compile, and then click the Upload button.

Open the Serial Monitor under Arduino IDE's Tools menu. If you've done everything properly, commands and status messages from any X10 hardware you've plugged into your power line will appear in the Serial Monitor (Figure L).

We have reached the really fun part, where we exercise the Arduino assembly as a true X10 macro module. To begin, replace the loop function in the checkout example with new code that listens for macro triggers on the power line and responds by sending a sequence of X10 control commands. Figure M shows an example, and you can download the full C++ sketch at makezine.com/projects/x10-macro-module.

```
                    /dev/tty.usbserial-A6008iAW

x10 receive/send testSC-1110 HOUSE-A UNIT-1 CMND5 (ON)
SC-1110 HOUSE-A UNIT-1 CMND7 (OFF)
SC-1110 HOUSE-A UNIT-2 CMND5 (ON)
SC-1110 HOUSE-A UNIT-2 CMND5 (ON)
SC-1110 HOUSE-A UNIT-2 CMND7 (OFF)
SC-1110 HOUSE-A UNIT-7 CMND5 (ON)
SC-1110 HOUSE-D UNIT-1 CMND7 (OFF)
SC-1110 HOUSE-A UNIT-6 CMND5 (ON)
SC-1110 HOUSE-D UNIT-1 CMND5 (ON)
SC-1110 HOUSE-C UNIT-1 CMND5 (ON)
SC-1110 HOUSE-B UNIT-3 CMND5 (ON)
SC-1110 HOUSE-C UNIT-5 CMND7 (OFF)
SC-1110 HOUSE-A UNIT-4 CMND7 (OFF)
SC-1110 HOUSE-A UNIT-7 CMND7 (OFF)
SC-1110 HOUSE-D UNIT-1 CMND7 (OFF)
SC-1110 HOUSE-A UNIT-7 CMND7 (OFF)
```

L

```
    void loop(){          // Macro module - sniffs out X10 commands, and sends a new sequence
      if (SX10.received()){          // received a new command
        SX10.debug();                // print out the received command
        SX10.reset();
    if (SX10.houseCode() == 'A')
      if (SX10.unitCode() == 1)              // Outdoor motion detector
        if (SX10.cmndCode() == ON){
          SX10.write(HOUSE_B,UNIT_1,RPT_SEND);   // Turn on lamp
          SX10.write(HOUSE_B,ON,RPT_SEND);
          SX10.write(HOUSE_E,UNIT_1,RPT_SEND);   // Turn on Zone 1 sprinklers
          SX10.write(HOUSE_E,ON,RPT_SEND);
          SX10.write(HOUSE_D,UNIT_1,RPT_SEND);   // Sound chime module
          SX10.write(HOUSE_D,ON,RPT_SEND);
          SX10.write(HOUSE_C,UNIT_1,RPT_SEND);   // Flash the porch light
          SX10.write(HOUSE_C,UNIT_1,RPT_SEND);
          SX10.write(HOUSE_C,OFF,RPT_SEND);
          SX10.write(HOUSE_C,UNIT_1,RPT_SEND);
          SX10.write(HOUSE_C,ON,RPT_SEND);
          SX10.write(HOUSE_E,UNIT_12,RPT_SEND);  // Turn on Zone 1 & 2 sprinklers together
          SX10.write(HOUSE_E,ON,RPT_SEND);
          SX10.write(HOUSE_D,UNIT_1,RPT_SEND);   // Sound chime module again
          SX10.write(HOUSE_D,ON,RPT_SEND);
          delay(30000);                          // wait, with time in milliseconds
          SX10.write(HOUSE_E,UNIT_12,RPT_SEND);  // Sprinklers off
          SX10.write(HOUSE_E,OFF,RPT_SEND);
          SX10.write(HOUSE_B,UNIT_1,RPT_SEND);   // Turn lamp off
          SX10.write(HOUSE_B,OFF,RPT_SEND);
        }
      }
    }
```

M

N

MAXIMIZE YOUR MACROS

If you've done any programming, even if you're not a C++ whiz, you should be able to understand the event catcher loop code in Figure M, following the conditional phrase if (SX10.received()), and alter it to suit your needs. You will typically use if and == conditionals to check the X10 events received, the SX10.write function to send the messages that result, and the delay function to invoke wait periods. But of course, you have the entire C++ language at your disposal.

For instance, you can have the macro module sniff the power line for an indoor motion detector signal, and then turn on a series of lights that illuminates a path to the bathroom. This example is also included in the project code at makezine.com/projects/x10-macro-module, showing how multiple macros can coexist in the same Arduino sketch.

That's about it. Reading the project code will give you a sense of how easy this really is. The possibilities for automation are limited only by your imagination. ∎

Open, compile, and upload it (or a variant) to your board from your computer as you did with the X10_receive code earlier, then reconnect the module to power and the PSC05. The macro module will begin code execution immediately, waiting for the proper X10 trigger before performing its magic.

This powerful macro sequence controls an outdoor motion detector (set to address A1), lamp module (set to B1), light switch module (C1), PowerHorn siren (D1), and IrrMaster sprinkler controller (E12). You can buy the sprinkler controller from homecontrols.com and everything else from X10.com.

The macro example waits for a signal from the motion detector, and then responds by turning on a lamp, sounding a siren, flashing the porch light, and activating sprinklers — to scare away raccoons and other intruders. After 30 seconds, the sprinklers and the lamp are turned off.

Jim Newell is an engineer/physicist at the NASA Jet Propulsion Laboratory in Pasadena, Calif.

MODERATE

ADVANCED ARDUINO SOUND SYNTHESIS

From "bit banging" to morphing and fading.

Written by **Jon Thompson**

Gunther Kirsch

THE ARDUINO IS AN AMAZING PLATFORM for all kinds of projects, but when it comes to generating sound, many users struggle to get beyond simple beeps. With a deeper understanding of the hardware, you can use Arduino to generate any waveform you can imagine, and manipulate it in real time.

Basic Sound Output

"Bit banging" is the most basic method of producing sound from an Arduino. Just connect a digital output pin to a small speaker and then rapidly and repeatedly flip the pin between high and low. This is how the Arduino's `tone()` statement works. The output pins can even drive a small (4cm or less) 8-ohm speaker connected directly between the pin and ground without any amplification.

Cycling the pin once from low to high and back again creates a single square wave (**Figure A**). Time spent in the high state is called *mark time* and time spent low, *space time*. Varying the ratio between mark and space times, aka the *duty cycle*, without changing the frequency of the wave, will change the quality or "timbre" of the sound.

The Arduino's `analogWrite()` function, which outputs a square wave at a fixed frequency of 490Hz, is handy to illustrate the concept. Connect your speaker to pin D9 and ground (**Figure B**) and run this sketch:

```
void setup() {
   pinMode(9,OUTPUT);
}

void loop() {
   for (int i=0; i<255; i++) {
     analogWrite(9,i);
     delay(10);
   }
}
```

You should hear a tone of constant pitch, with a timbre slowly changing from thin and reedy (mostly space time) to round and fluty (equal mark and space time), and back to thin and reedy again (mostly mark time).

TOOLS & MATERIALS
» **Arduino Nano v3.0 microcontroller board**
» **Solderless breadboard**
» **Capacitor, 0.1µF ceramic**
» **Resistor, 10kΩ**
» **Speaker, 8Ω, approx. 4cm diameter**
» **NPN transistor, BC548 type** or similar
» **LED (optional)**
» **Computer running Arduino IDE software** free from arduino.cc
» **Mini-B USB cable**
» **Oscilloscope** If you don't have access to a hardware oscilloscope, check out Christian Zeitnitz's Soundcard Scope software.

» **Code listings 1–6** Each is a complete running Arduino sketch. Download them from makezine.com/projects/advanced-arduino-sound-synthesis.

EXTRA RESOURCES:
» **OCR2A frequency table** from makezine.com/projects/advanced-arduino-sound-synthesis
» **Wave table spreadsheet** from makezine.com/projects/advanced-arduino-sound-synthesis
» **Soundcard Scope software:** zeitnitz.de/Christian/scope_en

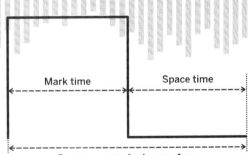

Mark time | Space time

Square wave: single waveform

A

B

A square wave with a variable duty cycle is properly called a *pulse-width modulated (PWM)* wave. Altering the duty cycle to change timbre may serve very basic sound functions, but to produce more complex output, you'll need a more advanced approach.

From Digital to Analog

PWM waves are strictly digital, either high or low. For analog waves, we need to generate voltage levels that lie between these 2 extremes. This is the job of a *digital-to-analog converter (DAC)*.

There are several types of DAC. The simplest is probably the R-2R ladder (**Figure C**). In this example, we have 4 digital inputs, marked D0–D3. D0 is the least significant bit and D3 the most significant.

If you set D0 high, its current has to pass through a large resistance of $2R + R + R + R = 5R$ to reach the output. Some of the current also leaks to ground through the relatively small resistance $2R$. Thus a high voltage at D0 produces a much smaller output voltage than a high voltage at D3, which faces a small resistance of only $2R$ to reach the output, and a large resistance of $5R$ to leak to ground.

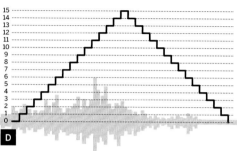
D

distorting the waveform. The jagged "steps" must also be smoothed, using a low-pass filter, to prevent a discordant metallic sound. Finally, an R-2R DAC uses up more output pins than are strictly necessary.

Though a bit harder to understand, the "1-bit" DAC produces very smooth, high quality waveforms using just one output pin with a single resistor and capacitor as a filter. It also leaves the Arduino free to do other things while the sound is playing.

One-Bit DAC Theory

If you replace the speaker from the bit-banging sketch with an LED, you will see it increase in brightness as the duty cycle increases from 0 to 100%. Between these 2 extremes, the LED is really flashing at around 490Hz, but we see these flashes as a continuous brightness.

This "smoothing" phenomenon is called "persistence of vision," and it can be thought of as a visual analogy to the low-pass filter circuit shown in **Figure E**. You can use this filter to smooth the output from a 1-bit DAC.

The mark time of the incoming PWM wave determines the voltage at V_{out} from moment to moment. For example, a mark/space ratio of 50:50 outputs 50% of the high voltage of the

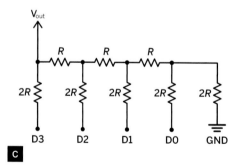
C

Setting D0–D3 to binary values from 0000 to 1111 (0–15 decimal), and then back down to 000 in quick succession, ought to output the triangle wave shown in **Figure D**. To produce other waveforms, in theory, we must simply present the right sequence of binary numbers to D0–D3 at the right rate.

Unfortunately, there are drawbacks to using an R-2R DAC, foremost probably that it requires very precise resistor values to prevent compound errors from adding up and

An LED subjected to "bit banging" in place of the speaker will steadily increase in brightness.

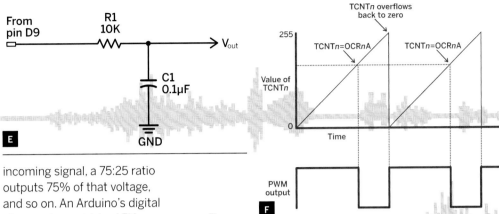

incoming signal, a 75:25 ratio outputs 75% of that voltage, and so on. An Arduino's digital pins produce a high of 5V, so a 50% duty cycle, for example, would give 2.5V at V_{out}.

For best sound quality, the frequency of the PWM signal should be as high as possible. Luckily, the Arduino can produce fast PWM waves up to 62.5KHz. The hardware also provides a handy mechanism for updating the mark time from a lookup table at absolutely regular intervals, while leaving the Arduino free to do other things.

The Arduino 1-Bit DAC

The ATmega328 chip at the heart of the Arduino Nano 3 contains 3 hardware timers. Each timer includes a counter that increments at each clock tick, automatically overflowing back to 0 at the end of its range. The counters are named TCNT*n*, where *n* is the number of the timer in question.

Timer0 and timer2 are 8-bit timers, so TCNT0 and TCNT2 repeatedly count from 0 to 255. Timer1 is a 16-bit timer, so TCNT1 repeatedly counts from 0 to 65535, and can also be made to work in 8-bit mode. In fact, each timer has a few different modes. The one we need

Listing 1

```
#include <avr/interrupt.h> // Use timer interrupt library

/******** Sine wave parameters ********/
#define PI2      6.283185 // 2*PI saves calculation later
#define AMP      127 // Scaling factor for sine wave
#define OFFSET   128 // Offset shifts wave to all >0 values

/******** Lookup table ********/
#define LENGTH   256  // Length of the wave lookup table
byte wave[LENGTH];   // Storage for waveform

void setup() {

  /* Populate the waveform table with a sine wave */
  for (int i=0; i<LENGTH; i++) { // Step across wave table
      float v = (AMP*sin((PI2/LENGTH)*i));  // Compute value
      wave[i] = int(v+OFFSET); // Store value as integer
  }

  /****Set timer1 for 8-bit fast PWM output ****/
    pinMode(9, OUTPUT);  // Make timer's PWM pin an output
    TCCR1B  = (1 << CS10); // Set prescaler to full 16MHz
    TCCR1A |= (1 << COM1A1);  // Pin low when TCNT1=OCR1A
    TCCR1A |= (1 << WGM10);   // Use 8-bit fast PWM mode
    TCCR1B |= (1 << WGM12);

  /******** Set up timer2 to call ISR ********/
    TCCR2A = 0; // No options in control register A
    TCCR2B = (1 << CS21); // Set prescaler to divide by 8
    TIMSK2 = (1 << OCIE2A); // Call ISR when TCNT2 = OCRA2
    OCR2A = 32;  // Set frequency of generated wave
    sei();  // Enable interrupts to generate waveform!
}

void loop() {  // Nothing to do!
}

/******** Called every time TCNT2 = OCR2A ********/
ISR(TIMER2_COMPA_vect) {  // Called when TCNT2 == OCR2A
  static byte index=0;  // Points to each table entry
  OCR1AL = wave[index++]; // Update the PWM output
  asm("NOP;NOP");  // Fine tuning
  TCNT2 = 6;  // Timing to compensate for ISR run time
}
```

is called "fast PWM," which is only available on timer1.

In this mode, whenever TCNT1 overflows to zero, the output goes high to mark the start of the next cycle. To set the mark time, timer1 contains a register called OCR1A. When TCNT1 has counted up to the value stored in OCR1A, the output goes low, ending the cycle's mark time and beginning its space time. TCNT1 keeps on incrementing until it overflows, and the process begins again.

This process is represented graphically in **Figure F,** previous page. The higher we set OCR1A, the longer the mark time of the PWM output, and the higher the voltage at V_{out}. By updating OCR1A at regular intervals from a pre-calculated lookup table, we can generate any waveform we like.

Basic Wave Table Playback

Listing 1 (previous page) contains a sketch that uses a lookup table, fast PWM mode, and a 1-bit DAC to generate a sine wave.

First we calculate the waveform and store it in an array as a series of bytes. These will be loaded directly into OCR1A at the appropriate time. We then start timer1 generating a fast PWM wave. Because timer1 is 16-bit by default, we also have to set it to 8-bit mode.

We use timer2 to regularly interrupt the CPU and call a special function to load OCR1A with the next value in the waveform. This function is called an *interrupt service routine* (*ISR*), and is called by timer2 whenever TCNT2 becomes to equal OCR2A. The ISR itself is written just like any other function, except that it has no return type.

The Arduino Nano's system clock runs at 16MHz, which will cause timer2 to call the ISR far too quickly. We must slow it down by engaging the "prescaler" hardware, which divides the frequency of system clock pulses before letting them increment TCNT2. We'll set the prescaler to divide by 8, which makes TCNT2 update at 2MHz.

To control the frequency of the generated waveform, we simply set OCR2A. To calculate the frequency of the resulting wave, divide the rate at which TCNT2 is updated (2MHz) by the

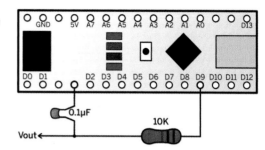

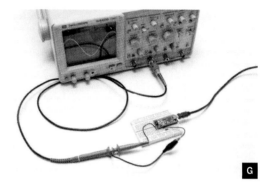

G

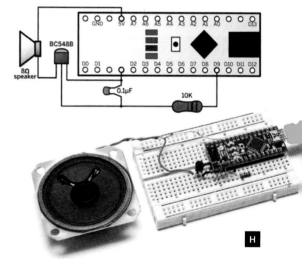

H

value of OCR2A, and divide the result by the length of the lookup table. Setting OCR2A to 128, for example, gives a frequency of:

$$\frac{\text{TCNT2 rate}}{\text{OCR2A value} \times \text{wavetable length}} = \frac{2{,}000{,}000\text{Hz}}{128 \times 256} = 61.04\text{Hz}$$

which is roughly the B that's 2 octaves below middle C. For a table of values giving standard musical notes, see makezine.com/projects/advanced-arduino-sound-synthesis.

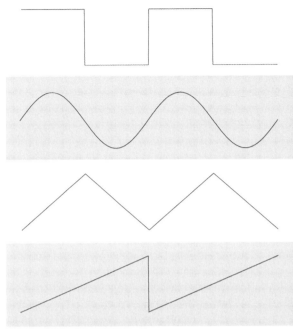

A SQUARE waveform can have a timbre ranging from round and fluty to thin and reedy, depending on its duty cycle. From our initial experiments with bit banging, we already know that a 50% duty cycle gives a flute-like sound.

A SINE waveform consists of a smoothly oscillating curved signal representing the graph of the trigonometric function. It produces a clear, glassy tone that sounds very much like a tuning fork.

A TRIANGLE waveform looks a little like a sine wave with sharp points and straight lines, or 2 ramp waveforms squashed back to back. It has a more interesting, rounded sound than a sine wave, a little like an oboe.

A RAMP waveform consists of steadily increasing values, starting at zero. At the end of the waveform the value suddenly drops to zero again to begin the next cycle. The result is a sound that's bright and brassy.

A RANDOM waveform can sound like anything, in theory, but usually sounds like noisy static. The Arduino produces only pseudorandom numbers, so a particular randomSeed() value always gives the same "random" waveform.

The ISR takes some time to run, for which we compensate by setting TCNT2 to 6, rather than 0, just before returning. To further tighten the timing, I've added the instruction asm("NOP;NOP"), executing 2 "no operation" instructions using one clock cycle each.

Run the sketch and connect a resistor and capacitor (**Figure G**). You should see a smooth sine wave on connecting an oscilloscope to V_{out}. If you want to hear the output through a small speaker, add a transistor to boost the signal (**Figure H**).

Programming Simple Waves

Once you know how to "play" a wave from a lookup table, creating any sound you want is as easy as storing the right values in the table beforehand. Your only limits are the Arduino's relatively low speed and memory capacity.

Listing 2 (available at makezine.com/ projects/advanced-arduino-sound-synthesis) contains a waveform() function to prepopulate the table with simple waveforms: SQUARE, SINE, TRIANGLE, RAMP, and RANDOM. Play them to see how they sound (**Figure I**).

The RANDOM function just fills the table with pseudorandom integers based on a seed value. Changing the seed value using the randomSeed() function allows us to generate different pseudorandom sequences and see what they sound like. Some sound thin and weedy, others more organic. These random waveforms are interesting but noisy. We need a better way of shaping complex waves.

Additive Synthesis

In the 19th century, Joseph Fourier showed that we can reproduce, or *synthesize*, any waveform by combining enough sine waves of different amplitudes and frequencies. These sine waves are called *partials* or *harmonics*. The lowest-frequency harmonic is called the first harmonic or *fundamental*. The process of combining harmonics to create new waveforms is called *additive synthesis*.

Given a complex wave, we can synthesize it

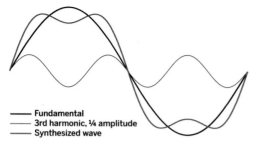

Fundamental
3rd harmonic, ¼ amplitude
Synthesized wave

Fundamental
Various harmonics
Synthesized wave

```
harmonic[PARTIALS] =
{1,3,5,7,9,11,13,15};

attenuate[PARTIALS] =
{1,3,5,7,9,11,13,15};
```

J

Adding the third harmonic creates a waveform that has a distinctly square look and sound, though still very rounded.

K

Adding the first 8 odd harmonics gives a fairly good approximation of a square wave. Individual sine waves appear as little ripples. Combining more partials would reduce the size of these ripples.

roughly by combining a small number of harmonics. The more harmonics we include, the more accurate our synthesis.

Professional additive synthesizers can combine over 100 harmonics this way, and adjust their amplitudes in real time to create dramatic timbre changes. This is beyond the power of Arduino, but we can still do enough to load our wave table with interesting sounds.

Consider the loop in **Listing 1** that calculates a sine wave. Call that the fundamental. To add, say, the third harmonic at ¼ amplitude (**Figure J**), we add a new step:

```
for (int i=0; i<LENGTH; i++) { //
Step across table
   float v = (AMP*sin((PI2/LENGTH)*i));
// Fundamental
   v += (AMP/4*sin((PI2/
LENGTH)*(i*3))); // New step
   wave[i]=int(v+OFFSET);    // Store
as integer
   }
```

In this new step, we multiply the loop counter by 3 to generate the third harmonic, and divide by an "attenuation" factor of 4 to reduce its amplitude.

Listing 3 (available at makezine.com/projects/advanced-arduino-sound-synthesis) contains a general version of this function. It includes 2 arrays listing the harmonics we want to combine (including 1, the fundamental) and their attenuation factors.

To change the timbre of the sound loaded into the lookup table, we just alter the values in the 2 arrays. A 0 attenuation means the corresponding harmonic is ignored. The arrays in **Listing 3**, as written, produce a fairly good square wave (**Figure K**). Experiment with the arrays and see what sounds result.

Morphing Waveforms

Professional synthesizers contain circuits or programs to "filter" sound for special effects. For instance, most have a low-pass filter (LPF) that gives a certain "waa" to the start and "yeow" to the end of sounds. Basically, an LPF gradually filters out the higher partials. Computationally, true filtering is too much for Arduino, but there are things we can do to the sound, while it's playing, to give similar effects.

Listing 4 (available at makezine.com/projects/advanced-arduino-sound-synthesis) includes a function that compares each value in the wave table to a corresponding value in a second "filter" table. When the values differ, the function nudges the wave value toward the filter value, slowly "morphing" the sound as it plays.

Using a sine wave as the "filter" approximates true low-pass filtering. The harmonics are gradually removed, adding an "oww" to the end. If we morph the other way — by loading the wave table with a sine wave and the "filter" table with a more complex wave — we add a "waa" quality to the start. You can load the 2 tables with any waves you like.

Creating Notes

What if we want to make a sound fade away, like a real instrument that's been plucked, strummed, or struck?

Listing 5 (available at makezine.com/projects/advanced-arduino-sound-synthesis) contains a function that "decays" the sound to silence by steadily nudging the wave table values back toward a flat line. It steps across the wave table, checking each value — if it's more than 127, it's decremented, and if less, incremented. The rate of decay is governed by the `delay()` function, called at the end of each sweep across the table.

Once the wave is "squashed," running the ISR just ties up the CPU without making sound; the `cli()` function clears the interrupt flag set in setup by `sei()`, switching it off.

Using Program Memory

The Arduino's Atmel processor is based on the "Harvard" architecture, which separates program memory from variable memory, which in turn is split into volatile and nonvolatile areas. The Nano only has 2KB of variable space, but a (relatively) whopping 30KB, or so, of usable program space.

It is possible to store wave data in this space, greatly expanding our repetoire of playable sounds. Data stored in program space is read only, but we can store a lot of it and load it into RAM to manipulate during playback.

Listing 6 (available at makezine.com/projects/advanced-arduino-sound-synthesis) demonstrates this technique, loading a sine wave from an array stored in program space into the wave table. We must include the *pgmspace.h* library at the top of the sketch and use the keyword PROGMEM in our array declaration:

```
prog_char ref[256] PROGMEM =
{128,131,134,…};
```

Prog_char is defined in *pgmspace.h* and is the same as the familiar "byte" data type.

If we try to access the `ref[]` array normally, the program will look in variable space. We must use the built-in function `pgm_read_byte`

Generating Wave Table Data

To read wave table data from program memory, you have to hard-code it into your sketch and can't generate it during runtime. So where does it come from? One method is to generate wave table data in a spreadsheet and paste it into your sketch. We've created a spreadsheet that will allow you to generate wave tables using additive synthesis, to see the shape of the resulting waves, and to copy out raw wave table data to insert into your sketch.

Download it from makezine.com/projects/advanced-arduino-sound-synthesis.

instead. It takes as an argument the address of the array you want to access, plus an offset pointing to individual array entries.

If you want to store more than one waveform this way, you can access the array in `pgm_read_byte` like a normal two-dimensional array. If the array has, say, dimensions of [10] [256], you'd use `pgm_read_byte(&ref[4] [i])` in the loop to access waveform 4. Don't forget the & sign before the name of the array!

Going Further

Audio feedback is an important way of indicating conditions inside a running program, such as errors, key presses, and sensor events.

Sounds produced by your Arduino can be recorded into and manipulated by a software sampler package and used in music projects.

Morphing between stored waveforms, either in sequence, randomly, or under the influence of performance parameters, could be useful in interactive art installations.

If we upgrade to the Arduino Due, things get really exciting. At 84MHz, the Due is more than 5 times faster than the Nano and can handle many more and higher-frequency partials in fast PWM mode. In theory, the Due could even calculate partials in real time, creating a true additive synthesis engine. ◼

Jon Thompson is a UK-based freelance technology writer. Among other things, he spends his time making strange props for magicians.

Jeffrey Braverman

Raspberry Pirate Radio

Broadcast your own go-anywhere FM station with this amazingly simple Pi hack. Written by Sam Freeman and Wynter Woods

THIS SIMPLE HACK TURNS YOUR RASPBERRY PI INTO A POWERFUL FM TRANSMITTER with enough range to cover your home or dorm, a DIY drive-in movie, a high school ball game, or even a bike parade (depending on the stragglers). It's the coolest Pi device we've ever seen with so few materials.

You'll start with the absolute minimum you need to run a Raspberry Pi — an SD card, a power source, and the board itself — and add a single piece of wire. The PiFM software cleverly uses hardware that's meant to generate spread-spectrum clock signals on the GPIO pins to output FM radio energy instead.

PiFM was originally created by Oliver Mattos and Oskar Weigl, and revised by Ryan Grassel. MAKE's contribution, the *PirateRadio.py* Python script, now enables automatic playback without using the command line and handles all the most common music file formats. It was written here in the MAKE Labs by engineering intern Wynter Woods.

1. Make the antenna

Stick a wire into GPIO pin 4 of your Raspberry Pi. That's it! We used 40cm of fat 12 AWG solid copper wire, soldered to a female jumper. We covered the connection with heat-shrink tubing and reinforced it with a gob of hot glue.

SAM FREEMAN
Raised in the galactic capital of Earth, Sam was destined to work as the MAKE Labs manager — testing, designing, and breaking projects for MAKE.

WYNTER WOODS
is a MAKE engineering intern and a programmer with one too many interests, ranging from hardware hacks to audio processing to 3D visualization of chemical sample data.

MATERIALS:

- » **Raspberry Pi single-board computer** Maker Shed #MKRPI2, or get the Raspberry Pi Starter Kit, #MSRPIK2, at makershed.com
- » **SD Card, 4GB or more** An 8GB card is in the Starter Kit.
- » **USB wall charger, 2A, with USB cable** also in the Starter Kit
- » **Hookup wire, solid-core, 12 AWG, 40cm length**
- » **Female jumper wire**
- » **Heat-shrink tubing**
- » **FM radio**
- » **Battery pack with USB socket (optional)** for portable operation, such as the Smart Power Base, a rechargeable 5V 1A lithium-ion power pack that works with many popular development boards.

TOOLS

- » **Computer**
- » **Wire cutters / strippers**
- » **Soldering iron**
- » **Hot glue gun**

2. Flash the SD card and add music

MAKE Labs created a disk image that runs the *PirateRadio.py* script on startup, so your music starts broadcasting immediately. It handles MP3, WAV, FLAC, AAC, M4A, and WMA files automatically. Download it from the online project page and flash it to the SD card. It's easy; check the project page for more advice.

Then just drag any music files, or artist or album folders, to the root of the Pirate Radio partition. Your music files can be nested within these folders, so there's no need to dump all your music into one mess on the root directory.

3. Edit the *config* file

Set **frequency** to the station you want to broadcast on. Useable FM frequencies are typically from 87.5MHz to 108.0MHz.

Set **shuffle** to **True** to shuffle files, or **False** to play them alphabetically. Set **repeat_all** to **True** if you want to loop forever through your playlist.

4. Start it up!

Tune an FM radio to your frequency and plug in the Raspberry Pi. In about 15 seconds you'll hear your music loud and clear!

How It Works

The PiFM software manipulates the frequency of the Raspberry Pi's internal PLLD clock (500MHz) using a fractional divider. For a target broadcast frequency of 100MHz, for example, the frequency is modulated between 100.025Mhz and 99.975Mhz. That's how FM radio transmits an audio signal.

The Python code defaults to 87.9 FM with shuffle and repeat turned off. It scans the SD card for music files and builds a playlist based on the options in the *config* file. It then passes each file along to a decoder based on the file type. Each file is then re-encoded into a format the PiFM radio can handle.

> **NOTE:**
> The Pi's broadcast frequency can range FROM 1Mhz to 250Mhz, which may interfere with government bands. Limit your transmissions to unoccupied portions of the FM band of 87.5MHz–108.0MHz.

Going Further

Tuck everything into the plastic case from the Raspberry Pi Starter Kit, or 3D-print this awesome radio tower enclosure, drawn up by MAKE Labs manager Sam Freeman. Download it at thingiverse.com/make.

Then add a USB battery pack so you can carry your station wherever you need to take over the airwaves. (It fits in the radio tower, too.) ●

Get complete step-by-step instructions and download the Pirate Radio code at makezine.com/projects/raspberry-pirate-radio

Share it: #rasppirateradio

HOW TO BAKE AN ONION Pi

Written by Limor Fried and Phillip Torrone

Hack your Raspberry Pi into an anonymizing Tor proxy!

⚡ TIME: 1–2 HOURS   **⚡ COST:** $90–$130

Feel like someone is snooping on you? Browse the web anonymously anywhere you go with the Onion Pi Tor proxy. This is a cool weekend project that uses a Raspberry Pi mini computer, USB wi-fi adapter, and Ethernet cable to create a small, low-power, and portable privacy Pi.

Using it is easy-as-pie. First, plug the Ethernet cable into any internet connection in your home, work, hotel, or conference/event. Next, power up the Pi with the Micro-USB cable connected to your laptop, or with a wall adapter. The Pi will boot up and create a new secure wireless access point. Connecting to that access point will then automatically route any web browsing from your computer through the anonymizing Tor network. Your tracks are swept clean.

Nate Van Dyke

MATERIALS:

» **Raspberry Pi Starter Kit #MSRPIK2 from Maker Shed, maker-shed.com. Our kit is the best way to get started using your Raspberry Pi. Includes Raspberry Pi Model B, 4GB SD Card, 5V 2A power supply, Micro-USB and HDMI cables, custom MAKE: Pi enclosure, Adafruit's Cobbler GPIO (General Purpose Input/Output) break-out, a breadboard for electronics prototyping, a selection of common components, and a copy of our bestselling book, Getting Started with Raspberry Pi.**
» **Mini USB wi-fi module Maker Shed #MKAD55**

—OR—

» **Onion Pi Bundle (Tor Router) w/Mini Wi-Fi Adafruit #1410 (adafruit .com). For more experienced users who want to build a dedicated wireless Tor proxy.**

—OR—

» **Raspberry Pi Model B** Ethernet is required.
» **Raspberry Pi case (optional)**
» **Ethernet cable**
» **USB wi-fi adapter that supports the RTL-8192CU chipset**
» **SD card 4GB or more**
» **5V Micro-USB power supply rated at least 700mA**

TOOLS:

» **Computer Windows, Mac, or Linux**
» **Router with working internet connection**
» **USB keyboard**
» **Display with HDMI or composite video-in**

What Is Tor?

Tor is an "onion routing" service: Internet traffic is wrapped in layers of encryption and sent through a random circuit of relays before reaching its destination. This makes it much harder for the server you're accessing (or anyone snooping on your internet use) to figure out who and where you are. It's an excellent way for people who are blocked from accessing websites to get around those restrictions. Journalists, activists, businesspeople, law enforcement agents, and even military intelligence operatives use Tor to protect their privacy and security online.

Why Use a Proxy?

You may have a guest or friend who wants to use Tor but doesn't have the ability or time to set it up on their computer. You may not want to, or may not be able to, install Tor on your work laptop or "loaner" computer. You may want to browse anonymously on a netbook, tablet, phone, or other mobile or console device that cannot run Tor and does not have an Ethernet connection. There are lots of reasons you may want to build and use an Onion Pi, not least of which is that it is an interesting way to learn about Raspberry Pi, network interfaces, and the Linux command line.

1. Prepare your SD card.

When you buy a Raspberry Pi, it may or may not come with an SD card. The SD card is important because this is where Raspberry Pi keeps its operating system and it's also where you'll store your documents and programs. Even if your Pi came with an SD card with the operating system already installed, it's a good idea to update it to the latest version, as improvements and bug fixes are going in all the time.

WARNING!
Before you start using your proxy, remember that there are a lot of ways to identify you, even if your IP address is "randomized." So delete and block your browser cache, history, and cookies — some browsers even allow "anonymous sessions." Do not log into existing accounts with personally identifying information (unless you're sure that's what you want to do). Use SSL whenever available to encrypt your communication end-to-end. And visit torproject.org for more info on how to use Tor in a smart, safe way.

This tutorial is a great way to make something fun and useful with your Raspberry Pi, but we can't guarantee it's 100% anonymous and secure. Be smart and paranoid about your Tor usage.

RASPBIAN
2012-10-28

OCCI
012
2012-11-19

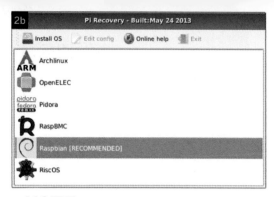

Install OS   Edit config   Online help   Exit

ARM Archlinux

OpenELEC

pidora fedora Pidora

RaspBMC

Raspbian [RECOMMENDED]

RiscOS

2. Boot and configure.

If you want to mount your Pi in a case, now's a good time.

2a. Insert the SD card you just prepared into the Pi's drive slot, being careful to note the correct orientation of the contacts. Connect your display and keyboard before plugging in the Micro-USB power cable. The Pi should boot automatically.

2b. Install Raspbian. From the NOOBS boot screen, select Raspbian, press Enter, and con-firm that you want to overwrite the disk. When installation is complete, press Enter again to dismiss the notice, and your Pi should reboot automatically.

NOTE : This tutorial assumes you'll be using Raspbian, and may not work exactly as written with other Linux distributions.

Experienced users have many options for preparing an SD card. We recommend new users visit raspberrypi.org and follow the instructions in their Quick Start Guide for for-matting an SD card and installing the official New Out Of Box Software (NOOBS) package. Briefly, the steps are:

1a. Format the card. The Raspberry Pi Founda-tion recommends using the SD card founda-tion's official formatting tool, SD Formatter, which is available for Windows, Mac, and Linux. The settings may vary depending on your OS. Refer to the Quick Start Guide for details.

1b. Download NOOBS. You can get the .ZIP archive directly from the Raspberry Pi web-site, one of several mirror servers, or through BitTorrent.

1c. Extract the NOOBS archive to your SD card. The contents of the archive, including the file *bootcode.bin* and the *images* and *slides* folders, should be in the top-level directory.

2c. After a lot of scrolling text, you'll arrive at the `raspi-config` options screen. Using the arrow keys to navigate and Enter to select, first update the default password ("rasp-berry") for the default user account ("pi") to a secure phrase known only to you.

TIP : You may notice a short lag between selecting options or entering commands and the system's response. This is normal. Be patient.

2d. Select Internationalisation Options and set the time zone, language, and keyboard layout options to match your preference. Then select Finish and press Enter.

3. Connect Ethernet/Wi-Fi.

For most home networks, you should also be able to connect to the internet through the Ethernet connection via your router without any further configuration. After `raspi-config`

```
┤ Raspberry Pi Software Configuration Tool (raspi-config) ├
Setup Options

    1 Expand Filesystem              Ensures that all of the SD card storage is available to the OS
    2 Change User Password           Change password for the default user (pi)
    3 Enable Boot to Desktop         Choose whether to boot into a desktop environment or the command-line
    4 Internationalisation Options   Set up language and regional settings to match your location
    5 Enable Camera                  Enable this Pi to work with the Raspberry Pi Camera
    6 Add to Rastrack                Add this Pi to the online Raspberry Pi Map (Rastrack)
    7 Overclock                      Configure overclocking for your Pi
    8 Advanced Options               Configure advanced settings
    9 About raspi-config             Information about this configuration tool


              <Select>                                        <Finish>

2c
```

3a

MAKE ME A SANDWICH.

SUDO MAKE ME
A SANDWICH.

WHAT? MAKE
IT YOURSELF.

OKAY.

Classic *xkcd* webcomic #149, "Sandwich."

Randall Munroe

exits, you'll be presented with the Raspbian command prompt: `pi@raspberrypi ~ $_`

3a. When you see the prompt, connect your Pi to your router using a standard network cable. As soon as you plug your Pi in, you should see its network LEDs start to flicker.

3b. At the Raspbian command line, type in: `sudo wget makezine.com/go/onionpi` The Linux command `sudo` allows one user to assume the security privileges of another, commonly the superuser or root. (Think: "<u>s</u>uper<u>u</u>ser <u>do</u>.") The next command, `wget`, will not run correctly unless preceded by `sudo`.

NOTE : Linux user rights and privileges can get pretty complicated, but as a general rule, you'll need to `sudo` any commands that involve making changes to the disk. Read-only commands, like listing directories or displaying (without modifying) the contents of files, can usually be executed without `sudo`.

The command `wget` instructs the operating system to retrieve a file from the web, and takes as argument the web address of the file to be retrieved. In this case, we're grabbing a pair of *shell scripts* that will automate much of the fiddly typing for configuring your Pi as a wireless access point.

TIP : If you get tired of typing `sudo` all the time, the command `sudo  su` allows you to become the superuser as long as you want.

When you understand what the command is supposed to do, press Enter to execute it. If your Ethernet connection is working, you'll shortly be notified that the file has been saved.

If your Ethernet connection is *not* working, you'll see an error message (such as `failed: Name or service not known`). Make sure that your Pi is correctly connected to your router, the network cable is good, and your router is correctly

configured for DHCP (Dynamic Host Configuration Protocol).

3c. Don't plug in your wi-fi adapter yet — you'll crash the Pi and corrupt the SD card. First, turn off your Pi by entering `sudo halt`. After shutdown, plug in the wi-fi adapter. Now restart your Pi by cycling the power.

4. Set up the "PiFi" access point.

Now we'll set up the Pi to broadcast a wi-fi service and route wireless internet traffic through the Ethernet cable. One of the great things about Linux is that every little detail of a system's configuration can be easily modified to suit your application by typing in commands or modifying the contents of text files.

The tradeoff is that the details can get pretty complicated, and you have to know what you're doing to understand exactly what needs to be changed, and how.

To make the process easier, we've prepared a script (which you just downloaded with `wget`) that will automatically make these changes for you (**Figure 4**). If you just want to get it working, all you have to do is run the script, as explained below.

4a. After your Pi reboots, you'll be prompted to log in. Enter the default user ID "pi" followed by the password you set from `raspi-config`.

4b. At the Raspbian command prompt, enter these commands to extract the shell scripts:
```
sudo unzip onionpi
sudo bash pifi.sh
```

We just made friends with `sudo`; now it's time to meet `bash`, the Linux *command-line interpreter*. In fact, you've already been introduced: Whenever you enter text at the command prompt, you are interacting with `bash`, which is the program that processes what you've typed and figures out what to do with it. `bash` runs automatically whenever you're working from the Linux command line, but can also be called as a command, itself, to execute a script file.

In this case, we're telling `bash` to read through the script `pifi.sh` and execute each line of text as if it had been typed in at the command prompt.

4c. Press Enter and you'll soon see the script splash screen, with the option to start the script or abort. Press Enter again to start.

4d. When prompted, enter the name (SSID) for your new wireless network, and the password required to access it.

When the script is complete, your Pi should reboot automatically, after which you should be able to detect your new "PiFi" network from nearby computers, smartphones, and other wi-fi appliances. Log on to the wireless network using the password you just set, open a web browser, and navigate to your favorite web page to verify that everything is working properly.

If you just want to configure your Pi as a wireless access point, you're done! You shouldn't even have to log in to Raspbian

NOTE: Both network name and password can be updated later by editing the config file with any text editor.

```
4  pi@raspberrypi ~ $ sudo bash pifi.sh




                    Raspberry PiFi

This script will configure your Raspberry Pi as a wireless access point.
Press [Enter] to begin, [Ctrl-C] to abort...
```

NOTE: For a slower and more instructive experience, we recommend opening the *pifi.sh* script (which is just a text file) in another computer and typing in the commands by hand, to get a feel for what each one does and how the system responds. The script file also contains comments that explain each step in more technical detail, for those who are interested.

again; the Pi will now automatically function as a wireless router whenever it's on.

5. Install tor.
To continue setting up your Pi to anonymize your wi-fi traffic with Tor, log in to Linux again and run the second script with:
sudo bash tor.sh (Figure 5)
This script is less complicated. Basically, it installs and configures the Tor software, then updates your IP tables to route everything through it. As always, it's a good idea to read through the commands and comments in the script file before running it. More technical detail is available there.

The Pi will automatically reboot again when the script is done. Your Tor proxy may not work until the reboot is complete.

6. Browse anonymously.
When your Pi has finished rebooting, log on to your "PiFi" wireless network from a nearby computer, smartphone, or other wi-fi appliance. Then open your favorite internet browser and visit check.torproject.org. If your Onion Pi is working correctly, you should see something like **Figure 6**.

Going Further
We use Ethernet because it requires no configuration or passwords — just click the cable to get DHCP. But if you want, it's not too hard to set up a wi-fi-to-wi-fi proxy. You'll need to use two wi-fi adapters and edit the settings in /etc/networks/interfaces to add the wlan1 interface with SSID and password to match your internet provider. See makezine. com/go/pifi2wifi for more details.

It's also pretty easy to configure Tor to give you a presence in any country you choose. For example, here's a *torrc* configuration file that sets up a Pi at IP address 192.168.0.178 to appear "present" in Great Britain:

```
Log notice file /var/log/tor/notices.log
SocksListenAddress 192.168.0.178
ExitNodes {GB}
StrictNodes 1
```

You'll also need to configure your browser to use a SOCKS5 proxy on 192.168.0.178 (or whatever your Pi's IP address may be), port 9050.

If you like using Tor, you can help make it faster by joining as a relay, or increase its effectiveness by becoming an exit node. Check out torproject.org for details.

Finally, if you want to support Tor but can't run your own relay or exit node, please consider donating to the project to help cover development, equipment, and other expenses. Your donation is even tax-deductible if you live in the United States. ∎

Limor Fried is owner of Adafruit Industries, an open-source hardware and electronics kit company based in New York City.

Phillip Torrone is an editor at large of MAKE magazine and creative director at Adafruit.

3D Printing AND CNC Fabrication

PART 4 >>>>>

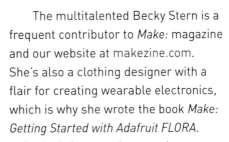

3d printing and CNC fabrication have been around for decades, but their popularity with home users has increased dramatically in recent years because of improved technologies, lower prices, and the availability of shared resources online and in community maker spaces. 3D printing is an *additive* process, meaning that shapes are created by laying down and fusing material. By contrast, CNC processes such as milling and laser cutting are *subtractive*, meaning they create shapes by cutting away excess material. Both use computer guidance to create precise pieces.

Make: book editor, Anna Kaziunas France (who also happens to be dean of the Providence Fab Academy) introduces you to CNC fabrication with a great-looking, versatile workbench for your workshop or office. The CAD file for the bench was created by a pair of award-winning architects and can be modified to suit your needs. When you're satisfied with the design, you can cut it yourself or send it out to a service bureau.

Tyler Worman explains how CNC end mills work and provides helpful graphics and tips on what to buy to get you started.

Dan Spangler's CNC air raid siren is *guaranteed* to attract attention. You'll find his CAD files for the project on our website. Then all you need is an AC motor to spin some air. Dan discovered that the Maker Labs bench grinder was all he needed to create an awesome wail.

The multitalented Becky Stern is a frequent contributor to *Make:* magazine and our website at makezine.com. She's also a clothing designer with a flair for creating wearable electronics, which is why she wrote the book *Make: Getting Started with Adafruit FLORA.*

Becky is here to show you how to transform flexible 3D-printed spikes into a luminous fashion accessory.

Have you ever heard of a pinhole camera? It's one of the oldest forms of camera, dating all the way back to the 17th century. Pinhole cameras don't have lenses; instead, they let light in through a small hole and can produce hauntingly beautiful photos. Todd Schlemmer has long been fascinated by pinhole photography. At first he built his cameras by hand, but eventually he began designing them with CAD software and creating them with a 3D printer. Here, Todd shares one of his designs with you and provides construction guidelines. Discover what kinds of artistic images *you* can produce!

CNC Maker Bench

Create custom, open-source CNC tables for your workshop using AtFAB's parametric program — or just download and fabricate MAKE's design.

Written by Anna Kaziunas France

Time Required:
2 Weekends
Cost:
$100-$180

ANNA KAZIUNAS FRANCE is the digital fabrication editor at MAKE. She's also the dean of students for the Global Fab Academy program, the co-author of *Getting Started with MakerBot*, and the editor of the book *Make: 3D Printing*.

Marie Kaziunas

WE HAD JUST MOVED INTO MAKE'S OFFICE IN PROVIDENCE, AND WE NEEDED FURNITURE. I decided to create a set of standing-height plywood workbenches to house our 3D printers and other CNC machines. I have access to a large CNC router, and I wanted to design custom tables tailored to my measurements. And I prefer to stand when I'm wrenching on machines, rather than sitting in a desk chair.

So I used AtFAB's parametric, open-source table configuration software (atfab.co) to create a table to my personalized ergonomic dimensions. I then adjusted the files in a CAD program, programmed the toolpaths, and cut the plywood on a ShopBot CNC router. Finally, the plywood was sanded and stained

to give it the look of reclaimed, weathered wood, and then the bench was assembled by hand.

Depending on the tools and supplies on hand in your shop, you can build this CNC Maker Bench for as little as $100 unfinished, and up to about $180 nicely finished.

1. Measure your workspace
Mark off the area with masking tape, and check that light switches and other fixtures will still be accessible.

2. Make your ergonomic and design decisions
Where do your arms rest when you're standing? For good ergonomics, this table's height should be at or just below your bent elbow height.

WHERE TO GET CNC ACCESS?

All over the world, there are FabLabs, makerspaces, hackerspaces, and TechShops where you can access a large CNC router. Costs vary widely; at AS220 Labs, where I cut these files, 2 hours of machine time was just $25. You can also have independent fabricators like *Fabhub* and *100kGarages* cut the files for you.

Find a machine near you: makezine.com/where-to-get-digital-fabrication-tool-access

Determine the dimensions of your table and record them. You'll need to enter them into the AtFAB parametric app. My table is 600mm wide, 1,520mm long, and 1,042mm high.

3. Procure your materials

I optimized my files for 18.5mm (¾" nominal) plywood. Buy the nice veneer; it looks better and resists tearout during routing. Take care to get 2 straight sheets, without veneer gaps or "voids," from the same pallet if possible. And bring your calipers — if thickness varies radically (by say, a full millimeter), choose sheets that are closer to the same dimensions. For more tips, see the project page at makezine.com/projects/cnc-maker-bench.

4. Measure your plywood (again)

Record each sheet's thickness at several points along its length and width. To ensure that your joints fit together, you'll input your maximum thickness measurement into the parametric app.

5. Create the CAD files

Now you'll actually design your table using computer-aided design (CAD) software. I used AtFAB's Parametric "One to Several" Table program, which runs on the desktop in Processing. (It's slated to be available soon as an online app at the AtFAB site.)

This table, created by award-winning architects Filson and Rohrbacher (filson-rohrbacher.com) can be configured into many different variations on the Processing app (**Figures 5a** and **5b**). You can also buy it preconfigured from atfab.co, along with 5 other furniture designs, or download those files for free from opendesk.cc/atfab.

Open the "One to Several" sketch in Processing, enter your values into the AtFAB parametric

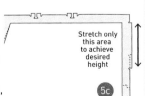

Materials

- » Plywood, ¾" (18.5mm), 4'×8' sheets (2) with a decent-looking veneer. I used Home Depot Pure Bond Plywood in birch ($48) and poplar ($40).
- » Sandpaper, 80 and 100 grit
- » Tack cloth
- » Dropcloth
- » Rags
- » Work gloves
- » Wood screws You can use dowels and glue for assembly, but our MAKE design uses #6 × 1⅝" drywall screws, Grip Rite #158SDDW1.
- » Dowels (optional)
- » Wood glue, clear (optional) such as Loctite Power Grab

For staining (optional):

- » Wood stain I used Minwax Oil-Based Ebony Wood Finish Interior Stain.
- » Pre-stain treatment I used Minwax 1-Qt. Pre-Stain Wood Conditioner.
- » Stain application pad
- » Gloves, latex or nitrile

Tools

- » CNC router, minimum 4'×6' cutting area 4'×8' preferred.
- » Calipers, digital
- » Drill with 5/16" bit and Phillips driver bit (optional) if you use screws for assembly
- » Clamps (optional) if you use glue. They're cheap at Harbor Freight (harborfreight.com).

- » Computer with software:
  - » CAD or drawing software that can manipulate vectors and open and export DXF files, such as Rhino, Illustrator, AutoCAD, or Inkscape
  - » Processing, version 2.2.1 processing.org/download/?processing
  - » ControlP5 library, version 1.5.2 code.google.com/p/controlp5/downloads
  - » AtFAB "One to Several Table" parametric design program github.com/akaziuna/cnc-standing-height-workbench

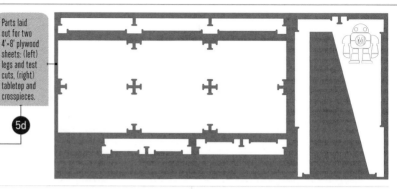

Parts laid out for two 4'×8' plywood sheets: (left) legs and test cuts, (right) tabletop and crosspieces.

5d

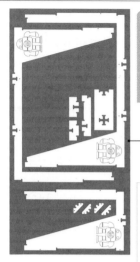

design app (see **Chart A**), and Save your custom design to a DXF file.

Lengthen the legs. The parametric app is awesome, but it doesn't lengthen the table legs to standing height. Use your favorite CAD program to lengthen them to your chosen height. Stretch only the straight middle portions of the legs, leaving the joints and feet intact (**Figure 5c**, preceding page).

Join vectors. Ensure that the vectors for each individual part are joined into one continuous shape. You'll need closed shapes to generate the toolpaths for machining. You can do this in your CAD program or use the Join tool in PartWorks or V-Carve Pro CAM software.

Rearrange the parts. Set your canvas size (sheet size) to 1,219.2mm × 2,438.4mm to avoid any possible resizing issues when importing into CAM software (DXFs exported from the app are in millimeters). You'll probably need to use screws to secure the wood to the router bed, so use an offset tool to create a 25.4mm border inside the canvas to avoid hitting these screws.

Then rearrange the parts to fit onto two 4'×8' sheets of plywood (**Figure 5d**). I used downspiral bits that pack the sawdust into the kerf, so I didn't need to create tabs to hold the parts in place while cutting. You can grab my final files at github.com/akaziuna/cnc-standing-height-workbench.

6. Program the toolpaths

You can cut your workbench with just 2 toolpaths: one inside cut (the "cross" or "plus" notches on the tabletop) and one *outside* cut for the rest of the file. (Inside and outside cuts refer to what side of the vector the router bit cuts on.) But if you're adding drill holes, or a decorative image etched into the surface of the wood (like the MAKE robot in this project), you'll need to create additional toolpaths. Before you start, download the machining and assembly instructions from opendesk.cc/atfab/one-to-several-table.

Plan toolpaths. If you're using dowels to assemble your table, you'll need to drill holes for them. I used screws instead and marked their locations with shallow 6.5mm (¼") holes.

Toolpaths must be cut in the proper order: first etching or "pocketing," then drilling, inside cuts, and finally, outside cuts. (You don't want to cut out a part and then try to drill or etch the loose part!)

Select bits. Preview your toolpaths in your CAM software, and make sure your router bit is small enough to cut any small features (**Chart B**).

Feeds and speeds. I've provided feeds and speeds settings for different sizes of 2-flute bits that worked well for me (**Chart C**). However, these numbers will

---

A VALUES ENTERED INTO ATFAB APP

FIELD	INPUT
Table width	600mm
Table length	1520mm
Constant ratio	No
Lock proportion	No
Dowel holes	Yes
Sniglet rows	5.0
Material thickness	18.5mm
Dowel diameter	6.5mm

B BIT SIZES FOR TOOL PATHS

TOOL PATH	BIT SIZE*
Test cuts	¼"
Robot pocket	⅛"
Drill	¼"
Inside profile	¼"
Outside profile	¼"

*All bits used were 2-flute downcut bits.

C FEEDS AND SPEEDS FOR SHOPBOT PRS STANDARD

SETTING	FOR 1/4" ENDMILL	FOR 1/8" ENDMILL
Stepover	0.125	0.125
Spindle speed	12,000 rpm	14,000 rpm
Feed rate	3.2"/sec	3.27"/sec
Plunge rate	1.0"/sec	1.1"/sec

For more on bit selection and chip load, see the full tutorial at makezine.com/projects/cnc-maker-bench.

vary according to your CNC's capabilities, the tooling used, and material machined. You want to move the tool as fast as the "chip load" for your bit will allow, without breaking the bit or sacrificing finish quality. If you move too slowly, the tool will heat up, wear out faster, and possibly burn the wood.

Create toolpaths. To do it in PartWorks, follow ShopBot's fantastic tutorial at shopbottools.com/msupport/ tutorials.htm.

7. Make some test cuts Cut
your "Test cuts" toolpaths and slot your test pieces together. Ideally, you should be able to fit 1–3 business cards through the assembled joints, although mine were much tighter. You may need to adjust your files to get your parts to fit properly; for tips, check the OpenDesk machining and assembly instructions.

8. Cut the files This is the fun part,
time to let the sawdust fly! Remember to wear eye and ear protection.

Load and secure the plywood, zero your axes, warm up your machine, load the appropriate bit, and cut your toolpaths in the correct order. (Drill or pocket cuts first, then profile cuts.) To see how I routed the MAKE robot, check out the videos on the project page.

9. Finish the wood
(optional) I sanded and stained my tables for a dark, weathered look with visible grain. For more details, see the project page online.

10. Assembly Assemble the
workbench in the following order, or it

won't fit together properly: crosspieces, back, side, front, other side, top.

If the joints are a little too tight, give them some encouragement with a mallet and the plywood will give. If your screw heads are too big to fit inside the ¼" drill holes, chuck a ⁵⁄₁₆" bit into your drill to countersink them.

Then, making sure the parts are aligned, screw everything together. Plywood tends to bow a bit, so I used clamps to tightly align the leg joints when adding the screws.

You're done. Stand back and admire your work. Then put your machines on it!

Share Your Design. We want to see your parametric table variations! Send your designs and stories to anna@ makermedia.com. ◐

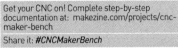

Get your CNC on! Complete step-by-step documentation at: makezine.com/projects/cnc-maker-bench
Share it: #CNCMakerBench

EASY

THE SKINNY ON
End Mills

Written by Tyler Worman ■ Illustrated by Rob Nance
Photos by Anna Kaziunas France

Don't be intimidated by CNC routing. Take the first step into machining by gaining a solid understanding of subtractive tooling basics.

INTERESTED IN CNC ROUTING BUT CLUELESS ABOUT TOOLING? Can't tell an end mill from a drill bit? Here's an overview of end mill anatomy, some basic cutter types, and tips on how to choose the correct tooling for basic wood or plastic jobs.

Drill Bits vs. End Mills

CNC machining is a subtractive process that uses rotational cutting tools called "end mills" to remove material. An end mill, while similar in appearance to a drill bit, is far more versatile.

However, in practice the terms "bit" and "end mill" are often used interchangeably. Here's the key difference. Drill bits are designed to plunge directly into material, cutting axially and creating cylindrical holes. End mills are typically used for horizontal carving and cut laterally.

Additionally, most mills are "center-cutting," meaning they are able to cut both axially and laterally. This is due to cutting flutes that extend to — and protrude from — the end face and enable plunge cutting.

To minimize tool breakage and stress on the material being cut, most CNC software will "ramp" the end mill slowly into lateral cuts.

The project type, material being cut, and desired surface finish determines the tool geometry. Key tooling features include the diameter, shank, flutes, teeth, tip shape, center cutting capability, helix angle, helix direction, length of cut, and overall tool length.

Tip Shapes and Applications

Each tip shape is designed for a particular purpose. Drill bits have a pointed center tip, while common end mill tip geometries include: fish tail, ballnose, straight, surface planing and v-bits.

Fish tail cutters will produce a flat surface, while ballnose mills produce a rounded pass and are ideal for 3D contour work. V-bits produce a "V" shaped pass and are used for engraving, particularly for making signs.

Flutes and Chipload

Flutes are the helical grooves that wrap around the sides of the end mill. Each flute has a single tooth with a sharp cutting edge (although there can be more than one) that runs along the edge of the flute.

As the tooth cuts into the wood, each

TYLER WORMAN
I'm a maker and software developer living in Ann Arbor, Michigan. I love playing with new microcontrollers and experimenting with rapid prototyping tools.

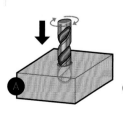

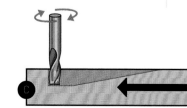

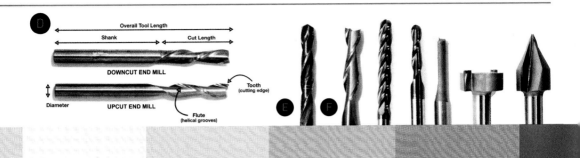

Overall Tool Length

Shank | Cut Length

DOWNCUT END MILL

Diameter | UPCUT END MILL

Tooth (cutting edge)

Flute (helical grooves)

flute whisks away a small section or "chip." The fewer the flutes, the more material that is ejected with each tool rotation.

Chipload is the thickness of a machined chip as cut by a specific tool type. More flutes create a smoother surface finish, while fewer flutes remove material fastest, but make rougher cuts.

See our full tutorial (makezine.com/endmills) for more detail, but proper chipload is important because chips dissipate heat. Hot cutters can lead to suboptimal results, including burned wood, a poor edge finish and dull tooling.

If you're machining a material like HDPE plastic, you want to use an "O" or single flute bit to clear the chips away as quickly as possible or heat will build up and melt the plastic, which can "reweld" to the tool.

Helical Direction, Chip Ejection and Surfaces Produced

A CNC router spins a cutter clockwise. The helical direction of the flutes as they wrap around the tool determine if chips are ejected towards the top or bottom of the workpiece.

True to their name, upcut mills eject chips towards the top of the workpiece, producing a cleanly cut bottom surface. The downside is possible surface splintering or "tearout" on the top surface as the chips are ejected upwards.

Downcut tools do the opposite, producing a smooth upper surface. They are ideal for pieces that have been previously engraved or v-carved and cannot be flipped to hide tearout. In addition, as downcut mills pack the chips into the cut path, they can be used instead of tabs to hold down a workpiece and keep it from moving.

Which Cutters to Buy First?

If you are looking to purchase a great wood and plastic starter set, consider picking up a few of the following carbide tool types in ¼" and ⅛" diameters:

- 2 flute upcut and downcut end mills (great for hardwood and plywood)
- 2 or 4 flute ballnose mill (great for 3D contours)
- Single or "O" flute mill (great for plastics like HDPE and acrylic)
- 60° or 90° v-bit (great for cutting hardwood signs)

The quality of your work can be significantly improved by selecting the right tooling for your project and materials —plus you'll spend less time on hand-finishing.

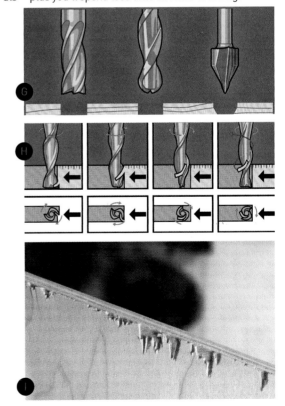

Cyberpunk Spikes

3D-print these soft, flexible
spikes and light them up with
full-color programmable LEDs

Written by Becky Stern and Phillip Burgess

BECKY STERN
(adafruit.com/becky
stern) is a DIY guru
and director of wear-
able electronics at
Adafruit. She pub-
lishes a new project
video every week
and hosts a live
show on YouTube.

Time Required:
A Few Hours
Cost:
$70–$85

Materials

» NeoPixel RGB
LED strip, 60 per
meter, individually
addressable Adafruit
Industries part #1138,
adafruit.com
» NinjaFlex flexible 3D
printing filament, Snow
White Adafruit #1691
» Adafruit Gemma
microcontroller Maker
Shed item #MKAD71,
makershed.com, or
Adafruit #1222
» Slide switch, SPDT, 0.1"
pin spacing Adafruit
#805
» Battery, LiPo, 500mAh
Adafruit #1578
» Battery extension
cable, JST male-female
Adafruit #1131
» Rare earth magnets (6)
Adafruit #9
» Safety pins or needle
and thread
» Silicone adhesive,
Permatex 66B
» Heat-shrink tubing
» Tape, nonconductive

Tools

» 3D printer, fused-
filament type
» Computer running
Arduino IDE software
free download from
arduino.cc/en/main/
software
» Soldering iron
» Solder, rosin core,
60/40
» Scissors
» Wire cutters / strippers

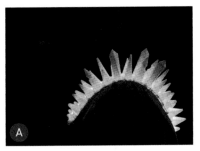

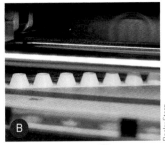

Becky Stern

MAKE YOUR OWN FLEXIBLE, SPIKY, GLOWING ACCESSORY USING NEOPIXEL LED STRIP lights diffused by NinjaFlex flexible 3D printing filament! Magnets let you attach the spikes to anything in your wardrobe. The soft flexible enclosure holds Gemma, the tiny microcontroller that animates the LEDs, and a rechargeable lithium polymer battery.

We designed 2 styles of spike strip — one with regular round spikes and one crystal-inspired statement piece (Figure A). Whichever you choose, it'll get you noticed!

1. 3D-PRINT THE SPIKES AND ENCLOSURE

Download whichever spikes you like from thingiverse.com/thing:262494 and print them in NinjaFlex filament at 225°F with a nonheated build plate (Figure B). For more tips on working with NinjaFlex, check out the guide by the Ruiz Brothers at learn.adafruit.com/3d-printing-with-ninjaflex.

Also download and print the 2 pieces of the flexible enclosure for the Gemma microcontroller and battery, from thingiverse.com/thing:262522. Since it's printed in NinjaFlex, the enclosure is soft and flexible, yet firm enough to protect your components (Figure C). The enclosure shape includes tabs for pinning or sewing to your garment.

2. PREPARE THE NEOPIXEL STRIP

Prepare the input end of your NeoPixel strip by tinning the pads with solder. The strip won't work if you solder wires to the wrong end, so be sure the arrows on the PCB point away from the end you're wiring.

Solder 3 stranded wires, about 8" long, to the tinned pads of the NeoPixel strip. To prevent the solder joints from being too cramped, solder the center pad's wire on the reverse

side of the PCB as shown: 2 on top, one on bottom (Figure D).

Wrap 3 rare-earth magnets in tape to prevent short circuits (Figure E), and slide them into the NeoPixel strip sheathing on the underside of the PCB (Figure F). Our spike strip is 16 pixels long, and we used 3 magnets evenly spaced (one at each end and one in the center).

Prepare a protected work surface in an area with good ventilation.

Use Permatex 66B silicone adhesive to affix the 3D-printed spikes to the NeoPixel strip (Figure G). Apply adhesive to both the strip's silicone sheathing and the NinjaFlex strip of spikes, using a toothpick to spread it around if necessary (Figure H).

Squish a bit of silicone adhesive into the ends of the NeoPixel strip sheathing to provide water resistance and strain relief (Figure I). Allow adhesive to dry overnight.

3. ASSEMBLE THE CIRCUIT

Route your NeoPixel strip's wires through the hole at the top of the enclosure (Figure J, following page), and solder them up to Gemma as follows: NeoPixel GND to Gemma GND; NeoPixel + to Gemma Vout; and NeoPixel signal to Gemma D1 (Figure K, following page).

Seat Gemma into the round outline inside the enclosure, with the USB port facing its opening at the bottom end of the enclosure (Figure L, following page).

Use a JST extension and slide switch to make this tiny adapter (Figure M, following page). Solder the

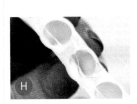

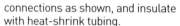

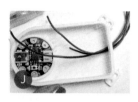

connections as shown, and insulate with heat-shrink tubing.

The slide switch fits into the opening in the enclosure (Figure Ⓝ). Now you can easily power up your circuit while still making it easy to disconnect the battery for recharging.

Connect the battery, fit everything neatly in the enclosure (Figure Ⓞ), and press on the lid.

4. LOAD THE CODE

Download the NeoPixel library from github.com/adafruit/Adafruit_NeoPixel. Rename the folder (containing the *Adafruit_NeoPixel.h* and *.cpp* files) to *Adafruit_NeoPixel* (with the underscore and everything), and place it alongside your other Arduino libraries, typically in your *[home folder]/Documents/Arduino/Libraries* folder.

Now open the *strandtest.ino* sketch from the *Examples* sub-folder, and upload it to the Gemma using the Arduino IDE.

Did that sound like gibberish to you? It's easy — if it's your first time, just read the "Introducing Gemma" and "NeoPixel" guides at learn.adafruit.com before you start.

The code is well commented to guide you through each part of the sketch and what it does. Let's have a look:

DECLARING AN OBJECT

All NeoPixel sketches begin by including the header file:

```
#include <Adafruit_NeoPixel.h>
```

The next line of code assigns a number to the symbol **PIN** for later reference. (This isn't necessary, it just makes it easier if you want to change the microcontroller pin where the NeoPixels are connected without digging deeper into the code.) Your strip is connected to Gemma's pin 1:

```
#define PIN 1
```

The next line declares a NeoPixel object:

```
Adafruit_NeoPixel strip = Adafruit_NeoPixel(16, PIN, NEO_GRB + NEO_KHZ800);
```

We'll refer to this by name later to control the strip of pixels. There are 3 parameters or *arguments* in parentheses:
- » The number of sequential NeoPixels in the strip, in our case **16**. (Yours might be longer.)
- » The pin to which the NeoPixel strip is connected. Normally this would be a pin number, but we previously declared the symbol **PIN** to refer to it by name here.
- » A value indicating the type of NeoPixels that are connected. (You can leave this off; it's mainly needed for older NeoPixels.)

DEFINING COLORS AND BRIGHTNESS

The next block of code lets you define favorite colors, which the NeoPixel will call upon later:

```
// Here is where you can put in your
favorite colors that will appear!
// Just add new {nnn, nnn, nnn}, lines.
They will be picked out randomly
//                     R    G    B
uint8_t myColors[][3] = {{232, 100, 255},
// purple
                    {200, 200, 20},
// yellow
                    {30, 200, 200},
// blue
                    };
```

There are 2 ways to set the color of any pixel. The first is:

```
strip.setPixelColor(n, red, green, blue);
```

The first argument — **n** in this example — is the pixel number along the strip, starting from 0 closest to the Arduino. If you have a strip of 30 pixels, they're numbered 0 through 29. It's a computer thing. (You'll see various places in the code using a **for** loop, passing the loop counter variable as the pixel number to this function, to set the values of multiple pixels.)

The next 3 arguments are the pixel color, expressed as numerical brightness levels for red, green, and blue, where **0** is dimmest (off) and **255** is maximum brightness.

An alternate syntax has just 2 arguments:

```
strip.setPixelColor(n, color);
```

Here, color is a 32-bit type that merges the red, green, and blue values into a single number. This is sometimes easier or faster for programs to work with; you'll see the *strandtest* code uses both syntaxes in different places.

You can also convert separate red, green, and blue values into a single 32-bit type for later use:

```
uint32_t magenta = strip.Color(255, 0, 255);
```

Then later you can just pass **magenta** as an argument to **setPixelColor** rather than the separate red, green, and blue numbers each time.

The overall brightness of all the LEDs can be adjusted using `setBrightness()`. This takes a single argument, a number in the range 0 (off) to 255 (max brightness). For example, to set a strip to ¼ brightness, use:

```
strip.setBrightness(64);
```

ANIMATED EFFECTS

In the *strandtest* example, `loop()` doesn't set any pixel colors on its own — it calls other functions that create animated effects. So ignore it for now and look ahead, inside the individual functions, to see how the strip is controlled.

You'll see code blocks with parameters you can tweak to:
» change the rate of twinkling
» change the number of pixels to light at one time
» transition colors gradually through the whole spectrum
» display rainbow colors, static or animated
» flash or fade random pixels.

5. WEAR IT!

You can stitch or pin the 3D-printed enclosure to your garment wherever you'd like, using the mounting tabs. For permanent use, stitch a pocket for this enclosure inside your garment and route the wires inside (Figure **P**).
» Use a fluffy bun-maker hair accessory and tuck the enclosure under it to wear these spikes around your head (Figure **Q**)!
» Epaulets, two styles (Figure **R** and **S**).
» Around the collar (see page 126).
» Cyber dragon, anyone? Try the crystal-inspired spikes (Figure **A**, page 127).

How will you wear it? We'd love to see your variations! ●

See more photos, and share your spike builds and costume ideas at makezine.com/projects/cyberpunk-spikes. This tutorial originally appeared on the Adafruit Learning System at learn.adafruit.com/cyberpunk-spikes.

3D-Printed
Pinhole Camera

The fully functional P6*6 camera uses 120 roll film, comes in 35mm and 50mm lengths, and is printable without support even on the tiniest of print beds. Written and photographed by Todd Schlemmer

TODD SCHLEMMER is a firefighter/paramedic living in Seattle. He has studied paleontology and broadcasting, and has worked as a chef, a Birkenstock store manager, and a developer support engineer at Microsoft. His hobbies include 3D printing, photography, writing, computers, robotics, boatbuilding, auto mechanics, ham radio, and cooking. Todd is currently assembling a Shapeoko CNC router kit and lusts for a laser cutter.

THE P6*6 IS A 3D-PRINTED PINHOLE CAMERA, glued and fastened together with 3mm nuts and bolts. All of the individual parts print without support and fit on a 6-inch square print bed. The files are available for download from thingiverse.com/thing:157844.

The P6*6 comes in two focal lengths, 35mm and 50mm. It uses 120 roll film and makes an impressive 6cm square negative — roughly 4 times larger than a negative from a standard 35mm camera. 120 film is widely available and can be found at camera stores that cater to professional photographers or from internet vendors.

1. Print your camera parts

Download the 3D files from Thingiverse and print the camera parts (Figure ❶):

» **A** — Knob, used to advance the film
» **B** — Cap, snaps onto the body
» **C** — Baffle
» **D** — Winder, engages the take-up spool
» **E** — Film clip, keeps film tightly wound on the spool during unloading
» **F** — Frame slide, allows viewing of frame number on film backing
» **G** — Body
» **H/I** — Body clip and leveling spacer, prints as joined pieces

P6*6 SPECS:

120 film, 6×6 format

50mm focal length:
  f-stop of f/167 with 0.30mm pinhole
  62 degree vertical and horizontal angles of view

35mm focal length:
  f-stop of f/135 with 0.26mm pinhole
  77.4 degree vertical and horizontal angles of view

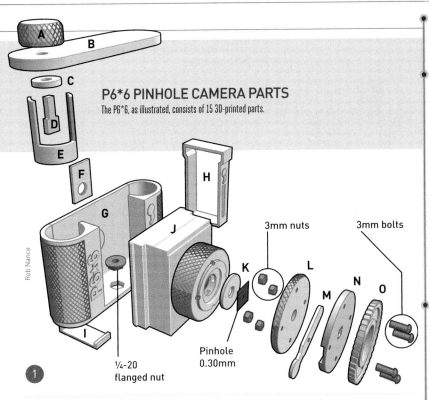

P6*6 PINHOLE CAMERA PARTS
The P6*6, as illustrated, consists of 15 3D-printed parts.

Rob Nance

A
B
C
D
E
F
G
H
I
J
K
L
M
N
O

3mm nuts
3mm bolts

¼-20
flanged nut

Pinhole
0.30mm

(1)

» **J** — Extension, 50 mm or 35 mm
length
» **K** — Pinhole disc, replaceable
pinhole mount
» **L** — Pinhole clamp
» **M** — Shutter blade
» **N** — Shutter clamp
» **O** — Trim ring

When preparing the STL files for
printing, use the following slicing
settings:
» 0.25mm layer height
» 2 perimeters (or "shells")
» 3 solid layers top and bottom
» 50% infill

2. Smooth and fit the printed parts

Every joint between parts in the P6*6
has a potential for photo-ruining light
leaks —unintended openings that allow
light into the camera. Careful attention
to fit will ensure awesome photos.
If necessary, use fine sandpaper or
a file to smooth mating surfaces.

Carefully enlarge bolt holes with
a ⅛" drill bit.

Pay special attention to the frame
surface, formed by the bottom of the

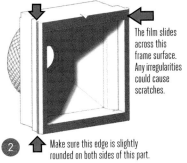

The film slides
across this
frame surface.
Any irregularities
could cause
scratches.

(2) Make sure this edge is slightly
rounded on both sides of this part.

extension — the film slides across this
surface when winding, and it forms the
margin of your photographs. Surface
irregularities could scratch the film,
and an unevenly trimmed inside
perimeter will be preserved as an
uneven border on every photograph you
make. Additionally, slightly round and
smooth the bevel edge of the frame to
avoid scratches on the film **(Figure ②)**.

Before proceeding, check the fit of all
mating parts. Refer to the exploded parts
diagram. All parts should fit together
without distorting. The cap should fit
the body securely. The shutter blade
should be slightly snug between the
pinhole clamp and the shutter clamp.

Time Required:
90 Minutes
Cost:
$12

Materials
» 3D-printed parts see Step 1
» Nuts, 3mm (4)
» Bolts, 3mm×15mm long (4)
» Washers (4) (optional)
» Flanged nut, ¼-20 for
tripod mount
» Adhesive-backed velvet or
similar to "trap light" that
would leak through joints in
the assembly
» Translucent red plastic,
15mm–18mm disc A cheap
plastic binder is a good
source.
» Thin sheet metal with
0.26mm or 0.30mm
pinhole
» Black permanent marker
for back of pinhole (no
internal reflections!)

Tools
» 3D printer (optional)
with black ABS or PLA
filament other colors will
require flat black paint on
interior surfaces. To find a
machine or service you can
use, see makezine.com/
where-to-get-digital-fabri-
cation-tool-access
» Flat files, large and small
» X-Acto / hobby knife
» Sandpaper, 500-1000 grit
» C-clamp or rubber bands
» Allen wrenches, small for
bolts, and for manipulating
tiny nuts inside the extension when assembling the
shutter
» Drill with ⅛" bit
» Epoxy, dark such as JB Weld
» Super glue aka cyanoacrylate (CA) glue
» ABS plumbing glue, black
for ABS only
» Plastruct "Plastic Weld" will
bond all manner of plastics

NOTE: RELAX —
The pinhole is not as
critical as it seems.
You can purchase a
precisely laser-drilled
pinhole on the Internet
or easily make your own
from brass shim stock,
a soda can, pie plate,
etc. (Aluminum foil is
too fragile.)

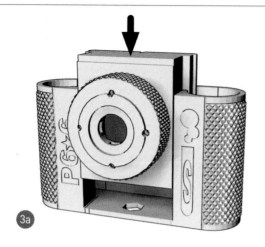

3a

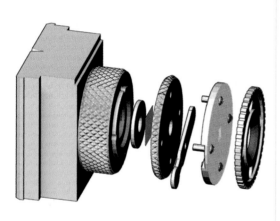

3b

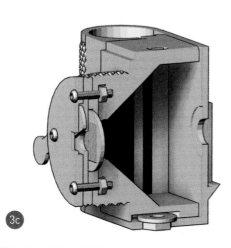

3c

3. Final assembly

TRIPOD MOUNT
A ¼-20 nut is the standard tripod attachment. Carefully bond the flanged nut in its hexagonal hole in the body (Figure 3a), flush with the bottom of the body, using a bit of epoxy on the inside.

CAP AND WINDER
Parts A, B, C, and D (see Figure 1).

The winder drive passes through the baffle and cap and into the knob. This is designed to be a friction fit. If the narrow part of the winder drive is slightly too large to fit through the baffle into the knob, enlarge the holes rather than reducing the size of the winder shaft.

EXTENSION AND PINHOLE/SHUTTER
Parts J, K, pinhole, L, M, N, O, nuts, bolts, and (optional) washers. See diagrams.

Everything should fit together tightly prior to fastening. The extension, pinhole clamp, and shutter clamp must fit without interference (Figure 3b).

Bolting all these parts together can be a bit fiddly, but it's important to assemble them before gluing the extension and body together. A small Allen wrench is handy to position the nuts in the nut traps (in the extension) during assembly (Figure 3c). The shutter should snap open and closed. It is easy to overtighten the bolts. Use super glue to mount the trim ring on the face of the shutter clamp.

VELVET LINING AND RED WINDOW
For best results, the inside back of the body can be lined with velvet behind the frame. The velvet provides a gentle friction that keeps the film in place and serves to reduce the effect of stray light from the frame index window. Lining the inside surface of the cap also minimizes light leaks (Figure 3d).

Cut a 15mm–18mm disc of transparent red plastic, and tack it in place in the recess inside the body with a couple tiny dabs of super glue. The hole in the middle of the adhesive-backed velvet will overlap the disc and secure it in place. Carefully use the tip of an X-Acto blade to slide the velvet into position when attaching it to the body and cap. It must be wrinkle free.

BODY/EXTENSION JOINT
"Dry-fit" the extension and body before gluing them together. They'll only fit one way — the "50" (or "35") marking will be visible. Any interference could mean light leaks. The tripod nut must fit without difficulty. Resolve any issues before you glue.

During gluing, space the frame surface about

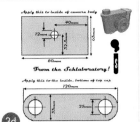

Apply this to inside of camera body
40mm
12mm
32.5mm
65mm
80mm

From the Schlaboratory!

Apply this to the inside, bottom of top cap.
120mm
33mm
25mm
87mm

3d
Download the self-adhesive velvet dimensions from thingiverse.com/thing:157844

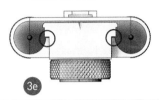

Mind the gaps! A proper fit looks like this.

3e

3f

0.50mm away from the velvet (Figure 3e). You can use 5 sheets of printer paper (0.10mm thick each).

GLUING

For ABS, plumbing cement works well (and comes in camera black). Work fast — the solvent evaporates quickly and the cement gets rubbery.

For bonding PLA, a dark epoxy is best, but gap-filling super glues or "Plastic Welder" type glues also give good results.

Follow the directions on the label. Too much glue will ooze out of the joint and muck up your lovely camera's appearance. Use a C-clamp or stout rubber bands to precisely clamp the 2 parts together.

Allow the glue to dry, load your new camera with 120 film (Figure 3f), slide the body clip onto the camera, and make some pinhole photographs (Figure 3g)! ✪

Get more photos and tips, and share your photos at makezine.com/projects/3d-printed-pinhole-camera

3g

3 FUN THINGS TO
3D Print
Written by Eric Chu

1

2

3

1. Low-Poly Bulbasaur by Agustin Flowalistik
thingiverse.com/thing:327753
This low-polygon representation of Bulbasaur from the Pokémon series really captures the essence of the little monster. Flowalistik is designing more, so follow him to catch (and print) them all.

2. Fuzzy Bear by Robo
(Remixed from Ice Bear by Virtumake)
thingiverse.com/thing:71156
A fuzzy, high-polygon texture was applied to this bear using Blender. It provides a tactile experience that feels great in the hand. Try printing with different materials to get different feels.

3. Lolly Box by Faberdasher
youmagine.com/designs/lolly-box
This 3-piece stash box in the form of a popsicle (or ice lolly, as it's known as in the UK and Ireland) is perfect for those sunny days.

CNC Air Raid Siren

It's loud, annoying, and fun. Cut the parts on a CNC router, then motorize them with a cheapo bench grinder! Written by Dan Spangler

**Time Required:
A Weekend
Cost:
$60-$75**

DAN SPANGLER is the fabricator for MAKE Labs and our resident retro technology connoisseur.

Learn more about CNC:

+SKILL BUILDER
CNC PANEL JOINERY
makezine.com/cnc-panel-joinery

Learn a bag of tricks to design and CNC cut clever joints in plywood, acrylic, and other sheet stock.

AIR RAID SIRENS FASCINATE ME, ESPECIALLY THE ONES FROM WORLD WAR II. The infamous wail indicated danger but also sounded the all-clear, inspiring both fear and relief. They're also just awesomely loud. Nowadays there are electronic sirens, but most civil-defense sirens are still the mechanical kind — basically blowers designed to make as much noise as possible.

I saw DIY sirens online and instantly thought of our ShopBot CNC router as an elegant solution — it could cut a perfectly balanced rotor every time. So I designed this siren for CNC cutters. Here's how I made it, and how you can too.

1. Motor. Fractional-horsepower AC motors cost $100 or more — so I used the MAKE Labs' crummy bench grinder. Get one at Harbor Freight ($45 brand new). It even looks like a WWII siren.

2. Cutting. I cut the plywood to 24"×18" to fit our ShopBot Desktop (Maker Shed item #DSSBDP, maker-shed.com). I fit all the parts for two rotor-stator assemblies on one sheet of ¼" ply and two sheets of ¾", and cutting went off without a hitch. Adjust them to fit the cutter of your choice.

3. Assembly. Building the rotors is a matter of glue, dowels, and gentle persuasion with a rubber mallet. Testing them was scary — would they hold together or explode catastrophically? — but they spooled up to full speed and blew a surprising amount of air with almost no vibration. Sweet!

Next I mounted the stators and adjusted them to eliminate any rubbing. It was finally time to see if this thing was going to work. As the  motor picked up speed, a faint wail began to emanate from the device, which quickly got louder and louder. By the time the motor got up to full speed the siren sounded so real and loud I had half a mind to duck and cover under my desk — it works! ⊘

Hear it wail, download the CAD files for CNC cutting, and get step-by-step build instructions at makezine.com/projects/cnc-air-raid-siren.

Robots
AND Drones   PART 5 >>>>>>

One of the most fun things you can do in electronics is to build robots. Once your electronics project begins to move around in the physical world, you're going to attract lots of attention! Along the way, you'll also learn a lot about how electronic and mechanical devices work together.

Jérôme Demers shows you how to get started quickly with a simple little robot he calls "Beetlebot." Despite the fact that it is so simple and inexpensive, and requires no programming, the Beetlebot can feel its way around rooms and avoid obstacles. It's a great project to do with kids.

Kris Magri developed her robot Makey when she was working as an intern at *Make:* magazine. The design you see here was her sixth. As of this writing, Kris is a Mechatronic Engineer working at Disneyland. Her recent animatronics projects for Disney include two of the seven dwarfs. And she gave Jack Sparrow a hand—literally!

You never know where an interest in hobby robotics will take you. It took Judy Aime' Castro to China. CoffeeBots started as an interactive project for an art show. Judy wrote up the project for *Make:* magazine, which led to her being invited to Maker Faire Shenzhen. There her robots won a blue ribbon, and one of them (CoffeeBot Zombie) was added to the permanent collection at Inno Park. You can learn how to build your own CoffeeBot and give it some personality by working through Judy's project, which was specially updated for this volume.

What could possibly be more fun than building robots? How about building and *flying* robots! That's one way to think of drones. They scale from simple toys up to sophisticated, programmable robots that can fly themselves, with applications ranging from consumer to professional to military. The editors of *Make:* introduce you to the anatomy of a drone with some sophisticated graphics that reveal all the details. Intel engineer Mikal Hart—inventor of the Reverse Geocache Puzzle—explains how to integrate GPS into your drone so it can find its way home.

Chad Kapper will get you off the ground with "The Handycopter UAV," a basic quadrotor that you can build mostly from parts found at your local hardware store. Chad is a video producer who is also passionate about r/c flight. You'll find great videos on his website FliteTest.com.

For a wealth of additional information about building and piloting a variety of r/c aircraft, be sure to check out the archive of Lucas Weakley's "Maker Hangar" video series on the makezine.com website. Lucas' project, which is reprinted here, introduces you to a different kind of multirotor drone—the tricopter. Read on to discover why the three-rotor configuration has some distinct advantages over the more common quad setup.

Photography by Jérôme Demers

BEETLEBOT

Ultra-simple bugbot navigates obstacles with feelers and switches. By Jérôme Demers

The Beetlebot is a very simple little robot that avoids obstacles on the floor without using any silicon chip — not even an op-amp, and certainly nothing programmable. Two motors propel the bugbot forward, and when one of its feelers hits an obstacle, the bot reverses its opposite motor to rotate around and avoid it. The project uses only 2 switches, 2 motors, and 1 battery holder, and it costs less than $10 in materials (or free, with some scrounging).

Beetlebot in 10 Easy Steps

1. Cut pieces of heat-shrink tubing and use a heat gun or other high-heat source to shrink them onto the motor shafts. Trim the tubing evenly, with a little bit running past the ends of the shafts. These will act as tires, improving traction (Figure B).

2. Glue the SPDT switches to the back of the battery holder, at the end with the wires. The switches should angle out at the 2 corners with their levers angled in toward each other, as shown in Figure C. Also, the contacts farthest from the buttons on each (the normally closed contacts) should touch. This will be the front end of our bugbot.

3. Cut the metal strip, mark enough length at each end to hold a motor, and bend each end in at about a 45° angle. This is your motor plate.

4. Examine or test your motors to determine their polarity. Tape the motors onto opposite ends of the motor plate so that their shafts point down and angle out. Orient their positive and negative contacts so that they'll spin in opposite directions.

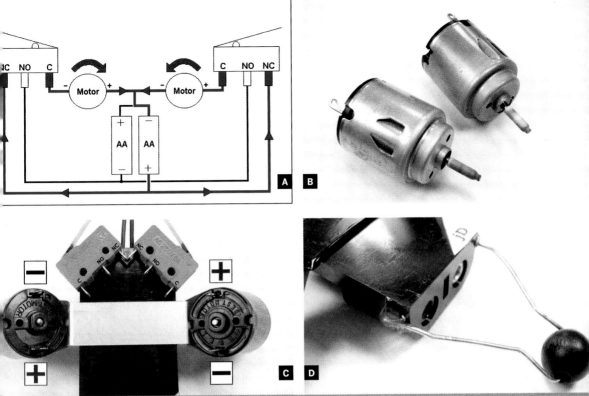

Fig. A: Diagram of Beetlebot running free, not bumped into anything; both motors draw current from the right battery only. Fig. B: Heat-shrink tubing acts as tires, giving traction to the motor shafts.

Fig. C: Switches and motor plate glued to the back of the battery holder. Fig. D: Bent paper clip threaded through a bead and glued to the rear end of the battery holder to make a rolling caster.

MATERIALS

1.5V motors (2) You can often scavenge these from toys, dollar store fans, etc.
SPDT (single pole double throw) momentary switches with metal tabs (2) You can scrounge these from an old VCR or mouse, or buy new ones for $1–$4 apiece.
Electrical wire around 22 gauge
AA batteries (2) You can also use AAAs.
AA battery holder
Spherical bead plastic or wood
Heat-shrink tubing to shrink to the widths of the motor shafts and the antennae connectors
Black electrical tape
Terminal connectors, spade type, small (2)
1"×3" piece of scrap metal plate I used aluminum.
Paper clips (4)
Cyanoacrylate (Super/Krazy) glue or epoxy
Soldering iron and solder
Toggle switch (optional) for on/off switch

FOR THE SHELL (OPTIONAL)

Round plastic lid I used a lid from a container of hair gel, but you can also use a peanut butter jar lid or anything similar.
Auto body filler putty or epoxy glue
Black and red enamel paint and primer
Clear varnish
Small, thin magnets (2) to attach shell to body

5. Use cyanoacrylate glue or epoxy to glue the motor plate down onto the back of the battery holder, just behind the switches (Figure C). Orient the motors so that the left motor spins counterclockwise as you view it from below, and the right one spins clockwise. For aesthetics, I then covered the plate with black electrical tape.

6. Unbend a paper clip, slip it through the bead, and bend it symmetrically on either side to make a caster (Figure D). Attach each end of the clip to the corners of the battery holder at the back. I used hot glue — not very professional. You could also try bending the clip ends under and soldering them to the battery connection tabs, but if you apply too much heat to the tabs, you might melt the plastic and ruin your battery holder. Beware!

Next we'll wire up the circuit, but first, an explanation: the key is that the 2 batteries work separately. Battery holders usually connect cells in series and combine their voltages, but with the Beetlebot, a wire soldered between the 2 puts them into separate subcircuits. The motors draw from only 1 battery at a time. Each switch's common connection (C) runs to a motor. The switches' normally open (NO) terminals connect together and run to the battery

E F

G H

Fig. E: Use pieces of paper clip and insulated wire to solder connections between the switches, motors, and battery holder. Fig. F: Complete wiring, with battery holder leads soldered to switch terminals.

Fig. G: Removable antennae made from paper clips use spade connectors to slip onto switch levers. Fig. H: The bare-bones Beetlebot, finished and working, but without any switch or decorative shell.

holder's negative lead, while the switches' normally closed (NC) legs run to the positive lead.

When the bot isn't hitting anything, voltage from the positive-side battery splits and runs through both motors via the NC terminals, and the negative-side battery is not used at all. But when a switch button is activated, it closes the circuit with the negative-side battery, through the NO terminal. This reverses the motor direction on that side while the unactivated side continues running forward, which results in a quick turn away from the obstacle.

When both switches activate, both motors momentarily run backward, and the bot backs away. (The feelers cross in front, so a bump on one side activates the button on the opposite side.) That's all there is to it. Now, back to the build.

7. Solder together the 2 switches' NC terminals that are close or touching. Then solder together their NO terminals, the middle legs. I use pieces of paper clip for short joins like this, since it's faster and stronger. Then connect the common leg of each switch to the front terminal of its nearest motor (Figure E, top).

8. Solder a wire between 2 motors' rear terminals. Connect another wire from either one to any contact point on the battery holder that's electrically in between the 2 batteries (Figure E, bottom). This is the Beetlebot's all-important "third connection."

9. Finish the wiring by soldering the battery holder's positive lead to the switches' NC terminals, and its negative lead to either of the switches' NO terminals (Figure F).

10. Remove the insulation from the 2 spade connectors, and unbend 2 paper clips. Slip the connectors over the paper clips, then squeeze them down with pliers and solder in place. Dress up the connection with some wide heat-shrink tubing (Figure G). These are the Beetlebot's feelers. The spade connectors clip onto the switch levers, which makes them easy to detach for packing, and prevents damage to the fragile SPDT switches. The long paper clips give sufficient leverage to activate the switches, even if they seem hard to trigger with your finger directly.

Your robot is finished (Figure H)! Add 2 batteries, and it should come to life. If it spins in a tight circle or runs backward, you need to reverse one or both of the motor connections. To change the bot's speed or to make it run straighter, bend the metal plate to adjust the motors' angles.

I J

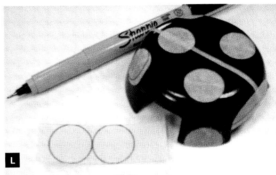

K L

Fig. I: The Beetlebot's on-off switch connects between the motors and batteries. Fig. J: Plastic lid cut to accommodate motors and antennae.

Fig. K: Building up and shaping the shell with auto body putty. Fig. L: Dime-sized masking tape circles give Beetlebot its spots.

For additional diagrams of how the circuit works, see makezine.com/projects/beetlebots.

Adding an On/Off Switch (Optional)

Every time you want to stop the robot, you need to remove the battery, which can get annoying. To solve this problem, splice a toggle switch onto the "third connection" wire between the motors and the batteries. Cut the wire, then solder in the switch and glue it to the edge of the battery holder. I neatened this connection up with more heat-shrink (Figure I).

Making the Shell (Optional)

Now here's the aesthetic part: adding the shell. I made mine out of the green plastic lid from a container of hair gel.

1. Fit the lid over the bot and cut holes in the sides to make room for the motors and the front switches/antennae (Figure J).

2. To make the shell more round, cover it with auto body putty (watch out — it cures pretty fast!) or epoxy glue, and then use files to shape and smooth it (Figure K). For final touch-up, I filled in any holes with a softer putty.

3. After sanding the lid smooth, give it a couple coats of primer, and then paint it. To make a ladybug beetle pattern, I started by painting the whole thing black (I also painted the antennae black). Then I used a dime as a template to cut round pieces of masking tape, which I applied to the lid along with a thin masking tape centerline (Figure L).

I painted glossy red over everything, and then removed the tape. For the final polish, I sanded the whole thing with very fine sandpaper and some water, which gives a glossier finish than sanding dry; this is a trick I learned from a friend who was restoring a guitar. I let everything dry and gave it 2 coats of clear varnish.

4. To connect the shell to your robot, you can glue it directly to the battery holder, or you can use magnets; glue one inside the lid and another in a matching position on the battery holder. This lets you remove the shell easily, to show your friends the insides of your biomech bug!

Jérôme Demers is a student in electronics engineering at the University of Sherbrooke in Québec. He is currently working on advanced sumo robots in both the 500g and 3kg categories.

MY ROBOT, MAKEY
By Kris Magri

Photograph by Sam Murphy

ROBOT ON, DUDE

I've made some fun robots, but I never liked the way they look, with their parts stacked up, and black tape and visible wires running everywhere. I decided to build a bot that combined tried-and-true workings with some Hollywood bling. I sketched some ideas, and this is the one that spoke to me.

Makey is an autonomous robot that I've programmed to follow objects around. It uses tank steering, aka differential drive, where separate DC motors power each of the 2 drive wheels. A servomotor moves its head, which carries a single ultrasonic rangefinder. Control comes from an Arduino microcontroller, which I've programmed to do object following and obstacle avoidance. With these behaviors, Makey constantly turns its head right and left to acquire differential ranging data, adding to its personality.

The chassis design is unusual in that there are no exposed components; I vowed to keep everything enclosed, which made the design harder, but it was worth it.

With different Arduino programming, the hardware would support mapping and other activities, and with a few hardware mods, Makey could also compete in Mini-Sumo, one of the most popular events in robotics competitions.

Kris Magri works at Disneyland as a Mechatronic Engineer. She is responsible for both the electrical and the mechanical issues for show effects and animatronics in Disney California's Adventure.

BRAINS AND BRAWN

Makey's Arduino microcontroller brain takes readings from the ultrasonic rangefinder on its head to "see" its distance to nearby objects. Using this data, the microcontroller determines where Makey should look and move next, by controlling the "neck" servo and wheel drive motors.

❶ **Rangefinder** Many bot builders use left and right IR sensors for ranging, but Makey uses a sonar unit that senses longer distances, up to 10 feet.

❷ **Servomotor** The servo turns Makey's neck, enabling its rangefinder to point in different directions.

❸ **Motor driver** The Arduino's output pins don't deliver enough current to power the drive motors directly, so the dual motor driver board takes the signal and uses it to route battery power to the motors.

❹ **Breadboard** Solderless breadboards are great for wiring up electronics without having to solder. Basically your best friend in electronics prototyping.

❺ **ProtoShield** Mounts on top of the Arduino, to give you a place to solder your circuitry.

❻ **Microcontroller** The Arduino reads the rangefinder input and controls the motors. Output pins set the drive motor speeds via pulse-width modulation (PWM) and direct the servomotor using pulse-width position servo (PWPS) pulses.

❼ **Skidder** Forms a stable triangular base with the 2 drive wheels. The ball bearing is smoother than casters.

❽ **Drive motors** Each wheel is powered by its own DC motor. This "tank drive" gives Makey maximum maneuverability. By running its 2 motors in opposite directions, the robot can turn in place. Capacitors between each motor's 2 contacts reduce its noise to the rest of the circuit.

❾ **Batteries** In single-battery robots, the motors can hog too much power and cause the microcontroller to reset. Makey avoids this problem by using one battery for brains, and another one for brawn.

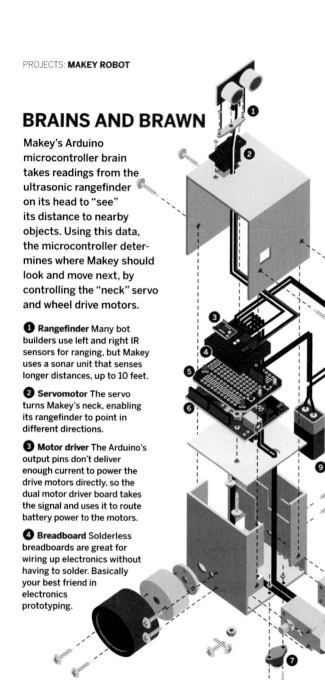

Illustration by Nik Schulz/L-dopa.com

HOW IT WORKS

For object-following behavior, Makey's code uses a strategy called "proportional control." Rangefinder readings are compared to an optimal distance. If they are equal, the drive motors stay still. Readings are taken in pairs, facing right and facing left, with each determining the speed on that side. If nothing is within sensor range, the motor runs at maximum speed.

An object at far left makes the left motor run at proportional speed while the right runs at maximum. This turns Makey left.

An object at dead center and closer than the optimal distance makes the motors run backward equally.

An object slightly to the right makes the left motor run slightly faster, turning Makey to the right.

SET UP.

MATERIALS

[A] Aluminum sheet,
0.032" thick, at least
9"×10", 5052 alloy

[B] 9V batteries (2)
[BB] 9V battery snaps (2)
Jameco #11280
(jameco.com)

**[C] Jumper wires, solid
core, 22 gauge** Maker Shed
#MKEE3 (makershed.com)
or Jameco #19290, or cut
your own from solid core wire

**[D] Du-Bro Mini E/Z
Connectors and Micro
Aileron System** Hobby-
Town USA #DUB845 and
#DUB850 (hobbytown.com)

[E] Paint I used Rust-Oleum
Painter's Touch Apple Red
Gloss #1966.

**[F] Arduino Duemilanove
microcontroller** The Diecimila version will also work.

**[G] Dual motor driver
board, 1A Dual TB6612FNG**
SparkFun #ROB-09457
(sparkfun.com)

[H] ProtoShield kit
SparkFun #DEV-07914

[I] Rigid plastic, around
3½"×3½"×1/16"

**[J] Lego tires, 49.6×28
VR (2)**

[K] Heat-shrink tubing

[L] FASTENERS
Machine screws:
1-72×¼" (2),
4-40×⅜" (2),
4-40×1" (8)

Sheet metal screws,
#6×¼" (4)

Nuts: 1-72 (2), 4-40 (6)

#4 lock washers (4)

⅛" pop rivets (2)

**[M] On/off switch, SPST
rocker** Jameco #316022

**[N] Servo extension
wire, 12"** HobbyTown USA
#EXRA115, to plug into the
rangefinder, not the servo

**[O] 3-pin right-angle male
headers (2)** Break apart a
10-pin header like Jameco
#103393.

**[P] Servomotor, Hitec
HS-55 sub-micro**
HobbyTown USA
#HRC31055S or ServoCity
#31055S (servocity.com)

[Q] 0.1μF capacitors (2)
Jameco #15229

[R] 10kΩ resistors (2)
Jameco #691104

[S] ⅜" metal ball caster
SparkFun #ROB-08909

**[T] 9V battery holder clips
(2)** Jameco #105794

**[U] Wire, solid core
insulated and stranded
insulated, 22 gauge,
red and black**

**[V] Scrap wood, 1×3 or
wider (¾" thick), 6" long**

**[W] GM2/3/8/9/17
gearmotor mounts (2)**
aka hubs, Solarbotics #GMW
(solarbotics.com)

[X] Mini breadboard
for ProtoShield, SparkFun
#PRT-12045 or Maker Shed
#MKKN1-B

**[Y] Drive motors,
Gearmotor 9 (2)**
Solarbotics #GM9

[Z] Ping rangefinder
Maker Shed #MKPX5

TOOLS

**9" bandsaw with
⅛" metal cutting blade**

18" bending brake Harbor
Freight (harborfreight.com)
#39103-8VGA, $40

**Drill press with vise clamp
and scrap wood block**

Fractional drill bit set

#43 drill bit

Step drill, aka unibit, Harbor
Freight #91616-0VGA

2" hole saw

4-40 tap and tap handle

Pop rivet tool

Small metal file

Drafting square

**Center punch and
small hammer**

Nibbler tool
Jameco #18810, $8

Handheld deburring tool

Soldering iron and solder

Screwdrivers for screws
and servo parts

Needlenose pliers

**Double-sided and regular
cellophane tape**

Double-sided foam tape

Black electrical tape

Small nail

Protective eyewear

**Computer with internet
connection and printer**

Photography by Ed Troxell

MAKE IT.

BUILD YOUR MAKEY ROBOT

START >>   Time: 2–3 Weekends Complexity: Difficult

1. MAKE THE BODY

The body consists of 2 pieces of sheet aluminum. You can cut, drill, and bend one piece at a time, or do both at once to minimize switching tool stations; see Step 1h.

1a. Download the 5 Makey templates from makezine.com/projects/my-robot-makey and print them out full-size. Cut out the base cutting template, and cut holes in the blank area of each panel. Securely tape the template to the aluminum sheet with double-stick tape on the back and regular tape over the holes.

1b. With a band saw, cut the aluminum roughly to size around the template, then cut the perimeter just outside the lines.

TIP: For inside corners, first cut a gradual curve close to the corner, then back up and cut into the corner's point from each direction.

1c. Use a center punch and small hammer to punch through the template at the 17 crosshairs (for drilling in the next step) and at the corners of the rectangles around the large holes.

1d. Drill the holes at the crosshairs following the sizes marked on the template. Remove (but keep) the paper first, to line up on the punch marks more accurately. Clamp the metal tightly onto scrap wood, and wherever possible use the unibit, which makes cleaner holes in thin metal than a twist drill. For the starter holes inside the rectangles, you may need to adjust the diameter smaller or larger to reach the rectangle edges.

1e. Finish the rectangular holes with the nibbler tool, cutting away until a rectangle appears. If you want, you can retape the template to see the rectangles more clearly. Then file the edges smooth.

1f. Use a handheld deburring tool to remove burrs from the metal's edges. To deburr the small holes, push the point of a larger drill bit over the holes and twist it by hand.

1g. Cut out the base bending template and attach it to the other side of the aluminum with double-stick tape, aligning the holes and rectangles. Insert the metal in the brake with the new template facing up, and make all indicated bends at 90°. For each bend, go up gradually, using a drafting square between small bends to check the angle. First bend the tabs on each long side of the metal, then bend up the sides of the body.

1h. For the body's top cover, repeat Steps 1a–1g using the top cutting and bending templates. Now you've got one sweet chassis any robot would be proud to wear!

---

2. ADD THE DRIVETRAIN

2a. Mount the drive motors in the base using 4-40×1" screws through the small holes. The motor shafts should poke out of the larger holes. Secure the screws with lock washers and nuts on the motor side. The base is small, so you may need needlenose pliers for tightening.

2b. Use a 2" hole saw in a drill press to cut wheels out of some scrap wood. I used 1×8 shelving and my finished wheels were ¾" thick by about 1.8" in diameter. Clamp the wood, and go slowly to avoid stalling the drill press.

2c. Center a wheel hub on each wooden wheel and use a small nail to mark 2 of the hole locations. Drill through the locations with a ⅛" drill.

2d. Paint the wheels. I love the Rust-Oleum red gloss; it is super thick, brightly colored, covers great, and cleans up easily. Try not to get too much paint in the mounting holes.

2e. Drill through 2 opposing holes in the hub with a #43 drill, then use a 4-40 tap to create threads in each hole.

2f. Use two 4-40×1" screws to attach the wheels to the hubs from the outside. Don't overtighten.

2g. Install tires on the wheels, orienting them with the larger diameter facing out. Then snap each wheel assembly into its motor shaft.

2h. Attach the skidder to the bottom of the base using the screws, nuts, and the thinner of the 2 spacers it comes with.

3. ADD THE POWER AND CONTROL

3a. Cut a plate out of hard plastic following the mounting plate template printed in Step 1a. Punch and drill it as indicated, then test-fit the plate into the robot body, resting on the motors, and file as needed for a snug fit. Use two 4-40×⅜" screws to fasten the Arduino board to the plate from the underside, securing it with nuts on top. The USB connector should line up with the notch in the tab.

3b. Pop-rivet the battery holders into the body through the holes in the left side tabs. Rivet from the outside so the ugly side of the rivet faces the battery.

3c. Solder together the ProtoShield following the manufacturer's instructions, linked at makezine.com/projects/my-robot-makey. Use the band saw to slice off the board's BlueSMiRF header, which connects to Bluetooth wireless modules. Wasn't that fun? The header won't fit in the robot and we don't use it. Stick the mini breadboard onto the ProtoShield and plug the ProtoShield onto the Arduino. If you are using a Diecimila, set its power jumper to EXT.

4. ADD THE SENSOR AND SERVO

4a. This project uses the shorter of the 2-arm horns that come with the HS-55 servo. Use a 1/16" bit to drill out the outermost holes in this horn.

4b. Press-fit the metal pieces of the 2 Du-Bro Mini E/Z Connectors into the servo horn holes from the front, and secure them in back with the black rubber pieces. Thread the control rods from the Du-Bro Aileron System through the connectors and screw them down using the included screws.

4c. Here's a tricky part. Plug the servo extension wire into the Ping sensor board. Bend the control rods from the horn in opposite directions 90°, to reach mounting holes at opposite corners of the board. The rods will point up from the servo, allowing room for the extension plug, and the sensor should face out. Slip the pushrod housing that came in the Du-Bro package over the rods to avoid short-circuiting the sensor, then secure the rods to the board using the connectors from the aileron control kit.

4d. Thread the wires from the servo and sensor down through the rectangular cutout in the body's top piece. Fit the servo in the cutout, and fasten using two 1-72×1/4" screws and nuts through the holes on either side. Clip the excess control rod length. Screw the horn onto the servo and use a small screwdriver to adjust it so that Makey's eyes face forward.

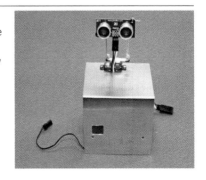

5. CONNECT AND TEST THE DRIVE MOTORS

The contact tabs on our inexpensive motors are fragile, so their connections must be strong and vibration-proof.

5a. Remove the motors and Arduino board from the robot body.

5b. Cut 2 red and 2 black 12" leads out of the stranded wire and strip 1" off an end of each. Without soldering, wrap each red/black pair around the round back end of the motor (for strain relief), then run the wires along the top and stick them on with a sandwich of double-stick foam tape. Don't cover any of the holes in the motor body, and leave room for the mounting nuts.

5c. Thread and solder the capacitor leads through the holes in each motor's connector tabs. This requires innovative bending with needlenose pliers. Then solder the motor wires to the capacitor leads, not the motor connectors, making a strong joint. Clip the extra lead length. Then cover the capacitor and the wrapped-around wires with black tape, and use more foam tape to cover the pointy bits.

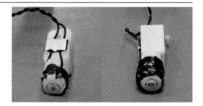

5d. Twist the free ends of the motor wire pairs together; this also reduces noise on the circuit. Mark the motors as Left and Right.

5e. Solder and heat-shrink short solid-core jumper wires to the drive motor and battery snap leads (this lets you plug them into the breadboard). Route the motor wires through the big holes in the plastic mounting plate.

5f. Plug the motor driver over the central trench of the breadboard and wire it to the drive motors and one battery, following the schematic. (Recall that on each side of the trench, holes in the same row are connected.) Use short jumpers to keep the wires close to the breadboard, as big loopy wires won't fit inside the robot.

5g. Download and install the Arduino software from arduino.cc and download the 5 project test programs from makezine.com/projects/my-robot-makey Hook the Arduino to your computer via USB, and if it's a Diecimila, move its power jumper to USB.

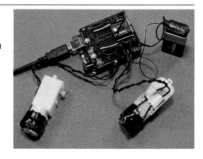

5h. To test the motors, run the program 01_Test_Motor_Rotation. The left motor should run forward and back, followed by the right motor. If not, check your wiring. Next, run 02_Test_Motor_Speed. The motors should start slow, speed up, and then reverse direction. Otherwise check wiring to pins D11 and D3.

6. CONNECT AND TEST THE SERVO AND SENSOR

6a. Replace the motors and Arduino assembly in the robot body. Plug a 3-pin right-angle header into the breadboard, plug the servomotor cable into it, and wire the servo: black to GND, red to +5V, and yellow to Arduino pin D10.

6b. Plug the other 3-pin header into the breadboard, then plug in and wire the rangefinder: black to GND, red to +5V, and white to Arduino pin D9.

6c. Run the program 03_Test_Servo_Center, which centers the servo, then unscrew and realign the servo horn as close to center as possible. You can't get it exactly in the middle, because the teeth on the shaft won't allow it, but we can nudge it later in software.

6d. Run 04_Test_Servo_Sweep, which should make the servo slowly rotate from one side to another.

6e. To test the sonar rangefinder, run 05_Test_Sensor_Distance and then click on the serial monitor icon in the Arduino software. You should see distance readings spitting out, and if you move your hand in front of the sensor, the readings should change. If your readings are stuck at 0cm or 255cm or otherwise incorrect, check your wiring, and make sure the sensor isn't plugged in backward.

7. CONNECT THE ARDUINO POWER

7a. Unpack the Arduino one last time. To add the on/off switch, solder the unused battery snap's red wire to one side of the switch and a solid red wire to the other. Also solder a solid black wire to the battery's black wire. Thread the wires out through the rectangular hole in the side of the robot body, and then press in the switch. Orient the switch with the "1" label at the top and fit it through the hole. It's a tight fit and you may need pliers.

7b. Wire the red lead from the switch to the RAW pin on the ProtoShield (which connects to Vin on the Arduino) and the black lead from the battery snap to the ProtoShield's GND pin. If you're using the Diecimila, move its Power jumper back to EXT.

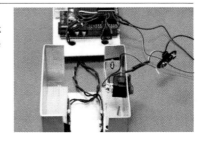

8. BUTTON IT UP

8a. Now that all your electronics are working, carefully put everything back into the body without knocking loose any wires. Install the batteries and prop the robot up on something so it doesn't run off the table. The USB programming jack should line up with the cutout in the body.

8b. Reload and run the test program 01_Test_Motor_Rotation, noting that the front of the robot is where the USB jack and skidder are. If the motors rotate the wrong way, check your wiring to pins AOut1, AOut2, BOut1, BOut2, AIn1, AIn2, BIn1, and BIn2. You may also need to reverse the motor connections.

8c. Rerun the other test programs to make sure all the wiring is still OK. When satisfied, fold up the servo and sensor wires and tuck them into the base. Slide the top cover on, and install 4 sheet metal screws to hold it on. You're done!

FINISH⊠

NOW GO USE IT »

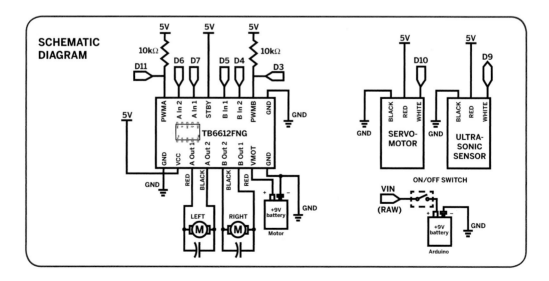

USE IT.

ROLL YOUR OWN ROBOT

ROBOT PROGRAMMING

Sometimes when you're done building, you're done building — but you're never done programming. This is where you get to be creative, think of new things you want the robot to do, then write or modify programs to implement the behavior.

In the code, you control the motors by using the digitalWrite and analogWrite functions to pass values to pins on the motor driver, 3 for each motor. One pin takes a number 0–255 and sets the current sent to the motor, which determines its speed. The other 2 pins take binary values that set each motor contact to either high or low voltage. This sets the motor's direction (when only one contact is high) or turns it off (when both are low). For example, here's a subroutine for moving the robot forward:

```
void Forward()
{
 digitalWrite(leftDir1, LOW);
 digitalWrite(leftDir2, HIGH);
 digitalWrite(rightDir1, LOW);
 digitalWrite(rightDir2, HIGH);
}
```

You can write similar routines for the more basic motions, such as Backward (both motors backward), Spin_Left (right wheel forward, left wheel back), Arc_Left (right wheel forward, left wheel stopped), and so on. The Arduino programming environment makes it easy to experiment with code and load new programs to your robot.

Makey's object-following behavior based on proportional control is described on page 142.

Another fun behavior is object avoidance, which runs the loop: Move forward a bit, then take a distance reading. If the object is too close, take evasive action, such as back up and turn. Repeat.

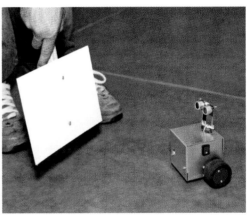

ROBOT MINI-SUMO

In a Mini-Sumo match, 2 autonomous robots are placed in a ring painted black with a white border. A robot wins when it pushes the other robot out of the ring (which means it has to find the other robot in the first place).

With narrower wheels, like the GM Series plastic wheels from Solarbotics (solarbotics.com/products/gmpw), Makey would fit within the maximum size and weight allowable for Mini-Sumo: a 10cm-square footprint and 500 grams. You would probably need another sensor pointed down to see the ring, but the Arduino has room for several more inputs.

RESOURCES

For more information about robot programming and behavior, I recommend *Robotics with the Boe-Bot* from Parallax, Inc. (parallax.com).

➕ For the Makey project schematic, templates, and code see makezine.com/projects/my-robot-makey.

🎥 To see videos of Makey in action, visit makezine.com/projects/my-robot-makey.

HOW TO BUILD CoffeeBots

Written by JUDY AIMÉ' CASTRO

Build a simple, program-mable robot with personality that's approachable for all ages.

Nate Van Dyke

Gunther Kirsch (materials); Judy Aimé Castro (various CoffeeBots)

Note: The 10KΩ resistor is matched to the photoresistor and the ambient light. The value of 10KΩ is suitable for a wide range of photoresistors and indoor lighting conditions. If your coffeebot doesn't seem responsive outdoors or in a bright room, replace the 10KΩ resistors with 1KΩ resistors. You may have to experiment with different values to suit your conditions.

⁄ TIME: 8–12 HOURS ⁄ COST: $$

MATERIALS

- » **Gearmotors with wheels (2)** Solarbotics #GM8 with GMPW wheel deal, solarbotics.com/product/gmpw_deal
- » **Transistors, logic level MOS-FET, 12N10L type (2)** such as Jameco #787798, jameco.com
- » **Photoresistors, CdS, 250mW, 12kΩ–1MΩ (2)** such as Jameco #120299
- » **Resistors, ¼W, 10kΩ (2)**
- » **Arduino Uno microcontroller board** Maker Shed item #MKSP11, makershed.com
- » **Battery holder with switch, 9V, with separate coax power plug** Maker Shed #MSBAT1
- » **Battery, 9V**
- » **Header, male, 36- or 40-pin** such as Jameco #68339 or #160882. You'll break off 20 pins, or 26 if you add an LED.
- » **Wire, solid core** Get 6 or 7 colors, or use colored tape to identify the wires. I use black, red, white, green, blue, yellow, and purple.
- » **Craft sticks, wood, 6"×¾": solid (6) and notched (2)**
- » **Wine corks (2)**
- » **Bottle caps, plastic (2)**
- » **Coffee can, 10oz or 12oz**
- » **Cable ties (10)** aka zip ties
- » **Decorative bits** Use recycled material, broken toys, googly eyes, and other fun bits to personalize your robot.

OPTIONAL COMPONENTS FOR ADDING LEDS:
- » **LED, jumbo, green** Jameco #2152104
- » **LED, jumbo, red** Jameco #2152112
- » **Resistors, ¼W, 220Ω (2)** such as Jameco #690700

I created CoffeeBots to observe chaotic behavior, by programming a bunch of them with subtle differences and a few choices — for example, how to respond to light sensors — and then turning them loose. To distinguish one from another, I named them after celebrities or historical figures and gave them each their own personality and style. I discovered that their interactions can create patterns that resemble our own social behavior!

TOOLS

- » **Hot glue gun with glue sticks**
- » **Soldering iron, fine, with solder**
- » **Wire strippers**
- » **Pliers**
- » **Helping hands tool**
- » **Magnifying glass**
- » **Multimeter**
- » **Battery, coin cell**

Wiring Diagram

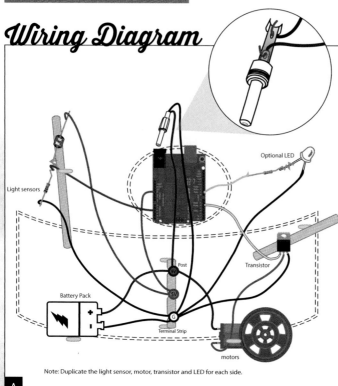

KEY

⬭ Represents hot glue

Wire Colors:

BLACK — All connections to the robot's ground terminal: battery pack ground, source pin transistor, Arduino ground, sensors ground.

RED — All connections to the robot's +9V terminal: the top terminal of each motor, the battery pack, and Arduino power.

WHITE — From the gate pin of each transistor to its respective Arduino digital I/O pin (pins 3 and 5).

BLUE — From the bottom terminal of each motor to the drain pin of its transistor.

YELLOW — All connections to the robot's +5V terminal: from one leg of each photoresistor, and from the Arduino's 5V pin.

GREEN — From the signal junction of each light sensor to its respective Arduino analog input pin (pins 0 and 2).

PURPLE (optional) — From the Arduino's pin 9 to the LED's resistor.

Note: Duplicate the light sensor, motor, transistor and LED for each side.

A

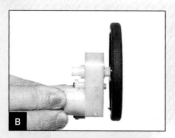

B

C

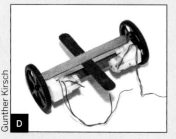

D

Gunther Kirsch

Each CoffeeBot is constructed from household materials that give it its own unique characteristics. Functionality comes from motors, sensors, wires, and an Arduino microcontroller for a brain. You can easily program your robot to detect light, to follow it or run away from it, to turn this way or that (by adjusting the timing of the motors), and to blink its LED light in whatever patterns you choose.

This robot was designed to get the most amount of robotic behavior for the least amount of coding, the least amount of money, and the fewest number of parts. And to be built by someone with the least amount of experience, in as little time as possible. You can definitely do this!

1. Wire the motors.

Attach a wheel to each motor shaft (**Figure B**).

Position the motor heads facing each other. Strip the ends of 2 red and 2 blue wires, and hook these into the copper terminals: a blue wire to the bottom and a red wire to the top terminal of each motor. Solder all 4 connections (**Figure C**). Hot-glue the wires to the motors for strain relief, as shown in the wiring diagram (**Figure A**).

Caution: Take care not to rest the hot soldering iron on the plastic motor body, and not to overheat the copper terminals. Also, it's wise to cover connections with electrical tape or heat-shrink tubing to avoid short circuits.

2. Build the chassis.

Hot-glue one 6" solid craft stick across the top of both motors, and one across the bottom, for structural support.

For front-and-back stability, hot-glue a third craft stick perpendicular to the first 2 (**Figure D**). At each end, hot-glue the wine corks and bottle caps as shown in the chassis diagram (**Figure E**).

Note: Make the front cork slightly shorter than the rear cork, and allow the chassis to teeter. This ensures that the wheels will always be in contact with the ground.

Place your favorite coffee can on top of the chassis. Glue additional pieces of craft sticks as necessary to strengthen the connection between the chassis and the can.

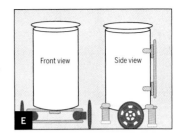

3. Mount the terminal strip.

This is where multiple connections are made to the positive voltages (+9V and +5V) and to ground. Build the terminal post by gluing 2 small pieces of cork between a notched craft stick and a solid craft stick, then glue the solid stick to the back of the coffee can as shown in (**Figure F**).

Label the notched stick for the 3 different terminal posts, from top to bottom: +9V, +5V, and Ground.

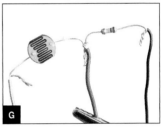

4. Make 2 light sensors.

Each sensor is made up of a photoresistor connected in series with a fixed resistor. Twist one leg of each component together, add a green wire, and solder the junction. The green wire will carry the signal from the light sensor.

Now solder the loose leg of the fixed resistor to a black wire, and the loose leg of the photoresistor to a yellow wire (**Figure G**).

Hot-glue each light sensor to a solid craft stick, then glue the stick to the can to make the robot's arms (**Figure H**).

Note: These resistors aren't polarized, so their orientation doesn't matter when you solder them together. But the assembled light sensor is polarized; you'll connect it in the correct orientation in Steps 8 and 9.

5. Wire the transistors.

Look at the component with the tab up, the 3 legs down, and the part numbers facing you.

The left leg of the transistor is called the *gate*. Strip a white wire, twist the stripped end tightly around the gate leg, and solder.

The middle leg is the *drain*. Strip, twist, and solder the blue wire coming from one of the motors.

The right leg is the *source*. Solder it to a black wire.

Note: Make sure the wires don't touch; you can bend the legs slightly to separate them (**Figure I**).

Wire the second transistor and motor the same way.

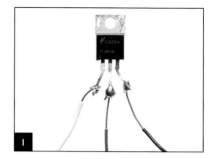

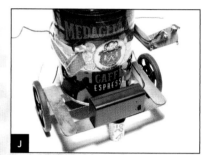

6. Mount the battery pack.

Depending on where you choose to mount the battery pack, you may need to extend its 2 wires to reach the terminal strip; they're typically not very long. Hot-glue the battery pack in place (**Figure J**).

Wrap the black wire a few times around the Ground terminal location you marked, then strip its end to create your Ground terminal post.

Repeat with the red wire to create your 9V terminal post.

Important: Don't put the battery in until you're done wiring the robot and programming the Arduino.

7. Solder the power plug.

Unscrew the DC power plug's cover and expose the 2 terminals. Solder a red wire to the inner terminal (+), and connect the other end to the +9V terminal post. Solder a black wire to the outer terminal (−) and connect it to the Ground terminal post (**Figure K**).

Caution: Put electrical tape between the 2 terminals to avoid having them touch and short circuit.

Important: Thread the wires through the plug's cover before soldering them to the robot's terminal strip, so that you can screw it back on.

8. Connect the terminal posts.

Wire the remaining terminal strip connections, following Figure A. At your 9V terminal post, solder together the red wire from the battery pack, the 2 red wires from the top of the motors, and the red wire from the Arduino's power plug.

For your 5V terminal, solder the 2 yellow wires from the light sensors, as well as a new yellow wire (which you'll connect to the Arduino's 5V pin in just a minute). You can wrap this new wire around the terminal strip to make it stay put.

And at the Ground terminal post, connect the black wire from the battery pack to the 2 black wires from the right legs of the transistors, the 2 black wires from the light sensors, and the black wire from the Arduino's power plug (**Figure L**).

Note: To avoid short circuits from exposed bare wires, only strip as much wire as is necessary to connect to your terminals.

Now you can tidy up your wires using the zip ties, and glue the transistors to the wooden arms if you wish.

9. Solder the headers for the Arduino.

Break off 3 strips of male header pins to fit the Arduino board's female headers for Power (6 pins), Analog In (6 pins), and Digital I/O pins 0–7 (8 pins). These male headers have one row of short pins, to which you'll solder your wires, and one row of longer pins, which you'll plug into the Arduino.

In the Power block, solder the free wire from the robot's +5V terminal post to your 5V header pin.

In the Digital block, solder the white wires from the 2 transistor gates to your Digital 3 and 5 header pins.

In the Analog block, solder the green signal wires from the 2 light sensors to your Analog 0 and 2 header pins (**Figure M**).

The schematic diagram (**Figure N**) shows how these connections work (for one side only).

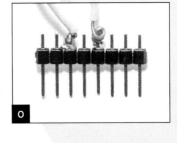

Tips: Solder the wires to the headers before you plug the headers into the Arduino. To avoid short circuits, make sure the wires don't touch adjacent pins; I have intentionally avoided using adjacent pins to give you more room to work.

It isn't necessary to twist the wires around the pins; it's best to strip only ⅛", apply solder separately to the wire and the pin, and finally, touch the wire to the pin and remelt the solder to form a good connection (**Figure O**).

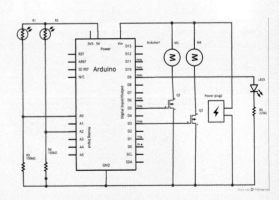

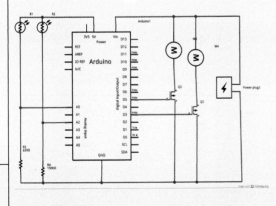

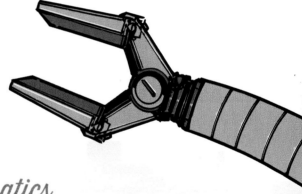

N CoffeeBot Schematics

P

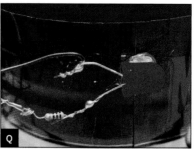
Q

R

10. Add an LED (optional).

Solder a black wire to the negative leg of the LED, typically indicated by a shorter leg and/or a flat spot on the LED's plastic dome.

Solder one end of a 220Ω resistor to the positive leg of the LED. Solder a purple wire to the loose end of the 220Ω resistor, forming a series circuit (**Figure P**).

Solder the other end of the wire to a male header pin in the Arduino's Digital block. You might wish to make an additional male header for Digital pins 8–13, and use pin 9 for the LED.

Solder the black (−) wire to the robot's Ground terminal post. Then glue the LED wherever you want it on the robot (**Figure Q**).

11. Program your Arduino.

The Arduino microcontroller will process all the inputs and outputs and control your CoffeeBot. Download and install the Arduino environment on your computer, following the instructions at arduino.cc/en/Guide/HomePage.

Then download the CoffeeBot code from bit.ly/coffeebot, and upload the correct version (with our without an LED) to your board.

12. Install the brain.

Use the coffee can's plastic lid as a tray for the Arduino, securing it with zip ties.

If your lid is metal, then make a tray from craft sticks, plastic, or other nonconductive material to avoid short-circuiting the board.

Plug your male headers into the corresponding female headers on the Arduino (**Figure R**). Now your CoffeeBot has a brain.

13. Give it nutty flavor.

CoffeeBots aren't just robots made with a coffee can and wires. Give them a personality you can see — a wheel that wobbles and drags a little to one side, or an LED that sort of hangs there like a weird antenna!

Now's the time to go nuts with hot glue and googly eyes, pirate flags, or whatever bits you'd like to add to make it unique and fun.

Use It

Put a battery in, plug the power plug into the Arduino, and switch on the battery pack.

The CoffeeBot Arduino program makes your robot either seek or avoid light, depending on how it's wired up. Shine a flashlight, or hold your hand over a light sensor, to change the robot's behavior.

If you added the optional LED, the program will make the robot's LED light up whenever the sensors measure equal brightness.

Try out this simple code first, and then start modifying it to give your robot the personality you desire. Position the LED so that other CoffeeBots can perceive the light, and see if they'll react to it according to their own programming!

I designed the CoffeeBot as a platform for experimentation by adding other sensors. Maybe you'd like to add bump sensors, or distance measuring sensors, or infrared LEDs and detectors.

With these additional sensors, and suitable programming, it would be possible for robots to identify other robots and to follow, lead, or communicate with them! ◼

BIO **Judy Aime' Castro** is a tinkerer, artist, and collaborator with Michael Shiloh at Teach Me to Make (teachmetomake.com). She can be reached at teachers@teachmetomake.com.

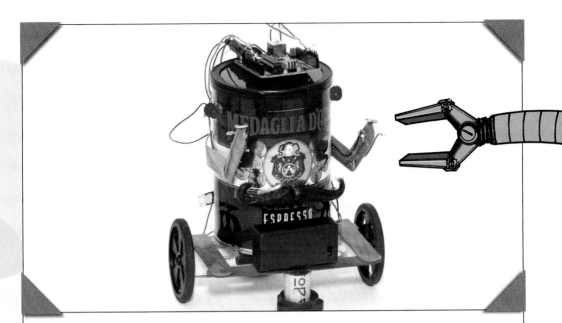

CoffeeBot Code

Here's the basic CoffeeBot code for your Arduino. If you installed the optional LED, get that version of the code at bit.ly/coffeebot. For component kits, visit bit.ly/CoffeeBot-store.

```
/*
This program makes your robot either seek light or avoid light, depending on how
it is wired up.
Try this program first, and then start modifying it to give your robot the
personality you desire.
Shine a flashlight, or hold your hand over a light sensor, to change the
behavior.
Other examples and suggestions are on our website.
*/
void setup()
{
 // Set the mode of the digital pins to outputs to drive the motors
 // (the analog inputs are automatically inputs and so don't need to be set)
 pinMode( 3, OUTPUT );
 pinMode( 5, OUTPUT );
}

void loop()
{
 // Compare the two light sensors
 if ( analogRead( 0 ) > analogRead( 2 ) )
 // If one light sensor has more light than the other ...
 {
 digitalWrite( 3, LOW ); // turn this motor off ...
 digitalWrite( 5, HIGH ); // and this motor on to turn in one direction
 }
 else // otherwise ...
 {
 digitalWrite( 3, HIGH ); // turn this motor on
 digitalWrite( 5, LOW ); // and this motor off to turn in the other direction
 }
}
```

ANATOMY OF A DRONE

Finding your way around a modern multirotor UAV.

Illustration by Rob Nance

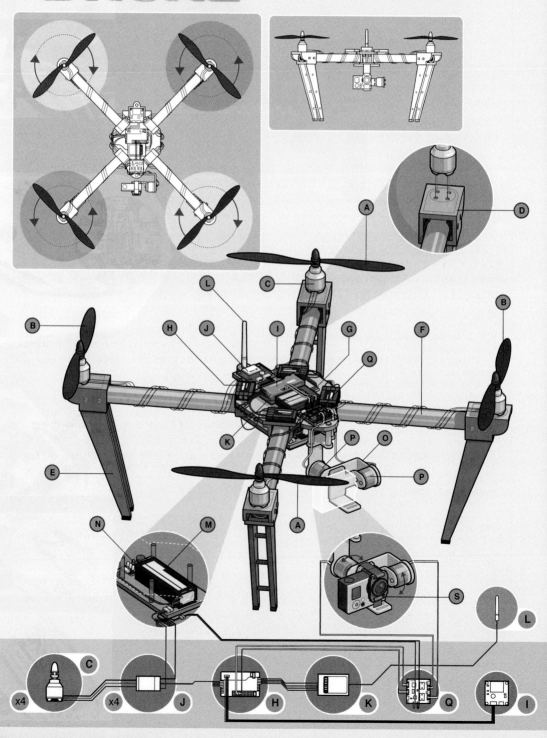

A. STANDARD PROP

The same "tractor" propeller used on standard front-engine R/C airplanes.

B. "PUSHER" PROP

These contra-rotating props exactly cancel out motor torques during stationary level flight. Opposite pitch gives downdraft.

C. MOTOR

Usually a brushless electric "outrunner" type, which is more efficient, more reliable, and quieter than a brushed motor.

D. MOTOR MOUNT

Sometimes built into combination fittings with landing struts.

E. LANDING GEAR

Designs that need high ground clearance may adopt helicopter-style skids mounted directly to the body, while designs with no hanging payload may omit landing gear altogether.

F. BOOM

Shorter booms increase maneuverability, while longer booms increase stability. Booms must be tough to hold up in a crash while interfering with prop downdraft as little as possible.

G. MAIN BODY

Central "hub" from which booms radiate like spokes on a wheel. Houses battery, avionics, cameras, and sensors.

H. ELECTRONIC SPEED CONTROLLER (ESC)

Converts DC battery power into 3-phase AC for driving brushless motors.

I. FLIGHT CONTROLLER

Interprets input from receiver, GPS module, battery monitor, and onboard sensors. Regulates motor speeds, via ESCs, to provide steering, as well as triggering cameras or other payloads. Controls autopilot and other autonomous functions.

J. GPS MODULE

Often combines GPS receiver and magnetometer to provide latitude, longitude, elevation, and compass heading from a single device.

K. RECEIVER

Often a standard R/C radio receiver unit. The minimum number of channels needed to control a quad is 4, but 5 is usually recommended.

L. ANTENNA

Depending on your receiver, may be a loose wire whip or helical "rubber ducky" type.

M. BATTERY

Lithium polymer (LiPo) batteries offer the best combination of energy density, power density, and lifetime on the market.

N. BATTERY MONITOR

Provides in-flight power level monitoring to flight controller.

O. GIMBAL

Pivoting mount that rotates about 1, 2, or 3 axes to provide stabilization and pointing of cameras or other sensors.

P. GIMBAL MOTOR

Brushless DC motors can be used for direct-drive angular positioning, too, which requires specially-wound coils and dedicated control circuitry that have only recently become commercially available.

Q. GIMBAL CONTROLLER

Allows control of direct-drive brushless gimbal motors as if they were standard hobby servos.

R. CAMERA

GoPro or other compact HD video unit with onboard storage. Real-time streaming is possible with special equipment.

SKILL BUILDER+

EASY

For makers, it has become quite cheap to incorporate high-quality geospatial data into electronics projects. And in the last few years, GPS receiver modules have grown much more diverse, powerful, and easy to integrate with development boards like Arduino, PIC, Teensy, and Raspberry Pi. If you've been thinking of building around GPS, you've picked a good time to get started.

FINDING YOUR WAY with GPS

Written by Mikal Hart

COUNTING STARS

One evening I built a little Arduino/GPS test gizmo to spy on the GPS satellite constellation. I was able to count all 32 distinct satellites over a 24-hour period, with as many as 13 visible at once. For more info:

makezine.com/gps

3 SINCE THE U.S. GLOBAL POSITIONING SYSTEM (GPS) HAS A PUBLISHED GOAL of being usable everywhere on Earth, the system must ensure that at least four satellites — preferably more — are visible at all times from every point on the globe. There are currently 32 GPS satellites performing a meticulously choreographed dance in a sparse cloud 20,000 kilometers high.

2 WHEN A GPS MESSAGE ARRIVES, the receiver first inspects its broadcast timestamp to see when it was sent. Because the speed of a radio wave in space is a known constant (*c*), the receiver can compare broadcast and receive times to determine the distance the signal has traveled. Once it has established its distance from four or more known satellites, calculating its own position is a fairly simple problem of 3D triangulation. But to do this quickly and accurately, the receiver must be able to nimbly crunch numbers from up to 20 data streams at once.

HOW IT WORKS

1 A GPS MODULE IS A TINY RADIO RECEIVER that processes signals broadcast on known frequencies by a fleet of satellites. These satellites whirl around the Earth in roughly circular orbits, transmitting extremely precise position and clock data to the ground below. If the earthbound receiver can "see" enough of these satellites, it can use them to calculate its own location and altitude.

FUN FACT:
GPS could not work without Einstein's theory of relativity, as compensation must be made for the 38 microseconds the orbiting atomic clocks gain each day from time dilation in Earth's gravitational field.

Time Required:
2 Hours
Cost:
$75–$150
A quick exercise in understanding and applying GPS data.

MIKAL HART
(arduiniana.org) is a senior software engineer at Intel Corp. in Austin, Texas. He is the inventor of the Reverse Geocache Puzzle and a founder of The Sundial Group. He has written about electronics development and prototyping for MAKE and for several books.

Whatever your project, GPS is simple to integrate. Most receiver modules communicate with a straightforward serial protocol, so if you can find a spare serial port on your controller board, it should take just a handful of wires to make the physical connection. And even if not, most controllers support an emulated "software" serial mode that you can use to connect to arbitrary pins.

For beginners, Adafruit's Ultimate GPS Breakout module is a good choice. There are a lot of competing products on the market, but the Ultimate is a solid performer at a reasonable price, with big through-holes that are easy to solder or connect to a breadboard.

The Adafruit Ultimate GPS Breakout board. A typical GPS interface, assuming both module and main board run at compatible voltages, is as simple as connecting four wires.

First, connect ground and power. In Arduino terms, this means connecting one of the microcontroller GND pins to the module's GND, and the +5V pin to the module's VIN.

To manage data transfer, you also need to connect the module's TX and RX pins to the Arduino. I'm going to arbitrarily select Arduino pins 2 (TX) and 3 (RX) for this purpose, even though pins 0 and 1 are specifically designed for use as a "hardware serial port" or UART.

Why? Because I don't want to waste the only UART these low-end AVR processors have. Arduino's UART is hard-wired to the onboard USB connector, and I like to keep it connected to my computer for debugging.

Materials

» **Arduino Uno** or compatible microcontroller / single-board computer. Maker Shed item #MKSP99, makershed.com
» **GPS module** such as Adafruit's Ultimate GPS Breakout. Maker Shed #MKAD47

Tools

» **Computer** PC, Mac, or Linux. Laptop preferred.
» **Soldering iron and solder** may be required to attach header pins to your GPS module
» **Solderless breadboard**
» **Jumper wires**

```
#include <SoftwareSerial.h>
#define RXPin 2
#define TXPin 3
#define GPSBaud 4800
#define ConsoleBaud 115200

// The serial connection to the GPS device
SoftwareSerial ss(RXPin, TXPin);

void setup()
{
  Serial.begin(ConsoleBaud);
  ss.begin(GPSBaud);

  Serial.println("GPS Example 1");
  Serial.println("Displaying the raw NMEA data transmitted by GPS module.");
  Serial.println("by Mikal Hart");
  Serial.println();
}

void loop()
{
if (ss.available() > 0) // As each character arrives...
  Serial.write(ss.read()); // ... write it to the console.
}
```

Sketch 1: A Toe in the Datastream

The instant you apply power, a GPS module begins sending chunks of text data on its TX line. It may not yet see a single satellite, much less have a "fix," but the data faucet comes on right away, and it's interesting to see what comes out. Our first simple sketch (**Figure 1**) does nothing but display this unprocessed data.

> ### NOTE:
> The sketch defines the receive pin (**RXPin**) as 2, even though we said earlier that the transmit (TX) pin would be connected to pin 2. This is a common source of confusion. **RXPin** is the receive pin (RX) *from the Arduino's point of view*. Naturally, it must be connected to the module's transmit (TX) pin, and vice versa.

Upload this sketch and open Serial Monitor at 115,200 baud. If everything's working, you should see a dense, endless stream of comma-separated text strings. Each will look something like **Figure 2**.

These distinctive strings are known as *NMEA sentences*, so called because the format was invented by the National Maritime Electronics Association. NMEA defines a number of these sentences for navigational data ranging from the essential (location and time), to the esoteric (satellite signal-to-noise ratio, magnetic variance, etc.). Manufacturers are inconsistent about which sentence types their receivers use, but *GPRMC* is essential. Once your module gets a fix, you should see a fair number of these GPRMC sentences.

Sketch 2: Finding Yourself

It's not trivial to convert the raw module output into information your program can actually use. Fortunately, there are some great libraries already available to do this for you. Limor Fried's popular *Adafruit GPS Library* is a convenient choice if you're using their Ultimate breakout. It's written to enable features unique to the Ultimate (like internal data logging) and adds some snazzy bells and whistles of its own.

My favorite parsing library, however — and here I am of course completely

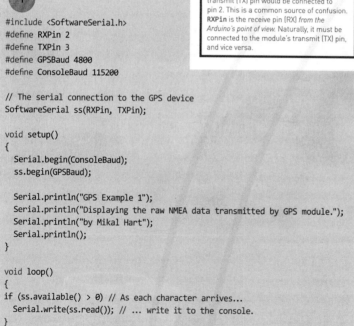

| "Talker" ID "GP" = GPS receiver | Time of fix 10:26:42 UTC | Receiver status A = OK | Latitude North | Longitude East | Heading 156.6705° | Magnetic variation 20.32° off true north | Magnetic variation East |

`$GPRMC,102642.03,A,4813.7943164,N,01621.5693035,E,7.158,156.6705,020713,020.32,E*5E`

Signature — Start of new sentence

| Sentence type ID "RMC" = GPS/Transit fix | Latitude 48° 13.7932164' | Longitude 16° 21.5693035' | Ground speed 7.158 knots | Date of fix February 7, 2013 | Checksum |

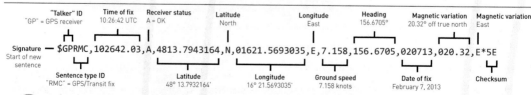

2 GPRMC is probably the most common NMEA sentence. It contains 12 comma-separated fields, followed by an asterisk (*) and a checksum to ensure data integrity.

unbiased — is the one I wrote called *TinyGPS++*. I designed it to be comprehensive, powerful, concise, and easy to use. Let's take it for a spin.

Our second application continually displays the receiver's location and altitude, using TinyGPS++ to help with parsing. In a real device, you might log this data to an SD card or display it on an LCD.

Grab the library and sketch *FindingYourSelf.ino* from makezine.com/gps. Install the library, as usual, in the Arduino *libraries* folder. Upload the sketch to your Arduino and open Serial Monitor at 115,200 baud. You should see your location and altitude updating in real time. To see exactly where you stand, paste some of the resulting latitude/longitude coordinates into Google Maps. Now hook up your laptop and go for a stroll or a drive. (But remember to keep your eyes on the road!)

Sketch 3: Finding Your Way

Our third and final application is the result of a personal challenge to write a readable TinyGPS++ sketch, in fewer than 100 lines of code, that would guide a user to a destination using simple text instructions like "keep straight" or "veer left." Every 5 seconds the code captures the user's

CODING WITH TINYGPS++

From the programmer's point of view, using TinyGPS++ is very simple:
1. Create an object **gps**.
2. Route each character that arrives from the module to the object using **gps.encode()**.
3. When you need to know your position or altitude or time or date, simply query the **gps** object.

location and *course* (direction of travel) and calculates the *bearing* (direction to the destination), using the TinyGPS++ **courseTo()** method. Comparing the two vectors generates a suggestion to keep going straight or turn, as shown in **Figure 3**.

Download the sketch *Finding YourWay.ino* from makezine.com/gps and open it in the Arduino IDE. Set a destination 1km or 2km away, upload the sketch to your Arduino, run it on your laptop, and see if it will guide you there. But more importantly, study the code and understand how it works.

Going Further

The creative potential of GPS is vast. One of the most satisfying things I ever made was a GPS-enabled puzzle box that opens only at one preprogrammed location. If your victim wants to get the treasure locked inside, she has to figure out where that secret location is and physically bring the box there. (See *The Reverse Geocache Puzzle*, MAKE Volume 25.)

A popular first project idea is some sort of logging device that records the minute-by-minute position and altitude of, say, a hiker walking the Trans-Pennine Trail. Or what about one of those sneaky magnetic trackers the DEA agents in *Breaking Bad* stick on the bad guys' cars?

Both are totally feasible, and would probably be fun to build, but I encourage you to think more expansively, beyond stuff you can already buy on Amazon. It's a big world out there. Roam as far and wide as you can. ◗

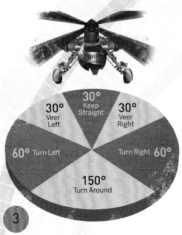

How *FindingYourself.ino* translates your course and bearing into text directions. The sketch can become the starting point for creating almost any type of autonomous or semi-autonomous vehicle.

THE FOURTH DIMENSION

Though we associate GPS with location in space, don't forget those satellites are transmitting time- and date-stamps, too. The average GPS clock is accurate to one ten-millionth of a second, and the theoretical limit is even higher. Even if you only need your project to keep track of time, a GPS module may still be the cheapest and easiest solution.

To turn *FindingYourself.ino* into a super-accurate clock, just change the last few lines like this:

```
if (gps.time.isUpdated()) {
    char buf[80];
    sprintf(buf, "The time is
%02d:%02d:%02d", gps.time.
hour(), gps.time.minute(),
gps.time.second());
    Serial.println(buf);
}
```

Find the codes and full steps at makezine.com/gps
Share it: **#makegps**

WRITTEN BY LUCAS WEAKLEY

BUILD YOUR FIRST TRICOPTER

LUCAS WEAKLEY is studying aeronautics engineering at Embry Riddle Aeronautical University. He also makes and sells aircraft kits at lucasweakley. com. He's a certified AutoCAD draftsman, an Eagle Scout, and the host of Make:'s Maker Hangar video series at makezine.com/go/makerhangar.

THEY FLY SMOOTHER AND MAKE BETTER VIDEOS THAN QUADS. BUILD THE MAKER HANGAR TRICOPTER AND SEE FOR YOURSELF!

SPECS
- Flight time: 12 minutes
- Frame weight: 325g
- Flight weight: 1kg
- Compatible with 8"–10" props
- Wire rope vibration absorber
- 22mm motor mounts

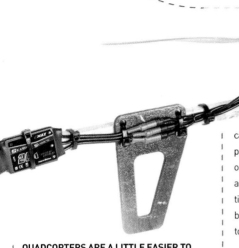

Time Required:
A Weekend
Cost:
$300–$400

Materials

Maker Hangar Tricopter Kit $85 from lucasweakley.com/product/maker-hangar-tricopter-kit, includes:
» Laser-cut plywood airframe parts
» 3D-printed tail assembly
» Carbon fiber hinge pieces
» Oak square dowels, $7/16$" × $7/16$"×12" (3) for the arms
» Bolts, stainless steel, M3: 25mm (8), 6mm (4), 10mm (16), and 22mm (8)
» Lock nuts, M3 (25)
» Washers: M3 (16) and M4 (2)
» Bolts, nylon, 6-32×$3/8$" (4)
» Nuts, nylon, 6-32 (4)
» Standoffs, 6-32×1½" (4)
» Cable ties (20)
» Push rods, 2½"×0.047" (2)
» Push rod connectors (2)
» Velcro straps (2)
» Wire rope, 3" lengths (4)

Electronics (not included) — see the kit web page for complete recommendations:
» **Flight controller board** see page 171
» **R/C receiver** to match your R/C transmitter
» **Motors, brushless outrunner, 900kV (3)** Emax GT2215/12
» **ESCs, 20A (3)** Emax Simon
» **Props, 10×4.7 (3)**
» **Batteries, LiPo, 3,300mAh (2)**
» Servo, micro
» Servo extension, 6"
» Wire, 16 gauge stranded
» Heat-shrink tubing
» Servo cable, male to male
» JST connector (optional)

Tools

» Drill and bits
» Pliers, needlenose
» Pliers, side cutting
» Wire cutters/strippers
» Hot glue gun
» Cyanoacrylate (CA) glue
» Screwdriver
» Hex driver set
» Adjustable wrench
» Sandpaper
» File
» Hobby knife
» Soldering iron and solder
» Heat gun or hair dryer
» Helping hands (optional)

QUADCOPTERS ARE A LITTLE EASIER TO BUILD, BUT TRICOPTERS HAVE ADVANTAGES that make them more exciting to fly — especially for shooting aerial video. I built my first one in 2010, inspired by David Windestal's beautiful aerial GoPro videos (rcexplorer.se/fpv-videos-setups). I didn't get many flights out of that first build, but I learned a lot. After building several more, I've developed an affordable kit that anyone can build — the Maker Hangar Tricopter.

WHY FLY TRI?

A tricopter's three motors are usually separated by 120°, not 90° like a quadcopter's. This makes them great for video because you can place the camera really close to the body and still have no propellers in view. And where quads must rely on counter-rotating propellers to handle torque and balance the aircraft, a tricopter can use identical props because it has a special servo in the back — a yaw servo — that twists the tail motor to counter torque (Figure A).

Tricopters fly differently too. With their dedicated motor for yaw (turning), they fly with more fluid, natural-looking movements — they can bank, pitch, and yaw like an airplane, but still hover like a helicopter. A quadcopter's flight is more robotic, as the controller board calculates the precise rotation for all four motors to create the proper torque and balance to yaw the aircraft. If you let go of the stick, a quad stops turning abruptly; for video work, this can be obvious and distracting. Let go of a tricopter's stick and the tilted tail motor takes a moment to return to a hovering position; this gives you a slow stop and even a little overshoot, as though a person were moving the camera.

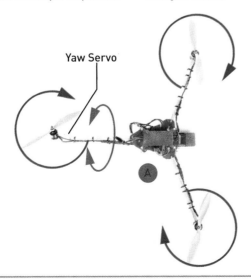

Yaw Servo

A

Hep Svadja

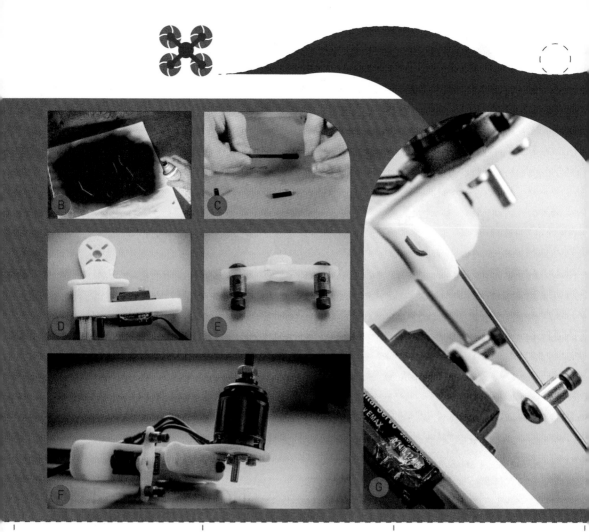

Finally, tricopters are a lot of fun to fly, especially for stunts and acrobatics. The tilting motor also gives you much higher yaw speeds — that means they turn faster.

A TRICOPTER FOR MAKERS

The Maker Hangar Tricopter is made of wood — hackable, easy to drill and cut, and a natural absorber of vibration, the enemy of aerial video. The airframe is big, with plenty of room for large controller boards, video transmitters, drop mechanisms, or whatever you can imagine. And we widened the front arms to about 150° so our tricopter is more agile.

The kit includes a 3D-printed tail assembly and all the hardware you'll need, plus a wire rope vibration absorber that will pretty much erase camera vibrations even if your propellers are unbalanced. A carbon-fiber hinge provides a strong, smooth connection between the tail motor and airframe.

Finally, like most tricopters, the two front arms lock in place for flight, then fold back neatly for transportation and storage.

It's a great kit for anyone wanting to get into multicopters or aerial photography. You can also build it totally from scratch: download the PDF plans, laser cutter layouts, 3D files for printing, flight controller settings, and watch the how-to video series at makezine.com/go/makerhangar.

1. SAND AND PAINT

Sand down any burrs or splinters in the wooden parts. If you wish, paint with a couple of light coats (Figure B).

2. ASSEMBLE THE HINGED TAIL

To build the hinge, glue the 2½" carbon rod flush into the ¾" carbon tube using CA glue (Figure C). Hot-glue this end into the 3D-printed motor mount. Also hot-glue the 1" carbon tube into the 3D-printed tail piece.

Now put it together: Slide onto the hinge rod an M4 washer, then the tail piece, then another washer. Finally, glue the ½" carbon tube to the end of the rod to capture the whole assembly.

Hot-glue the servo into the tail piece (Figure D) and install 2 "easy connectors" in ¹⁄₁₆" holes on the servo arm (Figure E). You can glue the hardwood tail arm into the tail piece now as well.

Bolt the tail motor into the motor mount with M3 washers (Figure F).

Finally, connect the servo linkages. Use

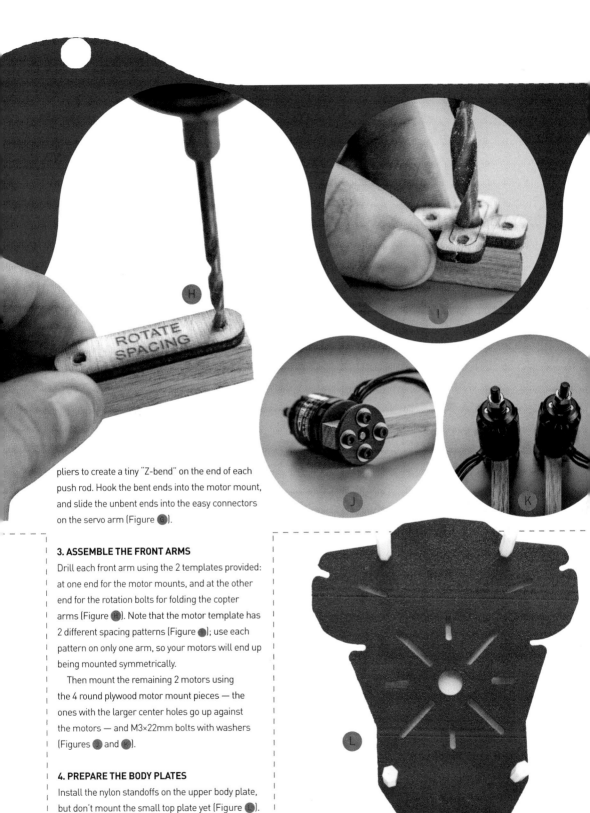

pliers to create a tiny "Z-bend" on the end of each push rod. Hook the bent ends into the motor mount, and slide the unbent ends into the easy connectors on the servo arm (Figure G).

3. ASSEMBLE THE FRONT ARMS

Drill each front arm using the 2 templates provided: at one end for the motor mounts, and at the other end for the rotation bolts for folding the copter arms (Figure H). Note that the motor template has 2 different spacing patterns (Figure I); use each pattern on only one arm, so your motors will end up being mounted symmetrically.

Then mount the remaining 2 motors using the 4 round plywood motor mount pieces — the ones with the larger center holes go up against the motors — and M3×22mm bolts with washers (Figures J and K).

4. PREPARE THE BODY PLATES

Install the nylon standoffs on the upper body plate, but don't mount the small top plate yet (Figure L).

Bolt 4 of the small plywood brackets to the lower body plate, and 4 to the camera/battery tray, using

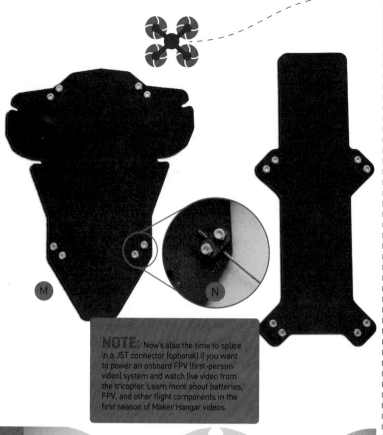

M3×10mm bolts and nuts (Figure **M**). This tray is optional (you could just velcro the battery to the bottom of the copter) but it's highly recommended for video because it's isolated from vibrations by short wire ropes. Clamp the 4 wire ropes into the brackets on the bottom plate, but don't connect the camera tray yet (Figure **N**).

NOTE: Now's also the time to splice in a JST connector (optional) if you want to power an onboard FPV (first-person video) system and watch live video from the tricopter. Learn more about batteries, FPV, and other flight components in the first season of Maker Hangar videos.

5. INSTALL THE ESCS

Connect the 3 electronic speed controllers (ESCs) to the motors and zip-tie them to the arms (Figure **O**).

Arrange the 3 arms in their folded configuration, then measure out enough wire to extend all the power and ground wires to meet at the back of the body (Figure **P**). Solder the extension wires and insulate connections with heat-shrink tubing. Strip the free ends and solder them into your battery connector (Figure **Q**).

6. ATTACH THE ARMS

Bolt the 2 front arms to the lower body plate through the outer mounting holes, using M3×25mm bolts and lock nuts. Place the upper body plate on top, then pass 2 more bolts through the locking slots and the inner arm holes, and secure with washers and lock nuts. Finally, clamp the tail arm between the body plates using 4 bolts (Figure **R**).

Test the folding action and loosen or tighten bolts until the arms fold smoothly and lock forward securely.

6. MOUNT THE LANDING GEAR

Zip-tie the 2 plywood landing struts to the front arms (Figure **S**).

7. SUSPEND THE CAMERA TRAY

Clamp the free ends of the wire ropes into the brackets on the camera tray. Make sure the camera platform faces forward (Figure **T**) and the bolt heads face outward; you'll need access to them to adjust the tray later. Strap the battery onto the tray with the velcro strap.

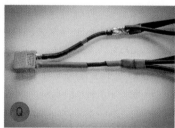

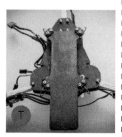

8. MOUNT THE RECEIVER AND FLIGHT CONTROLLER

Attach your flight controller to the upper body plate using hot glue, double-sided tape, or bolts through the mounting slots. (We used the Flip 1.5 MWC controller. You can download the settings at the Maker Hangar project page.)

Bind your R/C receiver to your transmitter (see Maker Hangar Season One, Episode 12), and then set the throttle ranges by plugging each of your ESCs, in turn, into the receiver's Throttle port (Season 2, Episode 4). Mount the receiver and plug it into the flight controller (Figure U). Center the yaw servo and tighten the linkages.

Finally, screw the top plate onto the standoffs to protect your electronics (Figures V and W), and your Maker Hangar Tricopter is complete! ✪

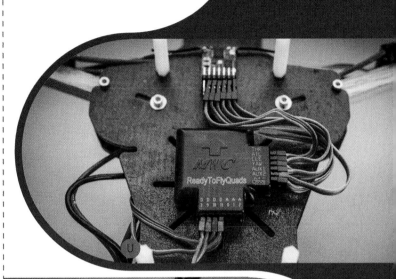

ABOUT FLIGHT CONTROLLERS

The flight controller board converts the signals from your transmitter into the motor speeds that move your tricopter. It also reads the aircraft's position and movements with its onboard gyros and accelerometers, and makes tiny changes to motor speeds to counter the wind, torque, and other forces that are trying to tip the copter over.

These are the boards I recommend for the Maker Hangar Tricopter:

- **OPENPILOT CC3D** — the best flight experience, easy setup, but tuning takes time
- **HOBBYKING KK2** — OK flight experience, fast tuning with onboard display, best for beginners
- **ARDUPILOT APM 2.6** — most powerful and expensive; programmable waypoint capabilities with GPS, compass, and barometer
- **FLIP 1.5 MULTI WII CONTROLLER (MWC)** — small, simple, and affordable, but powerful and flies well; optional barometer and compass

To learn more about flight controllers and how to fly your tricopter, watch the complete Maker Hangar how-to video series at makezine.com/go/makerhangar!

Time Required:
A Weekend
Cost:
Airframe: $30–$60
Avionics: $500–$800

THE HANDYCOPTER UAV

Written and photographed by Chad Kapper

Build your own quadrotor airframe from hardware store parts, then trick it out with stabilized onboard video and autonomous flight.

There are essentially two configurations for a quadcopter: the "+" frame and the "X" frame. Here we've chosen to build an X frame so your onboard camera can have a clear forward view. We'll take you all the way from building the airframe to adding autonomous flight capability with ArduPilot. Once you've got it working, you could program this drone, for instance, to automatically visit a series of landmarks or other waypoints and take pictures of them.

1. Fabricate the body

The copter's central hub consists of 2 polycarbonate plates. Download the cutting and drilling templates from makezine.com/projects/the-handy-copter-uav-2, print them full-size, and affix them temporarily to your polycarbonate sheet. Use a plastic cutter to score and snap each plate to shape, then drill out the holes with a ⅛" bit.

2. Cut and drill the booms

Saw 4 square dowel booms to 10"–11" each. Shorter booms will make your quad more agile, and longer booms will make it more stable. Drill two 3mm holes, one 6mm and one 26mm (on-centers) from the end of each boom.

3. Assemble the frame

Secure the booms between the hub plates using four M3×25mm screws through the inner holes and four M3×20mm screws through the outer holes. Once the booms are in place and you're happy with the fit, apply thread-locking compound to the *outer screws only*, add nuts, and tighten them down. Thread the inner nuts on just loosely, for now.

4. Wire the power hub

Six components will connect to the power hub — the 4 electronic speed controllers (ESCs), the power module, and the gimbal controller board. First, cut off the male XT60 connector from the APM power module cable. Then strip about ¼" of the insulation from each wire, red and black, on all 6 components, and tin the stripped ends. Saw a ⅜" ring from each

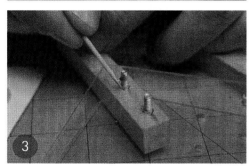

Materials

FOR THE AIRFRAME:

» Conduit clamps, 1½" (4)
» Square dowels, wood, ½"×36" (2)
» Machine screws: flat-head M3×6mm (8); M3×20mm (12); M3×25mm (4)
» Hex nuts, M3 (8)
» Flat washers, M3 (4)
» Thread-locking compound
» Liquid electrical tape
» Polycarbonate sheet, 0.093"×8"×10"
» Zip ties, 4" (100-pack)
» Flexible PVC coupler, 1¼" to 1¼"
» Aluminum bar, 1/8"×¾"×36"
» Hook/loop strap, ½"×8" (2)
» Hook/loop tape, ¾"×18"
» Weatherstrip tape, foam, 3/8"×12"
» Double-sided tape, 1"×5'
» Wire, stranded insulated, 12 AWG, 12" red and 12" black

FOR THE AVIONICS:

» Copper pipe reducer, 1" to ½"
» Gimbal motors (2) iPower 2208-80
» Gimbal controller iFlight V3.0
» Flight controller 3D Robotics ArduPilot Mega 2.6
» GPS module 3D Robotics LEA-6H
» R/C transmitter, 5+ channels
» R/C receiver, 5+ channels
» Motors, 850kV (4) AC2830
» Propellers, Turnigy 9047R SF (2)
» Propellers, Turnigy 9047L SF (2)
» Electronic speed controllers (4)
» M/M servo leads, 10cm (5)
» Camera GoPro Hero3 White Edition
» LiPo battery, 2,200mAh, 3S 20C
» Battery monitor APM Power Module with XT60 connectors

Tools

» Computer with printer
» Straightedge
» Plastic scoring knife
» Drill and bits: 1/8", 3/16", ¼", 5/16", 3/8"
» Wood saw
» Phillips screwdrivers: #1 and #2
» Pliers
» Wire cutters / strippers
» Hacksaw
» Soldering iron and solder
» Scissors
» Pencil
» File
» Hobby knife

Wiring the motors and electronic speed controllers together is tedious. Store-bought distribution boards are convenient, but cost space and weight. I prefer this homemade distribution hub made from 2 rings of nested copper pipe to keep things lean and tidy.

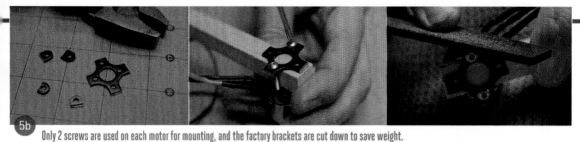

Only 2 screws are used on each motor for mounting, and the factory brackets are cut down to save weight.

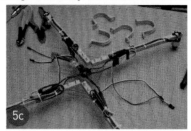

end of the copper reducer, and file off any rough edges. Solder each of the 6 red positive leads to the smaller ring, and the corresponding 6 black negative leads to the larger ring. Wrap the smaller ring in ⅜" foam weatherstripping tape and slip the outer ring over it. Finally, paint the entire hub with liquid electrical tape for insulation.

5c

7a

5. Add the motors and landing gear

Here we'll show you how to make your own landing struts from ordinary conduit clamps. You can also use inexpensive prefab combination landing gear / motor mounts that simplify the process quite a bit, and look better to boot. Please check out our product line at flitetest.com if you're interested in the prefab option.

5a. If you go the homemade route, you'll be mounting the motors directly to the booms. Mark and drill a shallow blind recess in each, so the shaft can spin freely. A ⁵⁄₁₆" bit works well for this.

5b. Cut down the bracket that came bundled with each motor and use two M3×20mm screws to clamp a motor to the end of each boom. Verify that each motor shaft spins freely when the screws are

5d

7b

fully tightened. If not, double-check that its boom is properly recessed underneath. Smooth any rough edges on the bracket with a file.

5c. Slip the power hub between the top and bottom body plates and route the ESC power leads out along the 4 booms. If you bought motors and ESCs from the same manufacturer, there's a good chance they came with preinstalled "bullet" connectors. In this case, simply plug the motor leads into the ESC leads and coil any slack under the boom. Or you can solder the motor wires directly to the ESC boards for a cleaner build. Secure the motor leads, the ESC power leads, and any leftover slack tightly against the booms with zip ties.

5d. Use wire cutters to snip off one side of each of 4 conduit clamps, leaving a J-shaped foot behind. Smooth the cut end with a file, then file or grind 2 small notches beside the remaining mounting hole as shown. Attach a foot to the end of each boom, just inside the motor mount, using a zip tie run through these notches.

6. Install the shock mounts

Remove the hose clamps from the flexible PVC coupler and save them for another project. Cut two ¾" rings from the coupler's rubber body with a sharp hobby knife. Align each ring across 2 of the frame's pro-

7c

7d

7e

truding inner screws and press down hard with your thumbs to mark 2 drilling spots. Drill ⅛"-diameter holes on the dents, through one side of the ring only. Install the rings over the frame screws with M3 flat washers and nuts. Secure with thread-locking compound when you're happy with the fit.

7. Build the camera/ battery mount

The gimbal and battery shelf are assembled from three simple L-shaped brackets. We refer to these as the *shelf*, *roll*, and *pitch* brackets.

7a. Saw a 36" length of ⅛"×¾" aluminum bar stock into two 18" sections, then saw one of those into two 9" sections, giving 3 pieces total. Make a right-angle bend in each section as indicated on the templates, working over a piece of wood or other scrap with a beveled edge to increase the bend radius to about ⅜". (Too sharp a bend can overstress and weaken the aluminum.) After you've made the bends, cut each bracket to final size per the templates.

7b. Accurately locate, mark, and drill a centered row of three ⅛"-diameter holes on the short leg of the shelf and pitch brackets, and on both legs of the roll bracket. In each case, the outermost hole should be 3mm from the bracket end on-center, and the holes themselves

8a

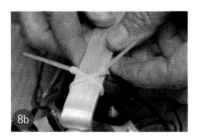

8b

9.5mm apart on-centers. Finally, step-drill the center hole in each row up to ⅜" to provide clearance for the motor shaft.

7c. Use two M3×6mm screws to attach the bottom of a gimbal motor to the shelf bracket, and then 2 more to attach the top of the motor to the longer arm of the roll bracket.

7d. Attach the bottom of the second motor to the free arm of the roll bracket, and its top to the pitch bracket, in just the same way.

> **TIP:** Though the GoPro is a tough camera, you may want to build a "dummy" version having the same weight, and approximately the same size, to mount during your maiden and subsequent shakedown flights.

9a

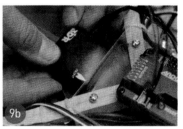

9b

9c

7e. For the gimbal motors to operate smoothly, the camera must be balanced along both axes. Weaken the adhesive on a piece of double-sided tape by sticking it to your shirt and peeling it off. Remove the backing and apply the exposed side to the pitch bracket, then use the weakened side to hold your GoPro in place while you adjust it to find the balance point. Once you've got it, use an elastic band or a velcro strap, in addition to the tape, to hold the camera securely in place.

The gimbal and battery shelf are attached via 2 shock mounts cut from thick flexible rubber tubing, which helps isolate the camera from propeller vibrations and adds a bit of space, above, to mount the gimbal controller board.

9d

9e

10

8. Mount the camera and battery

I designed this quad to balance properly with a 3S 2,200mAh LiPo battery and a GoPro Hero3 White. If you use other equipment be sure you keep the CG (center of gravity) in the middle of your airframe. Here's how to get it balanced.

8a. With the frame upside-down, balance the camera, brackets, and battery across the 2 shock mounts on the underside of the frame. Adjust the position of the whole assembly forward and backward along the frame until the entire quad balances evenly between your fingertips, centered on either side of the body.

8b. Once you've got the CG right, fix the shelf bracket to the shock mounts with 2 sets of crossed zip ties. Apply hook-and-loop tape on top of the shelf bracket and on the underside of the battery, and fix the battery in

place. Add a hook-and-loop strap around both bracket and battery as an added precaution.

9. Set up the avionics

Arrange your flight controller, receiver and other modules before attaching them to the airframe. Once you're happy with the layout, use double-sided tape to secure everything to the frame. Download the wiring diagram from makezine.com/projects/the-handycopter-uav-2 for a detailed list of all connections.

9a. Attach the flight controller. In this build we use 3D Robotics' ArduPilot Mega (APM) 2.6, which contains an accelerometer and must be oriented correctly with respect to the frame. Align the arrow on the APM case toward the front of the quad and fix it in place with double-sided tape.

9b. Add the GPS/compass module, which fits neatly on the rear extension of the bottom frame plate, and also must be aligned with the arrow forward. Tape the module in place and connect the cable to the APM's "GPS" port.

9c. Starting from the starboard-front position and proceeding clockwise (viewed from above), connect the ESC signal cables to APM outputs 1, 4, 2, and 3.

9d. Mount the receiver alongside the APM with double-sided tape, and connect channels 1–5 to the corresponding inputs on the APM.

9e. The gimbal controller consists of 2 boards: the larger controller board and the smaller IMU sensor unit. The controller board goes above the shelf bracket, in the space provided by the shock mounts.

Cover the top surface of the bracket with foam weatherstripping to keep the solder points from shorting against the bare aluminum, then fix the controller board to it with zip ties. The IMU detects the orientation of the camera and needs to be mounted in the same plane; fix it to the underside of the pitch bracket with double-sided tape, and run the connector cable back to the control board. Connect the 3 wires from each gimbal motor to the ports on the controller. Secure all wires with zip ties, leaving plenty of slack for the gimbal to rotate freely.

9f. The flight controller, ESCs, and gimbal controller all need to be calibrated and configured before flight. Refer to the bundled or online instructions that came with your equipment. Specific tutorials are available through makezine.com/projects/the-handycopter-uav-2.

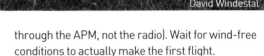

David Windestal

10. Add the props

Before you install the propellers, put bits of masking tape on the motor shafts to make it easy to see which way they are spinning. From above, motors 3 and 4 should spin clockwise, and motors 1 and 2 counterclockwise (see "Anatomy of a Drone," page 160). If a motor is reversed, simply swap any 2 of the 3 leads connecting it to the ESC.

The most important factor for steady flight is balanced props! There are lots of tricks for doing this, but the simplest involves sanding the heavier side of each blade until the prop balances level on a horizontal shaft. (Sand only the flat, not the leading or trailing edges.)

Once the props are balanced, install them on the shafts and tighten the nuts. You'll use 2 conventional airplane "tractor" props and 2 reverse-pitched "pusher" props. Motors 1 and 2 take tractor props, and motors 3 and 4 take pusher props. (If you're not using the APM flight controller, your prop configuration may be different.) Once you've got it right, mark the number and direction of rotation for each motor on its boom for easy reference.

The Maiden Flight

Make sure the props are balanced, the parts are securely fastened, and none of the props, gyros, or controls are reversed. Verify that all your radio trim settings are at zero (if you have to trim, do it through the APM, not the radio). Wait for wind-free conditions to actually make the first flight.

Don't expect your quad to fly perfectly the first time. You'll likely need to make some tweaks and adjustments before it flies well. If you've never flown a quad before, remember to work the controls gently, as most beginners tend to over-steer. Your first goal should be to hover about 24" off the ground for 1-2 seconds and then immediately land. Once you can do that consistently, try to take off, rise above the "ground effect" zone (3'-4'), and then land gently. Work your way up gradually to longer and higher flights.

It is likely that you will crash at some point, especially if this is your first multirotor. Keep a positive attitude, pay attention, and try to learn something every time. Crashing, learning, repairing, and improving your skills and your machine is part of the fun and challenge of the hobby. ⊘

Test Builders: Nick Parks, Brian Melani, and Sam Freeman, MAKE Labs

Chad Kapper is a veteran video producer with more than 18 years of experience. His passion for film and flight inspired him to create Flite Test (flitetest.com), one of the leading brands in the R/C flight industry. For more than 3 years, the team at Flite Test has been entertaining, educating and inspiring the world with R/C flight.

For complete parts spec, templates, wiring diagram, and ArduPilot tutorials, go to makezine.com/projects/the-handycopter-uav-2.

Share it: #handycopteruav

Music
AND Audio

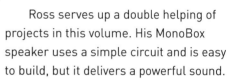

PART 6

Mark Frauenfelder's amazing range of accomplishments include having been the first editor-in-chief of *Make:* magazine, cofounder of the website Boing Boing, and author of the book *Made by Hand*. A dedicated evangelist of DIY culture, here Mark introduces you to the art of making cigar box guitars. The beauty of cigar box guitars is that you can make them from scrounged parts and materials found in any hardware store. With a few simple tools you probably already have in your workshop, you can make a musical instrument that will have a unique sound and provide entertainment for years to come.

IBM researcher Tom Zimmerman has contributed many projects to *Make:* magazine. He was drawn to DIY percussion instruments as a way to get kids interested in STEM through hands-on construction, as well as to capture the interest of less tech-oriented adults through art and music. "What's enticing about drums," he says, "is anyone can play them, and they are social; many people can play together like in a drum circle." Tom updated his "Electronic Drum Kit" project especially for this volume. The clever design uses foil pads stretched over PVC pipes to create a unique electronic drum set hack.

Audiophile and maker Ross Hershberger developed the Squelette amp especially for *Make:* readers. (Turn to the article to find out what the name means.) It's a cheap and simple chip-based amp that not only sounds great but also shows off your soldering skills at the same time with its see-through design. For this publication, Ross added some alternative cabinet designs that he created since the article's original appearance.

Ross serves up a double helping of projects in this volume. His MonoBox speaker uses a simple circuit and is easy to build, but it delivers a powerful sound. Plug it into the headphone jack of your iPod or smartphone, and you're ready to rock the house.

Rory Nugent created his solar xylophone for *Make:* several years ago. These days he is an engineer for littleBits Electronics, a company that is dedicated to making electronics fun and easy to understand for children. For his project, Rory drew inspiration from wind chimes, which draw natural energy from moving air to create music. Rory's chimes give a voice to the sun.

Software engineer Stephen Hobley hails from the UK but migrated to the US in the late 90s. His laser harp is undoubtedly the most visually arresting musical instrument. You can find videos of British electropop diva Little Boots playing one of his hand-built devices on YouTube. The version you will find here, simpler and using inexpensive laser pointers rather than more expensive scanning lasers, still creates a great visual effect and a wide array of sounds.

TRADITIONAL CIGAR BOX GUITAR

By Mark Frauenfelder

Photography by Mark Frauenfelder

HAND-ROLLED MUSIC

Several years ago, MAKE featured an electric Cigar Box Guitar project (*Volume 04, page 76*). The project's author, Ed Vogel, designed a simple instrument using only parts you'd find at a hardware store. I made one myself, and had a wonderful time playing it.

Recently, I decided I'd like to make a more traditional cigar box guitar. I soon found Cigar Box Nation (cigarboxnation.com), a fantastic online hangout for homemade stringed instrument enthusiasts. The photos, videos, and MP3s posted by these happy strummers and pluckers were inspiring, and the variety of guitars in the photo galleries was astounding.

I joined the group and was warmly welcomed by its members, who kindly answered my newbie questions about frets, choices of wood, and other aspects of guitar building. In a matter of days, I had built my first cigar box guitar (or CBG for short). I've now built more than a half dozen CBGs, and I guess you can say I'm hooked.

Because every CBG is built by hand, using different found and scrounged materials, no two sound alike. I love the suspense of not knowing what kind of "personality" a CBG is going to have until it's completed. Here's how to make a plain-vanilla, 3-string CBG that requires a minimum of tools and parts, yet sounds great.

Mark Frauenfelder is the founding editor-in-chief of *Make:* magazine and Wired Online, cofounder of Boing Boing, and editor of the Cool Tools website. He is the author of seven books.

A String, a Stick, and a Box

A cigar box guitar is much like a regular guitar except it usually has fewer strings. Before buying parts for one, rummage around in your junk drawers. A plastic comb can be cut and used as a bridge. An old cabinet hinge can serve as a tailpiece. A bolt makes for a dandy nut. The photo section of cigarboxnation.com shows ingenious and oftentimes humorous examples of CBG builders' resourcefulness.

A The cigar box serves as the instrument's **body** and resonating chamber. It doesn't necessarily have to be a cigar box. You can use a wooden craft box, a boxy tin can (like the kind turpentine comes in), or anything else with flat, light, solid surfaces that will push the air as they vibrate.

B At the lower end of the box, a simple cabinet hinge serves as the **tailpiece** that holds the strings' bottom ends. The strings pass through the screw holes on one face of the hinge, which holds back their end barrels. The other face of the hinge attaches to the end of the box.

C The strings stretch over a thin, blade-like **bridge** that transfers their vibrations to the guitar's body. For this guitar, I used a piece of wooden barbecue skewer, but I've also used a thin paintbrush handle with good results.

D The guitar's **neck** can be a piece of oak or maple. Don't use a softwood that bends easily or your neck will bow (especially if you use 4 or more strings or crank up the tension).

E The **scale length** (distance between the bridge and the nut) is up to you. Most guitars have a scale length somewhere between 24" and 25½". If you want to make a cigar box bass, try 30"–34".

F **Frets** determine the note a string plays, by restricting the length along which it vibrates. I made them using fret wire, but toothpicks or small nails with the tips and heads clipped off will also work. Or you can go fretless, which is great for slide guitar.

G The **nut** supports the upper ends of the strings so they can vibrate freely. I used another piece from the wooden barbecue skewer, but bolts also work nicely because their threads prevent the strings from sliding sideways.

H **Tuning pegs** and strings can be scavenged from an old guitar, or purchased at a music store or online.

Fret Spacing

A string's frequency is inversely proportional to its length; if you halve the string's vibrating length by fretting it at its midpoint, it vibrates twice as fast. Musically, this means it plays one octave higher. Western music divides the octave into 12 equal intervals, so you can determine the distance between adjacent frets by successively dividing each length by the 12th root of 2. This is an irrational number, and any rounding error accrues with each division, so you'll need to carry it out with a lot of digits. Using 1.05946309436 should be safe.

For example, starting with a 24" string gives you string lengths of 22.653", 21.382", 20.182", and so on. These lengths correspond to fret distances from the nut of 1.347", 2.618", and 3.818", and the 12th division takes the distance up to 12". Visit makezine.com/projects/cigar-box-guitar for links to online fret calculators that perform these operations instantly.

BONUS!
Illustrator Rob Nance created a sheet of Papercraft Guitar Picks to help you rock out in style with your new cigar box guitar! Get it at makezine.com/projects/cigar-box-guitar.

- Second fret/1.05946
- First fret/1.05946
- Scale length/1.05946
- Scale length

Illustration by Rob Nance

SET UP.

MATERIALS

[A] Cigar box I buy them at my local cigar store for $3 each. You can also find them on eBay.

[B] 1×2 oak or maple lumber, 3' length The actual dimensions are ¾"×1½". A 6' stick (enough for 2 necks) costs about $10. Pick the straightest, flattest, clearest (free of knotholes) piece you can find.

[C] Guitar strings Standard medium-gauge strings work well. CBGs typically use open G tuning. I use strings 5, 4, and 3 (A, D, G) and tune them to G-D-G.

[D] Tuning pegs Elderly Instruments (elderly.com) sells a set of 6 (enough to build two 3-stringers) for $10. Sometimes they're called "tuning machines."

[E] Fret wire $10 at elderly. com or cbgitty.com. You can also use flat toothpicks or go fretless.

[F] Cabinet hinge with 3 mounting holes on each side

[G] 1" wood screws Phillips head Grip-Rite Fas'ners work well.

[H] Bamboo barbecue skewer or other hard, thin rod for the bridge and nut. A ³⁄₁₆"×2" bolt also works well for the nut.

[I] Super glue (optional)

TOOLS

[J] Hole saw, ¾"

[K] Stanley Surform shaver from a hardware store

[L] Miter box hobby size

[M] Coping saw

[N] Wire cutters

[NOT SHOWN]

Yardstick A decimal inch or millimeter scale is best, as opposed to fractional inches.

Wood saw

Drill or drill press and bits for wood

Phillips head screwdrivers

Hammer

Marker or paint

Sanding block and sandpaper of various grits

Files

T square or carpenter's square

Jeweler's file (optional)

Pencil

Sharpie or paint

Magnifying glass

Utility knife

MAKE IT.

BUILD YOUR CIGAR BOX GUITAR

START >>>   Time: **An Afternoon** Complexity: **Easy**

1. MAKE THE NECK

To begin, we'll cut the neck to length, make the headstock (the part where the tuning pegs go), and saw off a rectangular slice so that the fretboard is flush with the cigar box lid. Making the neck and installing the frets are the most time-consuming parts of the build. Once you're finished preparing the neck, you'll be surprised by how fast the rest of the build goes!

1a. Using a wood saw, cut the oak or maple lumber to 36". You'll have to cut it a little shorter later on, but it's good to start out with more than enough.

1b. Saw off a rectangular slice from the lower end. This is the end that goes into the cigar box. Measure the length and thickness of the cigar box lid. I use the box itself as a guide, tracing along the oak stick with a pencil. Mark these dimensions on the wood, then use a saw to remove the part shaded red in the illustration.

Also make a pencil mark 2¼" from the end. This is where your bridge will go later on.

1c. Mark the lines for the nut and headstock. Starting from the pencil mark you just made for the bridge, make another mark indicating the scale length (I decided on a scale length of 24½"). This second mark is where the nut will go. Make a third mark ½" farther past the nut. Make a fourth and final pencil mark 3½" beyond the third mark.

1d. Cut out the headstock. Your third and fourth pencil marks indicate the beginning and the end of the headstock. Use your saw to cut away the material shaded in red in the illustration above. The headstock should be half as thick as the neck, or ⅜".

1e. Sand the fretboard. Now is a good time to sand the top surface of the neck so it's dead flat. Use a sanding block, starting with rough sandpaper and finishing with fine-grit sandpaper.

2. INSTALL THE FRETS

I used to be intimidated by the idea of frets. The process seemed mysterious and difficult. But it's really not. If you take your time and make careful measurements, you'll have no problem.

2a. Mark the fret locations. Enter your desired scale length into an online fret spacing program (see makezine.com/projects/cigar-box-guitar) and print out the table it generates. Using a yardstick and a square, make pencil marks along the length of the neck to indicate the location of the frets.

NOTE: If you don't want to install metal frets, you can glue flat toothpicks over the pencil marks. They work quite well, but will eventually wear out. If you want a fretless guitar (which also sounds great), go over the pencil marks with a Sharpie or with some paint. In either case, skip to Step 3.

2b. Cut the fret slots. About ¹⁄₁₆" should be deep enough. The saw blade should be thin enough so the fret tangs bite into the slots you cut. I buy medium-gauge fret wire and have had no problem with frets popping out. A coping saw and a hobbyist's miter box will help you keep the fret slots square with the neck.

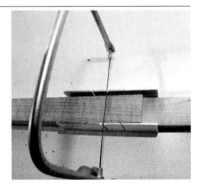

2c. Form the back of the neck. On the backside of the neck, shape the sharp 90° edges into soft curves so your fretting hand can easily slide up and down the neck. A Surform shaver tool will quickly rough out a rounded edge. Follow up with sandpaper until the wood is very smooth.

NOTE: Don't shave the headstock or the part that will fit into the cigar box — only work on the area under the frets and nut.

2d. Tap the frets into the slots. Fret wire usually comes pre-cut, and each piece is about an inch longer than the width of the neck. The wire's cross section is T-shaped, and the barbed center rail goes into the slot.

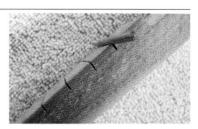

At each fret slot, align one end of the fret wire so it overhangs the side of the neck just a fraction of an inch. Press the fret wire into the slot, then place a thin block of wood on the fret and tap on the block with a hammer until the fret is all the way in.

NOTE: You can smear a tiny bead of super glue across the part of the fret that fits in the slot if you wish, but I usually skip it, because it's hard to keep the glue from getting onto the neck.

2e. Clip the fret wire. Cut it almost flush with the neck. Repeat Steps 2d and 2e until all frets are installed. I installed 21 frets on my cigar box guitar.

2f. File the ends of the frets. The cut ends of the fret wires are very jagged and would shred your hands if you attempted to play without filing them smooth. Use a file to form a gentle curve on both ends of each fret. (If you have a store-bought guitar handy, inspect it to see how the frets should look.) Run your hand up and down the neck. If your skin snags, you need to keep filing! Use a magnifying glass and look for any small burrs that need to be filed off with a jeweler's file.

3. INSTALL THE TUNING PEGS

Study the geometry of your tuning pegs and determine where the headstock holes need to be drilled so that the strings and pegs won't interfere with each other. Keep in mind the location of the mounting screws — they shouldn't be too close to the edge of the headstock, or they might split the wood.

3a. For each peg, drill a large hole for the post and 2 small pilot holes for the mounting screws. A drill press will make things easier, but if you use a handheld drill, try your best to drill straight down.

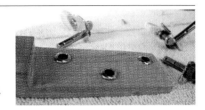

TIP: When you drill the holes for the posts, use drill bits made for wood. I used the wrong kind of bit and it tore out big splinters. (You can see a missing piece above the top tuning peg, inside the orange circle in the bottom photo.)

4. ATTACH THE NECK AND HARDWARE TO THE CIGAR BOX

Now that the neck is complete, the rest is smooth sailing!

4a. Cut a hole (a 3-sided notch) in one end of the box for the neck. Measure the cross section of the part of the neck that fits inside the cigar box, and draw a matching rectangle on the inside of the box. Use a coping saw to cut the 2 vertical lines, then use a utility knife to score the horizontal line several times until you can snap off the rectangle.

Insert the neck, close the lid, and make sure the fretboard is flush with (or a tiny bit higher than) the lid. If the fretboard is lower than the lid, sand down the cut-out part of the neck that comes into contact with the lid until the fretboard and lid are flush.

4b. Now we'll screw the neck to the box. I try to use as little glue as possible when I make a cigar box guitar because I don't like waiting for the glue to dry, and screws make it easy to take the guitar apart for repairs, modifications, or salvage.

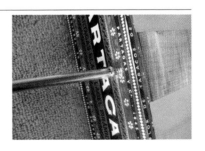

Drill a pilot hole in the far end of the box and drive a screw through the box into the neck. Close the lid and then pilot-drill and drive 2 more screws through the lid of the box into the neck (if you later want to install a pickup, you can remove these screws).

4c. Attach the tailpiece. Fold the cabinet hinge centered over the front lower edge of the cigar box, then drill pilot holes and screw it to the lower end of the box. The hinge will sit over the screw you inserted in the previous step, 4b.

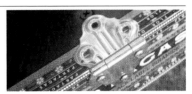

4d. Paint position marker dots on the neck. Use paint or a Sharpie to make dots above frets 3, 5, 7, 9, and 12.

4e. String the guitar. Thread the barrel ends of the strings through the hinge's unused mounting holes. Wind the other ends of the strings onto the tuning pegs, but not too tight yet. Here's a good video that will teach you how to wind a guitar string: makezine.com/projects/cigar-box-guitar.

NOTE: I inserted a screw to keep the middle string centered in the headstock. You might have to do this, too.

4f. Slip in a bridge and a nut under the strings. I used a wooden barbecue skewer to make the bridge and the nut. Snip 2 pieces to size and place one above the line you drew for the nut. Place the other under the strings on the cigar box at a distance equal to the scale length you chose; this is the bridge.

4g. Screw down the tailpiece. Drill a hole through the hinge and drive a screw through it into the lid and the neck. This will increase tension on the strings and prevent rattling.

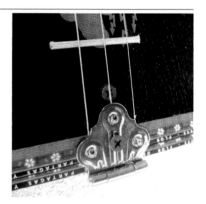

4h. Make a sound hole. Use a small hole saw (¾" diameter or so) to cut a sound hole in the top of the box. Make sure to position the hole so it doesn't cut into the neck. (I made this mistake when I made my first cigar box guitar!)

Guess what — you've built your guitar! In the next section, I'll explain tuning and playing, as well as direct you to other helpful cigar box guitar resources.

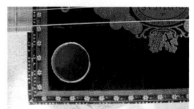

FINISH X

NOW GO USE IT »

TUNE UP
AND TURN ON

GET IN TUNE

The most popular tuning for cigar box guitars is called open G tuning. Many of the original blues guitar players used open G, and it's a favorite with Keith Richards of the Rolling Stones.

Visit makezine.com/ projects/cigar-box-guitar for an MP3 file of this tuning played string by string on a six-string guitar. For the CBG, you can ignore the first string that's plucked, and tune it to the 3 strings after that: G, D, and G.

FREE ONLINE LESSONS

Keni Lee Burgess, a well-known New York street musician, has posted a terrific series of cigar box guitar lessons on YouTube. Shane Speal, cofounder of cigarboxnation.com, also has fun lessons on YouTube that show you how to use a slide, and how to experiment with different tunings and scales. Visit the URL above for links to both of these series.

MAKE A BOTTLENECK SLIDE

Bottleneck slides sound great with open tuning, and both Burgess and Speal use them to enhance their playing.

To make one, take an empty wine bottle and score a ring around the neck with a Dremel cutting disc (wear eye protection). Wearing a pair of oven mitts, tap the score line with a spoon and snap off the neck (do this over a trash can to capture the

shards). Sand off the rough edges and you've got something far superior to a store-bought slide. YouTube has instructional videos on making bottleneck slides using different techniques.

TURN IT UP

You can electrify your cigar box guitar in 2 ways. The easiest is by adding a piezoelectric buzzer. Buy one at RadioShack or salvage one from a discarded smoke alarm. Carefully crack open the plastic housing, remove the metal disc, and sandwich it between the neck and lid of your guitar. Wire it to a patch cord jack and plug into an amp (if you don't have an amplifier, make our Cracker Box Amp featured in MAKE, Volume 09, page 104).

Another way to electrify a cigar box guitar is by adding an electromagnetic pickup. In MAKE Volume 15 we show you how to make your own.

SQUELETTE, THE BARE-BONES AMPLIFIER

By Ross Hershberger

Photograph by Sam Mur

SKELETAL SYSTEM

Squelette is a see-through amplifier that sounds ridiculously good while showing off your soldering (it looks nothing like a typical audio product). Build it with common materials and enjoy music played through a component you made yourself.

I love to build audio gear: speakers, turntables, preamplifiers, tube amps, transistor amps, and anything else that reproduces music. A few years ago I read rave reviews of an exotic audiophile amplifier based on a National Semiconductor amplifier chip. After more research, I learned that NatSemi's Overture series of audio amp ICs pack a ton of musical goodness into one robust and flexible chip, combining all the features of a well-designed transistor amp.

I decided to use their LM1875 chip to make a small, simple stereo amplifier. I dubbed the result Squelette (*skeleton* in French), after mechanical wristwatches that show their internal workings. Squelette is a chip amp in its simplest form: it has 2 source inputs and a volume control, and it puts out a useful 11 watts per channel. Building it requires no exotic parts; I made careful design choices and sourced everything except for the LM1875 from RadioShack and the plumbing aisle at the hardware store. Here's how you can build your own, from materials costing less than $50.

Ross Hershberger was working with hobby computers long before IBM coined the term PC. He worked as a mainframe systems analyst for 20 years before returning to technical electronics as a restorer of vintage audio gear.

Sound Design Practices

A good design makes every part earn its keep. Here's how the Squelette amplifier's common components team up to make a whole that's greater than its parts.

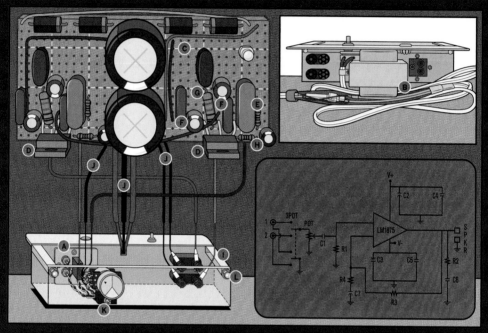

Illustration by Timmy Kucynda

(A) The input jacks connect to the audio source.

(B) The power cord connects to one side of a 25.2V 2A transformer. The LM1875 chip can handle larger transformers, but scaling the other components and the heat-sinking design to match would make the amp significantly bulkier, more complicated, and more expensive.

(C) The heart of the amp is a small, hand-built circuit board. The grid pattern of the board's perforations lends itself to easy, solder-by-numbers instructions. The power supply components (along the top and middle of the board as shown here) convert AC from the transformer into DC for the amplification circuits.

(D) Each of the 2 identical amplification channels is built around a NatSemi LM1875 amplifier chip.

(E) Capacitor C1 isolates the chips' pin 1 inputs from DC, and resistor R1 provides a ground reference.

(F) Capacitors C2, C3, C4 and C5 bypass the power supply to the V+ and V− pins of the LM1875, ensuring a low-impedance power source at high frequencies.

(G) Resistor R2 and cap C6 form a Zobel filter, presenting a low-impedance load to the output at very high frequencies. This loads down and damps out oscillations and spurious signals outside the audio band.

(H) Resistors R3 (on back of board) and R4, AC-coupled to ground through C7, form a global negative feedback loop from the chip's output pin back to its negative input signal terminal. This feedback sets the overall voltage gain, reduces the output impedance, and reduces internally generated noise and distortion.

(I) The circuit board is housed in a cabinet made from stock aluminum angle and plexiglass. The aluminum attaches to the LM1875 chips as a heat sink, as well as serving as the circuit ground. The plexiglass lets you see the circuitry.

(J) All ground connections run directly to one shared "star" ground, a large loop of wire centered on the back of the PCB that connects to the cabinet. This wiring prevents interactions between circuit features that could cause noise and instability.

(K) The 3P2T (triple-pole, double-throw) input selector switches both the signals and the grounds of 2 input devices, to completely isolate the deselected input device. This avoids ground currents and noise that can come from connecting different inputs' grounds together. Wired for "center off," the switch also lets you mute the amp without changing the volume setting.

(L) Two acorn nuts at the upper corners of the chassis let you expose the amp's innards to open air, and 5 more screws let you remove the Squelette's circuitry from the chassis completely, without unsoldering.

SET UP.

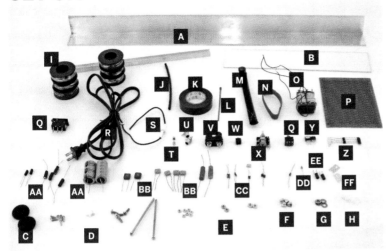

MATERIALS

**[A] Aluminum stock: 2"
angle, ¹⁄₁₆" thick, 28" long;
⁵⁄₈"×½" channel, ¼" inside
cavity, 13" long** Online
metals part #6063-T52;
National Mfg. #N258-525

**[B] Clear acrylic (plexi-
glass), ³⁄₁₆" thick, 13"×2⁷⁄₈"**

**[C] Plastic furniture feet,
about 1"×½" thick (2)**

**[D] Screws: #4-40×³⁄₈" (7);
nylon #6-32×½" (2);
#8-32×½" (4);
#8-32×3½" (2)**

**[E] Nuts: #4-40 (7);
#6-32 (2); #8-32 (2)**

**[F] Acorn nuts, #8-32,
chrome plated (4)**

**[G] Washers, #8, chrome
or nickel plated (4)**

[H] Nylon P-clip, ¼"

**[I] Insulated solid-core
wire: 18 gauge, black,
red, and green; 24 gauge,
various colors**

**[J] Heat-shrink tubing,
³⁄₁₆" diameter, at least 4"**

[K] Electrical tape

[L] Small zip tie

**[M] Tech Weave braided
cable sleeving, 7" long**

[N] Velcro tape, 2"

**[O] Transformer, 25.2V CT
2A** RadioShack #273-1512

**[P] Perf board with copper
pads on one side, 2"×3½"**
RadioShack #276-1395

**[Q] Panel-mountable
switches: power (any type)
and 3P2T toggle**

[R] Extension cord

[S] In-line fuse holder
RadioShack #270-1238

[T] 500mA fuse

[U] RCA jacks (2)
RadioShack #274-346

[V] Speaker terminals
RadioShack #274-718

**[W] ⅛" audio (headphone)
jack** RadioShack #274-249

**[X] 100kΩ dual potentiom-
eter** RadioShack #271-1732

**[Y] National Semiconduc-
tor LM1875 IC chips (2)**

**[Z] Rectifier diodes, ≥4A
and ≥40V (4)** Schottky
B540, B560, 1N5822

**[AA] Capacitors, electro-
lytic: 22µF (2), 100µF (4),
4,700µF (2)**

**[BB] Capacitors, film:
0.1µF (4), 0.22µF (2),
1µF (2)**

**[CC] Resistors: 1Ω, 1W;
830Ω, ½W; 1kΩ, ¼W (2);
22kΩ (4); 220kΩ (2)**

[DD] LED any color

**[EE] Crimp-on spade ter-
minals, 18–22 gauge (5)**

**[FF] TO-220 insulated chip
mounting pads (2)**

[NOT SHOWN]

Knob for potentiometer

TOOLS

Drill press or drill with bits:
⅛", ⁵⁄₃₂", ¹¹⁄₆₄", ¼", ⁵⁄₁₆",
⅜", ½" or use a step bit

Disk or belt sander

**Band saw or hacksaw with
fine-toothed metal blade**

Sheet metal nibbler

**Soldering iron, fine point,
with solder, flux paste, and
desoldering plunger**

**"Third hand" holder and
magnifying glass**

**Wire cutter/stripper,
needlenose pliers, and
crimping tool or pliers**

3" C-clamp

**Needle file, knife, or
Dremel tool with tapered
diamond bit**

Digital multimeter

Signal generator or
computer running a signal
generator app, or CD player
with a CD of sine waves

Clip-on heat sinks (2)

Resistors, 10W, 10–25Ω (2)

Alligator clip leads (3)

**Carpenter's square and
metal ruler, at least 12"**

Hammer and metal punch

Metal files, flat and round

Screwdrivers and wrenches

Reamer and cutting oil

**Masking tape, scrap card-
board, and Sharpie marker**

400-grit sandpaper

**Small brass wire brush and
disposable paintbrush**

Butane pocket lighter

Scotch-Brite scouring pad

**Rubbing alcohol, cotton
swabs, and paper towels**

**Eye protection and leather
shop gloves**

MAKE IT.

BUILD YOUR MINI AMPLIFIER

START ❖❖❖   Time: 2 Weekends   Complexity: Moderate

1. BUILD THE CIRCUIT BOARD

This board looks complex, but each LM1875 chip contains 46 transistors; imagine hand-wiring all of those!

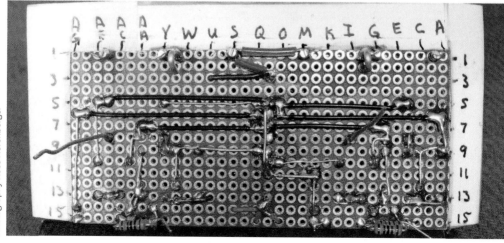

1a. Cut a 33×15-hole piece of perf board, then sand the edges. Scour the copper side with a Scotch-Brite pad and wipe with alcohol. Tape a cardboard frame around the edges and label the rows 1–15 and the columns A–AG, with the origin at upper left on the component (non-copper) side.

NOTE: You can download a schematic diagram for the Squelette at makezine.com/projects/build-bare-bones-skeleton-amplifier.

1b. Start with the power supply. Connect the following hole pairs with diodes, cathode side (marked with a line) listed second: B1 to G1, H1 to M1, X1 to S1, and AD1 to Y1.

On the copper side, jumper G1 to H1, X1 to Y1, and M1 to S1. Clip the leads at H1 and X1 off to ½" and bend them under in a tight loop. Clip all other leads off completely.

TIP: If any holes are too small, ream them from the component side with a needle file, a knife point, or (fastest) a Dremel tool with a tapered diamond bit.

1c. Strip some 24-gauge wire, thread it through A1, and solder it in back to bridge A1 and B1. Route it over the 4 diodes, and strip, thread, and solder the other end to connect AE1 and AD1.

1d. Stuff a 4,700μF cap into P3 (+) and P6 (−) and another into P10 (+) and P13 (−). Jumper S1 to P3. Electrolytic capacitors have the (−) terminal labeled.

1e. On the component side, pass a 24-gauge wire through AF1 and connect to AD1. Route it between rows 1 and 2, then down between columns S and T. Thread the other end through S14, insulated, then cut, strip, and connect only to P13.

1f. Connect P6 to P10 in back with a raised loop of bare 18-gauge wire. This is the star ground node, and a lot of wires attach here, so leave some room.

1g. To test this circuit, clip-lead (alligator jumper) the transformer's secondary center tap lead (black) to the 0V star ground at P6/P10, and its secondary (yellow) leads to X1 and H1. Temporarily solder one side of the fuse holder to one of the transformer's primary leads (black) and tape the joint. Insert a 500mA fuse. Split a power cord (not plugged in), connect it between the free side of the fuse and the free primary transformer lead, wrap with electrical tape, and use a switched power strip to apply power to the transformer.

If nothing sparks or pops, measure the DC voltages across both capacitors. You should get around −18V from P10 to P13 and around +18V from P10 to P3. If the voltages read 0V, check the fuse. If it's blown, there's a wiring error. If the voltages check out, turn it off and clip a 22K resistor between P3 and P13 to bleed off the charge on the caps.

1h. Stuff and solder the following left channel components:
• LM1875 into D13–H13, perpendicular to and facing the center of the board (pin 1 in H13, heat-sink side out)
  • 1K resistor into H15, K15
  • 22μF capacitor into L15 (+), L14 (−)
  • 22K resistor into J10, J13
  • 1μF cap into I8, I13.

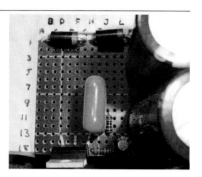

On the back of the board, solder K15 to L15 and clip off. Solder H15 to G15. Solder J13, I13, and H13 trogether and clip off. Clip the other leads except for I8, J10, and L14.

1i. Stuff and solder a 0.1μF cap into G10, G14; and a 100μF cap into F8 (+), F9 (−). On the back, bend lead F9 down and solder it to F13 and G14. Bend lead G10 up and solder it to F8. Clip the other leads, except for F8.

1j. Stuff and solder a 1Ω resistor into D9, E14; and a 0.22μF cap into D5, D8. In back, solder E14 to E15, and D8 to D9. Clip except for D5 and E15.

1k. Stuff and solder a 100μF cap into C12 (−) and C13 (+); and a 0.1μF cap into A9 and A13. Bend lead C12 up to A9 and solder, then solder A13, C13, and D13 together.

1l. Use separate jumpers to wire A9, D5, F8, J10, and L14 to the ground wire loop in back. Put a red 24-gauge jumper through L3 from the component side and connect it to P3. Pass the other end through B14 and connect it to D13. Put a green jumper through O15 from the component side and connect it to P13; pass the other end through F12 and connect it to F13.

Finally, connect a 20K resistor (or a 22K and 220K in parallel) on the copper side between G15 and E15, making sure it doesn't project past the edge of the board. This is the feedback resistor. Clip off its lead at G15 and leave its other lead in place.

1m. Stuff and solder the following right channel components (just like building the left channel):
• LM1875 chip into AC13, AB15, AA13, Z15, and Y13, perpendicular to and facing the center of the board (pin 1 in Y13, heat-sink side out). The chip should clear the bottom edge of the PCB.
  • 1K resistor into AC15, AF15
  • 22μF cap into AG14 (−), AG15 (+)
  • 22K resistor into AE10, AE13
  • 1μF cap into AD8, AD13.

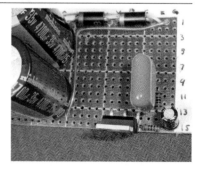

On the back of the board, solder AF15 to AG15 and clip off. Solder AC15 to AB15. Solder AE13, AD13, and AC13 together and clip off. Clip the other leads except for AD8, AE10, and AG14.

1n. Stuff and solder a 0.1μF cap into AB10 and AB14; and a 100μF into AA8 (+) and AA9 (−). On the back, bend lead AA9 down and solder it to AB13 and AA14. Bend lead AB10 up and solder it to AA8.

1o. Stuff and solder a 1Ω resistor into Z14 and Y9, and a 0.22μF cap into Y5 and Y8. In back, solder Z14 to Z15, and Y8 to Y9. Clip except for Y9 and Z15.

1p. Stuff and solder a 100μF cap into X12 (−) and X13 (+), and a 0.1μF cap into V9 and V13. Bend lead X12 up to V9 and solder it, then solder V13, X13, and Y13 together.

1q. Use separate jumpers to wire V9, Y5, AA8, AE10, and AG14 to the ground wire loop in back. Put a red 24-gauge jumper wire through U3 from the component side and connect it to P3; pass the other end through U13 and connect it to Y13. Put a green jumper through S14 from the component side and connect it to P13; pass the other end through AA12 and connect it to AA13.

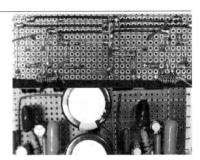

For the right channel feedback, connect a 20K resistor (or a 22K and 220K) on the back between AB15 and Z15. Clip off lead AB15.

2. TEST THE AMP

2a. For the power indicator light, stuff your favorite color LED into AC3 (anode, longer leg) and AD3 (cathode). Bend and solder the cathode lead to AD1. Stuff an 830Ω, ½W resistor into Z3 and V3, then solder V3 to the ground wire loop, and Z3 to AC3.

2b. The circuit board is done now, and the only unconnected leads remaining should be I8 (left channel input), E15 (left channel output), AD8 (right channel input), Z15 (right channel output), X1 and H1 (AC input), and the ground at P6–P10.

To set up testing, hook up the transformer to the power supply as in Step 1g. Clamp a heat sink to the tab of each LM1875 chip, making sure they don't touch anything conductive. For dummy loads, clip-lead a 10Ω–25Ω, 10W resistor from E15 to ground, and a second resistor from Z15 to ground. Connect AD8 to I8. Connect I8 to ground for the moment.

2c. Power up the circuit as in Step 1g. The power LED should come on. Measure the DC voltage across each dummy load resistor. If it's over 0.05 volts, power down and look for wiring errors.

2d. For signal testing, disconnect I8/AD8 from ground and connect them to any signal generator source with a volume control. Power the circuit back up and test across the load resistors, measuring AC voltage this time. Briefly increase the volume of the input, and look for the voltage to read up to about 9V AC output. But keep all signal testing brief, to avoid overheating the chips. If you have an oscilloscope, look for clipping at about 25V peak-to-peak. Reduce the signal to 1V AC output on one channel, and then test the other channel; it should also measure 1V AC. Switch the meter to DC and confirm that both channels measure 0V DC. Try a speaker in place of the load resistor to make sure the output signal sounds like clean sine wave. If it all checks out, your board is done and working.

3. MAKE THE CABINET

3a. Download and print the mechanical drawings at makezine.com/projects/build-bare-bones-skeleton-amplifier. Part A is the chassis front, B is the inside back, and C is the outside back. Parts D and E are rails that hold the plexiglass ends and top.

3b. Cut three 9" lengths of the 2" aluminum angle and four 3" lengths of the ⅝"×½" aluminum channel. Sand the edges smooth.

3c. To protect the aluminum and mark the patterns for drilling and cutting, cover the surfaces with masking tape. Transfer the markings from the drawings.

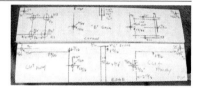

3d. On Part C, position the transformer and check the clearance between the switch and transformer underneath. Adjust the position of the switch hole to suit your switch, and mark it.

3e. Time to drill. Use a sharp punch and a hammer to dimple the center of each hole to be drilled. Drill the smallest holes first.

TIP: Some holes need to align for assembly: the row of four ⁵⁄₃₂" and ¹¹⁄₆₄" holes on parts A and B; the ⅛" holes in the D spacers and the bottom corners of A and B; and two ¹¹⁄₆₄" holes on parts B and C. For these, drill one part, clamp it up to the other part, and mark the hole locations with a Sharpie. Punch and drill on your marks, and the holes will match up. This is cheating but it works.

3f. For larger holes, drill a small hole first to make the bit easier to center. You can also cut large holes with a step bit (aka castle bit or unibit).

3g. Use a tapered reamer or round file to open the large holes up to their final size. Deburr the small holes with a large drill bit or chamfering tool. Deburr the large holes with a round file. Holes C at the bottom of part A must be completely smooth. Any burrs there might poke into and short out the LM1875 chips.

3h. For the square hole on part B, the input panel, you can drill a ⅝" hole and use a ¼" metal nibbler to cut the perimeter. Finish the hole with a flat file.

3i. Part C has 2 large areas to cut out. Saw these with a hacksaw or band saw and deburr. Make two 1"×¼" shims out of the cut-out material. These will fill a gap between parts B and D.

3j. For the long slots in parts B and C, drill a ¼" hole at the end, and saw up to the holes, making ¼" slots with round ends.

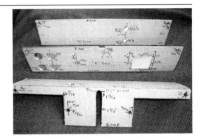

3k. Assemble the metal parts to check for hole alignment: two 3½" screws with acorn nuts at the tops of parts A and B, through rails E; #8-32×½" screws and nuts through the matching ¹¹⁄₆₄" holes on A and B, and B and C; and #4-40×3/8" screws securing rails D to the bottom corners of A and B. Adjust the positions of any off-center holes with a needle file.

3l. Cut ³⁄₁₆"-thick plexiglass into one piece 8½"×2⅞" (for the top) and 2 pieces 2⅞"×1½" (for the sides). Notch out the lower corners of the side pieces so they clear the nuts in the D rails. Test-fit these pieces and trim to fit.

3m. For the input terminal panel, cut a 1½"×2" piece of Formica or other thin, stiff plastic. Tape it inside the input terminal cutout on B, then transfer the screw hole positions and cutout shape to the plastic with a Sharpie.

Drill the 3 smaller holes for #4 screws (0.100") and the larger hole for the #8 screw (0.16") to mount the transformer. Find positions on the cutout area where the input jacks will not touch part B's metal, then drill the plastic and install the input connectors there.

4. CONNECT THE OFF-BOARD COMPONENTS

4a. Test-fit the volume knob onto the dual potentiometer shaft. Cut the shaft to length, being careful to keep metal bits from getting inside the potentiometer.

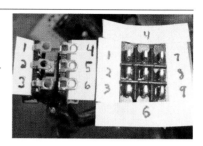

4b. Referring to the photo of the potentiometer and 3P2T switch at right (and the tip below), use 4" wires to connect the following input jack and switch terminals: input 1 left to switch 7, input 1 right to switch 9, input 1 ground to switch 8, input 2 left to switch 1, input 2 right to switch 3, and input 2 ground to switch 2.

Use 3" wires to connect switch 6 to volume pot 4, switch 5 to volume pot 3 and 6 (both), and switch 4 to volume pot 1.

With 6" wires, connect volume pot 3 and 6 to the star ground wire on the PCB, volume pot 5 to PCB lead AD8, and volume pot 2 to PCB I8.

TIP: Recall that with headphone plugs, the tip is the left channel, the ring is the right, and the body or "shield" is ground.

4c. Cut the socket end off your extension cord. Near the plug, cut out a section of the live wire (the wire connected to the plug's narrower prong). Splice in the inline fuse holder with enough slack so you can open it up, and insulate these solder joints with heat-shrink tubing. Write "½ A" on the fuse holder, or otherwise label it in case the fuse is lost.

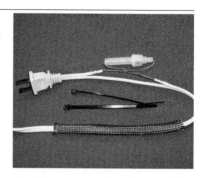

Slide cable sleeving up over the fuse holder and position it so it can be pushed back for changing the fuse. Fix one end of the sleeve to the cord with shrink wrap or a zip tie, and the other end with a loop of double-sided velcro.

4d. Install the speaker terminals, power switch, and transformer on parts B and C, running the transformer secondary (yellow and black) wires in through the slots. Temporarily screw the PCB to part B with screws through the LM1875 tabs. Bend the 2 yellow leads to the hooked diode leads H1 and X1, and carefully mark where they meet. Do the same for the 1 black center tap lead to connect to the ground wire loop at P6–P10. You'll connect these 3 wires later, but their lengths are critical because they support the upper edge of the PCB.

4e. Measure and note the distances from the PCB ground to the speaker terminal black posts, and from the speaker terminal red posts to the amplifier outputs at E15 and Z15. Remove the PCB from the chassis and set it aside.

4f. Test-fit the power line cord to the chassis with a nylon P-clip on the transformer mounting screw under the switch. Measure, cut, and solder the cord's neutral (wide prong) side to one transformer primary lead, and its live (narrow prong) side to one switch lead. Solder the other switch lead to the remaining transformer primary lead. Insulate all solder joints with heat-shrink, and refit the power cord and P-clip.

4g. Disassemble everything and then cut and strip the transformer wires (yellow and black) where you marked them in Step 4d. Flip everything over and solder the transformer leads to the PCB.

4h. Solder two 18-gauge stranded wires to the PCB's ground wire loop and run them to the speaker terminal black posts, trimming their lengths and connecting them to the posts with spade terminals crimped and soldered onto the wires. Similarly connect 2 more 18-gauge wires between the PCB's E15 and Z15 positions and the speaker terminal red posts. Be careful not to damage the feedback resistors soldered to those output terminals.

4i. Crimp and solder a spade lug to one end of a 2" length of 18-gauge stranded lead, and solder the other end to the PCB's star ground loop P6-P10. This will be the ground connection, screwed down at the transformer's right mounting screw. The circuit is now electrically complete. You can test it again to make sure it still works.

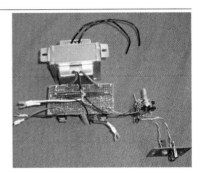

5. ASSEMBLE EVERYTHING

5a. Put two #8 screws through the plastic feet, pass them up through the ¹¹⁄₆₄" holes from underneath parts A and B, and secure them with nuts on top. Part A overlaps on top of part B. Install rails D as in Step 3k, filling the gap between B and D with the 1"×¼" shims you cut in Step 3i.

5b. Install the speaker binding posts.

5c. Place TO-220 mounting insulators over the ⁵⁄₃₂" holes on part A, where the chips will bolt down. Use thermal grease if they're mica insulators.

5d. Carefully place the assembled electronics into the A/B assembly. Secure the input switch and volume control to the front of A with nuts. Install the volume knob. Install the three #4 screws that hold the input panel to B, and nut them on the back.

5e. Slide part C between the transformer and part B. Install the power switch in part C. From the back, pass a #8×½" screw through the transformer mounting hole, parts C and B, and the input panel. Pass #8×½" screws through the P-clip holding the power cord, the transformer mounting hole, and parts C and B on the speaker side. Nut these 3 screws on the inside, trapping the PCB ground wire lug under one nut to ground the PCB to the chassis.

5f. Screw down the LM1875 chips with #6 nylon screws from below and metal nuts on top. Don't use metal screws. The chip tabs must be electrically insulated from the chassis.

5g. Connect the speaker ground leads from the PCB to the black speaker terminals. Connect the speaker signal leads to the red speaker terminals, right to lower and left to upper.

5h. Drop the plexiglass side panels in place in rails D. They're left loose for airflow.

5i. With washers, pass the #8×3½" screws through the upper corner holes in parts C, B, and A from the back. Install washers and acorn nuts on them at the front. Slide one rail E over the long screw from the side. Tighten the nut on the screw to hold the E rail in place. Slide one end of the plexi top into the E rail. Slide the E rail over the other end and tighten the long screw's nut to hold it in place.

5j. You're done. Hook it up and take it for a spin!

FINISH⊠

NOW GO USE IT »

USE IT.

ENJOY YOUR SIZZLIN' SQUELETTE

OPTIONS

Simple Power Amp If you plan to use a single signal source, you can omit the 3P2T source selector switch and wire your input connectors directly to the volume pot.

And if your source has a volume control of its own, you can build the Squelette as a plain power amp by replacing the 100K volume pot with a 100K resistor to ground for each channel.

Headphones There's no headphone jack in this build, but you can add one by connecting it to pin 4 of each chip through 150Ω resistors in series with the signal leads. This configuration limits current, preventing you from accidentally smoking your 'phones. Insulate the headphone jack body from the metal chassis and ground it to P6–P10.

Radio A good mini amp deserves a good mini source. For FM stereo, AM, and shortwave listening, I recommend Grundig's compact radios.

Speakers Quality speakers make a huge difference to music reproduction. I polled the speaker experts on audiokarma.org for suggestions on good-sounding compact speakers under $250. We recommend the JBL Control 1, the Paradigm Atom, the PSB Alpha Series, and the Sony SS-B1000. Any of these will provide clean, well-balanced sound from your Squelette.

The basic Squelette circuit can be used for other applications as well. Install a chip amp inside a speaker cabinet. Convert an old tube table radio to a powered iPod dock. Make an amplifier to match your customized Chumby.

Build your own steampunk amplifier, complete with brass meters, knife switches, a japanned steel cabinet, and a fly-ball governor for voltage regulation. Please! I so want to steampunk one of these, but I have always opted for minimalism.

➕ Visit makezine.com/projects/build-bare-bones-skeleton-amplifier for a schematic diagram and plan drawings for the Squelette's chassis.

ALTERNATIVE CABINETS FOR SQUELETTE

Squelette, the Bare-Bones Amplifier, can be built with a simpler cabinet to save time, materials, and labor. Two examples are shown here.

The blue box is a Radio Shack 6" × 4" × 2" plastic Project Box painted with Krylon Fusion paint for plastics. It comes with an optional aluminum bottom, which makes a good heat sink. To permit airflow, slots are cut in the edges of the aluminum, and holes are cut in the top of the box. The power transformer is housed separately, and the three transformer secondary wires connect to the amp box with an auto trailer plug and socket assembly. The transformer's power cord has the line fuse, and a lamp cord power switch installed in-line.

The silver box is an Eddystone 120mm × 95mm × 55mm diecast aluminum enclosure. Bud Industries makes similar boxes. It has been painted with appliance stainless steel spray paint for a matte pewter appearance. It is used upside down, with the bottom replaced by an acrylic panel so the insides can be viewed. Construction is similar to the original Squelette cabinetry except there's one aluminum piece instead of five, no measuring of aluminum stock, and no assembling the cabinet with numerous screws.

Many other cabinets will work, as long as they have at least 25 square inches of metal to dissipate the heat of the chips and some ventilation.

Laser
Harp

Play strings of light, using laser pointers, rangefinders, photo-cells, and Arduino.
By Stephen Hobley

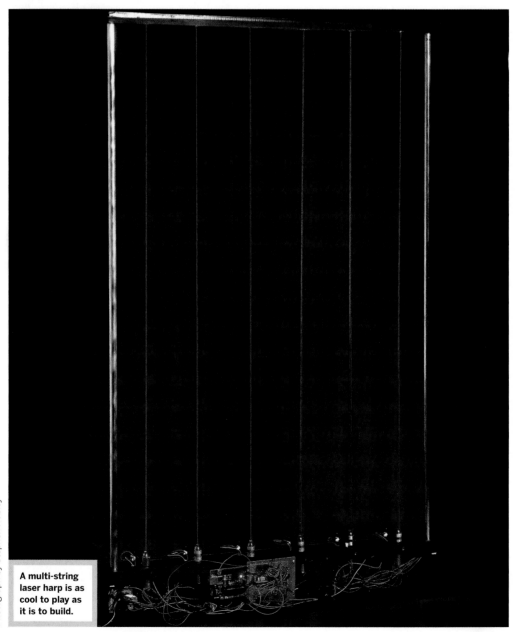

A multi-string laser harp is as cool to play as it is to build.

Photograph by Stephen Hobley

One of my most vivid concert memories is seeing Jean Michel Jarre perform in 1986 at the city of Houston's 150th birthday celebration. He played music by breaking laser beams with his hands. The beams came out of the stage and went off into space, and for a long time I thought it was a fake — I couldn't understand how this instrument could work without any sensors above. That started me researching and tinkering, and 22 years later, I figured it all out and built my own.

Now I have several versions of the laser harp. The one I perform with uses a powerful laser and a scanning mirror system, designed for professional lighting effects, that splits one beam into multiple beams that can fan out and move dramatically. This article describes a simpler harp I designed more recently, which uses inexpensive laser pointers and doesn't need the scanner.

The harp works as a MIDI controller, so it doesn't make sound itself, but generates a stream of MIDI data to drive an audio synthesizer. Each beam strikes a photocell, and when the player's hand interrupts it, the sensor prompts an Arduino microcontroller to send a MIDI "Note On" message. Additionally, a range sensor reads the position of the hand, which spawns MIDI controller messages that change the sound's qualities.

First I'll show how to make a single-beam laser theremin, which changes pitch with the position of your hand. Then we'll replicate the circuit and reprogram the Arduino to produce a multi-string harp, with each beam corresponding to a different note. The Arduino has 6 analog inputs, so this harp is limited to 6 beams, but at the end of the article I'll suggest ways to expand it.

MATERIALS

Arduino board I used an Arduino Diecimila, but more recent Arduino versions should work as well. Also consider the breadboard-pluggable Boarduino from Adafruit Technologies (adafruit.com).

Laser pointers (6) any color, but they need to have a decent IR filter to avoid confusing the range sensor. I bought 25 red pointers from eBay, where prices go as low as $1 each. Green is more visible and thus scores a higher coolness factor.

DC power source, switched, 8V–12V, 2–3 amps I used an old 8.5V camcorder charger.

Adjustable voltage regulator Trossen Robotics #P-VR-DE-SWADJ3, trossenrobotics.com

7805 voltage regulator, 5V

LM324 quad op-amp chips (2) RadioShack #276-1711; also available on eBay

Red LEDs (6) I used a 10-LED bar array, Jameco #1553686 (jameco.com).

Resistors, ¼-watt: 220Ω, 1.5kΩ (6), 3.9kΩ (6), 68kΩ, 1MΩ (6)

Capacitors: 0.1µF (3) and 300µF tantalum (6)

Photocell, 100mW (6) Jameco #202403

Sharp GP2D12 or GP2D120 IR range sensors (6) from Trossen Robotics

Tumbled rocks, translucent (6) craft or bead store

Potentiometer, 100kΩ

5-pin DIN (MIDI) connector

Blank circuit boards I used 1 dual mini and 1 medium, RadioShack #276-148 and #276-168.

24-gauge hookup wire various colors

Heat-shrink tubing

8-pin headers (5) (optional)

Aluminum tubes, ½" ×36" (2)

Wood and screws I used ½" fiberboard

Black paint

TOOLS

NOTE: I've developed my projects on a PC, so the software tools I use are PC-based, but there are equivalent tools for the Mac and Unix/Linux.

Computer

MIDI utility software to test output. I recommend MIDI-OX (midiox.com).

Software synthesizer I recommend Superwave P8 (home.btconnect.com/christopherg/main.htm).

USB-MIDI interface such as M-Audio Uno

Soldering equipment and solder

Insulated wire various colors

Wire cutters and strippers

Multimeter

Alligator leads (2)

Saw

Drill

Vise and clamps

Stephen Hobley, a photographer by trade, wants to continue tinkering with electronic instruments, but lately his brand-new role as "Dad" seems to take up most of his time.

Photograph by Jacques De Selliers

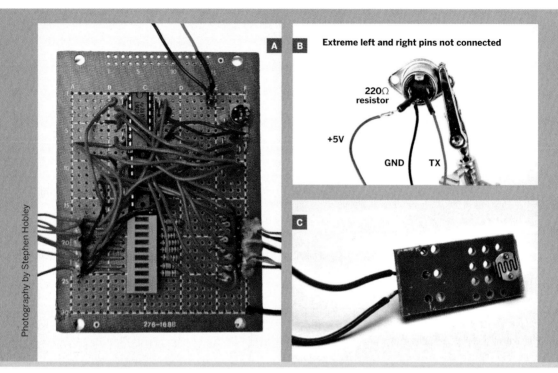

Fig. A: The detector board uses 2 quad op-amp chips to support 6 detector circuits (2 op-amps are unused).
Fig. B: MIDI jack wiring, with signal pin TX from Arduino.

Fig. C: A small piece of perf board holds the photo-detector for easy mounting.

Build the Pieces

We'll build the main electronic components first and then put them together. We'll start with the common power supply, light detector circuit board, and MIDI output jack. Then we'll add photocells, range sensors, and lasers — 1 of each for the theremin, and 6 of each for the harp.

POWER SUPPLY

I built the power supply on a mini circuit board following the schematic at makezine.com/projects/laser-harp. A 7805 regulator steps down the 8.5V from my camcorder charger to 5V for the range sensors. An adjustable voltage regulator lets you tune the power to the lasers to just above the detection threshold. The Arduino gets 8.5V directly, since it has its own onboard voltage regulator. To suck up any power spikes, I added a 0.1μF "bypass cap" across each of the 3 output voltages: 5V, variable, and 8.5V. These capacitors are optional. Finally, to neaten the connections out, I used two 8-pin headers for the outputs to the lasers and range sensors.

DETECTOR CIRCUITS

The photocells (aka light-dependent resistors or LDRs) are on their own little boards, across from the lasers. But I assembled the rest of the detector circuitry onto the larger circuit board (Figure A). The board needs to have 1 circuit for the theremin, or 6 circuits in parallel for the harp. Each of the two LM324 op-amp chips supports 4 detector circuits, and I went ahead and created 8 circuits, even though the harp only needs 6. The 68K and 100K resistors create a shared reference voltage, so we only need 1 of each. See the schematic online.

MIDI JACK

Wire the MIDI output jack by connecting pin 5 to the TX pin on the Arduino, pin 2 to circuit ground, and pin 4 through a 220Ω resistor to +5V. (MIDI jack pins are numbered 3, 5, 2, 4, 1, from left to right, facing the pins.) The outermost pins, 1 and 3, are not used for MIDI (Figure B).

LASERS AND PHOTOCELLS

The lasers connect in parallel to the variable voltage

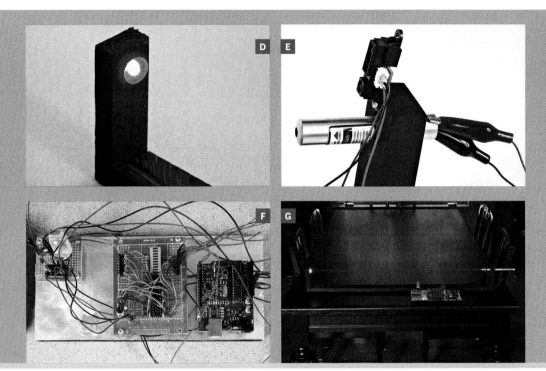

Fig. D: A tumbled translucent rock spreads laser light over the photodetector. Fig. E: The rangefinder works better when positioned vertically, away from the

reflective laser pointer barrel. Fig. F: The power supply board, detector board, and Arduino on a wooden base. Fig. G: A 1-string optical theremin with green laser.

on one side and to ground on the other. I soldered the photocells to small pieces of perf board for easier mounting (Figure C). They connect in parallel to +5V on one side and to the + input pins of the op-amps on the other.

RANGE SENSORS

Anyone who has played with a touchless D-Beam control on a Roland synthesizer will recognize these sensors immediately. The GP2D12/GP2D120 range sensors fire a pulse of IR light and measure distance by triangulating on the reflection.

For musical applications, I've found that the output from these sensors can be noisy, due to the constantly flashing IR drawing a lot of current every 40ms. You can smooth the output by connecting a capacitor between voltage (pin 3) and ground (pin 2); I used some 300µF tantalum caps.

You can also filter the signal with a dedicated filter circuit (see schematic online), or in the software, by averaging consecutive readings and using the average value.

Before connecting the range sensors, I removed

them from their plastic housings. They connect in parallel to +5V power, ground, and the Arduino's analog input pins 0–5.

The output from the range sensors is nonlinear, so the software converts output voltage into centimeters of distance using a simple equation, courtesy of Acroname Robotics (acroname.com). For the GP2D12 sensor:

Range [cm] = (6787 / (Voltage − 3)) − 4

And for the GP2D120:

R = (2914 / (V + 5)) − 1

Laser Theremin

Here's an optical version of a theremin, with 1 laser beam controlling both Note On/Off and pitch.

1. Download the Arduino programming software from arduino.cc. Upload the program *MAKE_MIDI_TEST.pde* from makezine.com/projects/laser-harp to your Arduino. This program lets the Arduino generate test MIDI messages. Set the baud rate of the Arduino to 31250.

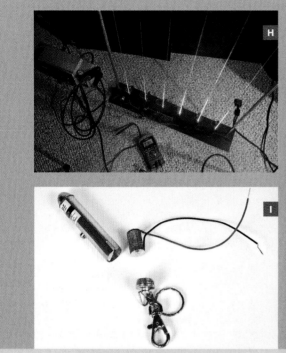

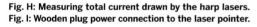

Fig. H: Measuring total current drawn by the harp lasers.
Fig. I: Wooden plug power connection to the laser pointer.

Fig. J: Author Stephen Hobley plays his full-sized laser harp.

2. Connect the MIDI jack to your computer with a USB-MIDI interface. Launch MIDI-OX (or equivalent software) and open that port. You should see Note On and Note Off messages in the MIDI inspector. If not, then test the +5V, ground, and TX pins for connectivity.

If the MIDI test is OK, upload the laser theremin program *MAKE_THEREMIN.pde* to the Arduino.

3. Take one of your laser pointers apart and measure the battery voltage. Adjust the variable regulator on your power supply board until its output matches this voltage. This lets you run the laser from your power supply board. One neat way to connect it is with alligator clips.

4. Now make the physical frame. I cut a long piece of scrap wood into 3 pieces: to make a base, a laser holder, and a detector holder.

Drill the laser holder piece for the laser to fit through horizontally, and drill a smaller perpendicular hole for a screw to hold down its power button.

5. Drill a hole through the detector holder where the laser will shine, tape the photosensor board to the outside with the sensor facing in, and glue a translucent tumbled rock over the hole in front (Figure D, previous page). The rock diffuses the light, which helps the sensor see it.

6. Next, attach the range sensor to the laser. I found that it worked better mounted vertically (Figure E) — when it was horizontal, I think the IR beam was reflecting off the laser pointer's barrel and causing misreadings. Having a rangefinder too close to a wall can also diminish accuracy.

7. Connect the photosensor circuit's output from the op-amp to pin 2 on the Arduino and connect the rangefinder's output to Analog In 0. Connect your computer back to the MIDI out and run MIDI-OX. Switch everything on (Figure G). Adjust the pot on the detector board so that the LED just comes on. At this point, breaking the laser beam with your hand should switch it off, and MIDI-OX should show you Note On, Note Off, and Pitch Bend messages as you move your hand in the beam.

That's it. Swap MIDI-OX for a soft-synth, or plug the MIDI jack into a hardware synthesizer, and you're playing!

Laser Harp

Now we'll expand on the theremin idea and create a 6-beam laser harp.

1. Build the frame. I made a wooden base to hold the lasers, rangefinders, and circuitry. Two metal tubes at either end support a top tube, which has 6 holes drilled through its underside to expose the photosensors. Space the lasers at least 4" apart, or else cross-talk between the range sensors can throw off their readings.

2. Wire the other lasers and photosensor/detector/ rangefinder loops in parallel with the first ones; see the schematic online. For neatness on the detector board, I used 8-pin headers for the photosensor and Arduino connections.

Also check that the variable regulator can handle the current drawn by the lasers: multiply the lasers' amperage by 6, and confirm that it's below the voltage regulator's rated max current. To make sure, you can also measure the current that comes into the regulator (Figure H).

3. Instead of messing with alligator clips, I made connector plugs for the laser pointers. I cut a slot in the back of each with a Dremel tool, made a wood plug to fit into the barrel, and thumbtacked a wire to each end (Figure I). Insert the plug, pass the wires through the slot, and screw on the back. The case contact in back will be ground. To keep the lasers switched on if the screws slip, wrap the barrels with electrical tape.

4. Connect the detector outputs from the op-amps to Arduino pins 2–7 and connect the range sensor outputs to the Arduino's analog input pins 0–5. Adjust the potentiometer until all 6 LEDs come on. You should now be able to turn them off individually by breaking the 6 beams. If ambient light becomes a problem, cut rings of narrow PVC pipe, paint them black, and attach one around each detector. If the lasers just miss the photosensor holes, glue on tumbled rocks as diffusers.

5. Upload the program *MAKE_HARP1_CTRL.PDE* to the Arduino, and start playing. The software assigns the MIDI note numbers 60, 62, 64, 65, 67, and 69 to the beams, but you can change this by editing the `notearray[]` structure. The controller messages from the range sensors are sent as note 74. With my synthesizer, this changes the filter sweep and creates a funky, retro synth sound.

You can also try *MAKE_HARP1_VEL.PDE*, a modified version of the code that maps your hand position to MIDI velocity, to mimic how hard you would strike a key on a keyboard.

Further Development

You're not limited to just playing notes. Ableton Live software allows you to MIDI-trigger drum loops, sequences, and other musical events. It's not free, but you can download a demo version at ableton.com.

If you want to go crazy and add more beams, you'll need to expand the digital and analog inputs of the Arduino using a multiplexer. There are a couple of neat, off-the-shelf ways to do this. I'll mention two here, and you can find more at the Arduino Playground (playground.arduino.cc).

One approach is to use an analog multiplexer like the R4 AIN MIDIbox module, which is based on the 4051 chip (kit available from AVI Showtech, avishowtech.com). This will support 32 inputs, for 32 harp strings.

With some clever programming, you should be able to bypass the detector board and read the laser harp through the analog multiplexer. To do this, connect the photocell array's outputs to the multiplexer's inputs, feed the multiplexed output to the Arduino, and detect which beams are broken in your software.

You can also use a digital multiplexer like the R5 DIN Module, another MIDIbox kit from AVI Showtech, which is based on the 74HC165 chip. With these, you can chain modules together to support an unlimited number of inputs.

If you want to tackle a full-sized scanning laser harp (Figure J), visit my website, stephenhobley.com.

➕ For project code, schematics, and further development resources, visit makezine.com/ projects/laser-harp.

Acknowledgment: This project is a testament to the collaborative power of the internet. I could not have done it without the help of many people who were good enough to answer the questions I posted on a variety of forums. I'd like to take this opportunity to pass on my gratitude!

SOLAR By Rory Nugent
XYLOPHONE

Photographs by Kay Canavino

MUSIC OF THE SPHERE

Solar cells gracefully link technology with the Earth's natural resources, bringing projects out of the dank, dusty workshop and giving them a sustainable home with the plants outside. This autonomous xylophone uses Solarengine circuits and pentatonic chimes to play in tune with that big nuclear power plant in the sky.

Wind chimes capture wind energy to move metal tubes that generate sound when they strike one another. They're simple, timeless, and beautiful. You can never predict the composition the chimes will play after the next gust of wind, which is what makes these inventions so compelling. The elegant overlay they add to our experiences brings us closer to nature.

I wanted to create a different kind of autonomous musical instrument that would, like wind chimes, generate tones from a natural resource. So I made this solar xylophone, which gives voice to the silent sun and takes the project (and ourselves) outside, where we belong. It uses eight simple, independent systems to strike its eight chimes in parallel. So you can lose the power cord and forget the batteries, but be sure to bring your suntan lotion.

Rory Nugent (prizepony.us) is a designer and engineer living in the New York City metro area. He currently works as a senior electronics engineer for littleBits.

X-PHONE SPEX

HOW IT WORKS

A regular xylophone fits inside of a solar-powered player box that holds a mallet over each of its 8 chime tubes. Each mallet is powered by a system that includes a solar cell, a simple Solarengine circuit, and a small motor. The systems work in parallel; the brighter the sunshine on each panel, the more frequently its corresponding tube will be struck.

The sun shines upon your project.

The 8V 44mA epoxy solar cell collects light energy, converts it to DC, and trickles it into the circuit. The more light it sees, the more power it delivers.

The circuit, a Miller Solarengine, collects the energy in a 4,700µF capacitor until its voltage exceeds 5V. Then a voltage trigger opens and discharges the capacitor into the motor. A smaller 1.0µF monolithic capacitor sets the discharge's duration.

With each pulse of voltage, the pager motor pulls its mallet down with a bent paper clip. The motor has a spring on the shaft that inhibits rotation beyond 60°. This means it gives one strong momentary nudge rather than rotating continuously. This is useful for our application.

A counterweight, made of plumbing solder, makes it easy for the motor to keep the mallet head lifted off the tube between strikes.

When the mallet is pulled down, it strikes its chime tube. The 2 balance strips that run along either side of the instrument hold the mallets and serve as a fulcrum for them to swing around.

The chimes are tuned to cover 1½ octaves using a pentatonic scale. This keeps the notes sounding pleasant and not dissonant in random combinations.

Illustration by Nik Schulz

SET UP.

MATERIALS

[A] Xylophone I used the 8-tube Pipedream from Woodstock Chimes. Check Amazon or eBay for others. To make it less expensive, you can use smaller models with fewer tubes, even just 1 or 2, if you adjust sizes and quantities accordingly.

[B] Mallets (8) Xylophones typically come with a pair, so you'll need to obtain 6 more. Buy cheap wooden ones, which are easier to drill through.

[C] Panasonic 1381U, 4.6V voltage triggers (8) Buy 10 for a discount from solarbotics.com.

[D] 2N3904 NPN transistors (8) RadioShack part #276-2016.

[E] 10kΩ single-turn trimpot potentiometers (8) RadioShack part #271-282

[F] 4,700µF electrolytic capacitors (8)

[G] 1.0µF monolithic capacitors (8)

[H] 1N914 silicon diodes (8) Buy 10 for a discount from solarbotics.com. RadioShack part #276-1122

[I] SCC3766 8V solar cells (8)

[J] GM10 geared pager motors (8)

[K] Dual mini perf boards (4) RadioShack part #276-148, 2 boards per package

[L] Paper clips (8)

[M] 10-penny finishing nails (8)

[N] 22-gauge solid wire, about 15'

[O] Plumbing solder, about 0.125" in diameter

[P] Hot glue Epoxy will also work nicely, but takes more time to mix and apply.

[Q] Wood glue

[NOT SHOWN] ⅛" plywood, at least 12"×31"

¾" wood board, at least 16"×26" You could go less than ¾" thick, but I wouldn't suggest anything less than ¼".

1¼"×5½" wood plank, at least 14" long You could go less than 1¼" thick, but I wouldn't suggest anything less than ½".

1"×1½" wood plank, at least 31" long

Female (16) and male (8) wire connection headers (optional) These will keep your components modular and will make testing much easier.

TOOLS

[R] Measuring tape

[S] Soldering iron and solder

[T] Wire stripper/cutter

[U] Hot glue gun unless you're using epoxy

[V] Needlenose pliers

[NOT SHOWN] Dremel tool

Drill press or drill and vise

Radial saw or other type of saw

Table saw or router and fence

MAKE IT.

BUILD YOUR
SOLAR XYLOPHONE

START ❯❯   **Time: A Weekend Complexity: Hard**

1. ASSEMBLE THE CIRCUITS

Putting the circuits together is the most straightforward step in this project, and it will ease us into the rest.

1a. Split your RadioShack perf board into single squares, or if you're using your own perf board, cut it into 2"×2" pieces.

1b. Collect and separate the parts needed to make each circuit: voltage trigger, transistor, trimpot, diode, 4,700µF capacitor, 1.0µF capacitor, motor, solar cell, a bit of wire, 2 female connection headers (optional), and a piece of perf board.

1c. Take one group of parts and solder a circuit together, following the Solarengine schematic at makezine.com/projects/solar-xylophone. This can be a bit tricky but it's good practice if you want to get better at electronics. I put the electrolytic capacitor and the trimpot at opposite ends, and included 2 female headers for connecting to the solar cell and motor at the edges.

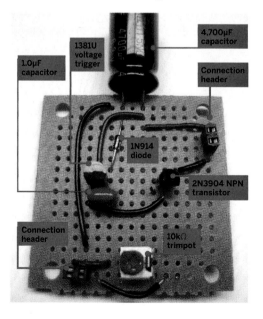

1d. After you finish one of the circuit boards, solder wire leads onto the positive and negative terminals on the solar cell. If you're using headers, solder male headers to the motor's delicate leads (we don't need these for the solar cell — the 22-gauge wire is thick enough).

1e. Test your first completed circuit board by taking it outside or putting the cell under a high-wattage incandescent bulb. You should see the motor move within 10 seconds. A compact fluorescent bulb or heavy-duty flashlight will probably also work but will take longer, and you might begin to think something is wrong with your circuit.

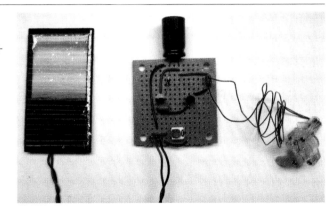

1f. If it doesn't work, adjust the trimpot using a screwdriver, by turning it all the way in one direction or the other. This controls the efficiency of the circuit, allowing more electricity to reach the capacitor, and thus, adjusting the timing of the motor. One side of the trimpot resists all the electricity while the other resists none.

1g. The circuit works correctly if the motor periodically turns about 60° and then quickly resets itself. If it works, pat yourself on the back, call up your friends for help, and build 7 more just like it (or however many your project calls for). Otherwise, check your solder connections, review the circuit, and use a multimeter on the solar cell and 4,700μF capacitor to see if electricity is flowing. Lastly, ask a friend to review the circuit.

2. BUILD THE PLATFORM AND STAND

I used a radial saw for cutting the wood and a table saw to put in some long grooves, but you can use other tools.

2a. Cut 2 pieces of ⅛" plywood to 5½"×15¼" each. These pieces will make the platforms that the mallet and motor mechanisms rest on.

2b. Cut 4 pieces of the 1¼" wood to 3¹⁄₁₆"×5½". These will be the legs that support the mallet platforms above the xylophone.

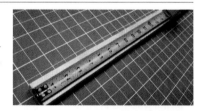

2c. Cut 1 piece of ¾" wood to 16"×25¾". This is the largest piece of wood and will support the entire structure.

2d. Take the 1"×1½" wood plank and cut 2 pieces 15¼" long each.

2e. This is the tricky part. Now we'll need to put a ½"-deep cut along the entire 1" edge of each 1"×1½" piece. A table saw is perfect for this. Place each piece on the table with the narrow edge down, adjust the fence so that the blade is centered on the wood, and run it all the way through the piece. Repeat for the second piece.

2f. Move the pieces back to the radial saw and cut four ⁵⁄₁₆" slits into each, 2¼" apart, on the same side as the long groove. These will be the notches that hold the mallets. Start at one end of the piece and measure 3³⁄₁₆" from the end. Mark off a section ⁵⁄₁₆" wide and ⅞" deep. Measure 2¼" from the marked section and repeat the marking 3 more times. Repeat this step for the second piece.

2g. Glue the 2 plywood platforms on top of the 4 leg pieces to make 2 U-shaped raised platforms. Then glue the platform tops onto the large piece of ¾" wood, running parallel and centered 2⅜" from each end. Let the glue dry if needed.

2h. Glue the 2 slotted strips on top of the platforms, running parallel and 1¼" in from the outside edges. You should now have a nice wooden structure that will house the xylophone inside, below the platforms. Each slotted wooden strip will act as a balancing point for 4 mallets.

3. INSTALL THE MALLETS

3a. Use a ¹⁄₁₆" drill bit to drill 2 holes into each mallet, one 5⅛" from the rubber ball end, and the other 4⅛" from the bare end. Slide a 10-penny nail into the hole closest to the bare end.

3b. Cut and bend your paper clips into the zigzag shape shown here using wire cutters and needlenose pliers. These wires will run through the holes closest to the mallets' striking ends and drive them from the motor.

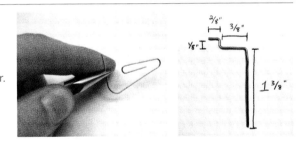

4. FINAL ASSEMBLY

This part is particularly tricky. We need to figure out where the motors should be placed and how much counterweight (plumbing solder) to spool around the mallets' ends.

4a. Slide the xylophone in place under the platforms. Use your eyes to center it.

4b. Place each mallet with its nail into its own groove on the balancing strips. The mallets should fall and rest on or near the center of each pipe. If not, adjust the placement of the xylophone underneath, or pull out and reverse the nail — the head can sometimes push a mallet off center. If that doesn't work, check the alignment of the strips and the placement of the platforms.

4c. Grab one of the mallets and wrap some plumbing solder around the wooden end. You want to wrap just enough to make the mallet balance nicely in the wooden groove and fall slowly down toward the xylophone, which requires trial and error.

4d. Place a properly counterweighted mallet into the groove, and then bring the motor from one of your circuits close to the mallet head. Fit a bent paper clip into position, connecting the motor to the mallet, with the short end through the eyehole of the motor and the long end through the unused hole in the mallet. Curve the long end of the wire up and around the mallet.

4e. Hold the motor slightly offset from the mallet, and make sure the motor is oriented so that it will pull down on the mallet.

4f. Test the circuit while holding the motor in place. This will give you an idea if you've found the right spot. If so, mark its position on the platform with a pencil. Grab your Dremel tool, attach a round wooden shaving tip, and grind out the line you marked so that the plastic tab on the bottom of the motor will fit in it. The motor will then sit flush with the platform.

4g. Repeat the installations and tests for all other mallet/motor/circuit combinations. After each motor placement mark has been ground out, do one last quick test and then hot-glue or epoxy the motor into place.

4h. Finally, drill a hole in the balancing strips next to each motor and thread the motor lead wires through. Hot-glue or epoxy the solar cells so that they rest at a 45° angle with the top touching the strip above the hole and the bottom touching the platform edge. Connect your solar cells and motors back to the circuit. You're done!

FINISH X

NOW GO USE IT »

USE IT.

PLAY SOME SOL MUSIC

FINE-TUNING THE PAPER CLIPS

Your solar xylophone will likely need some tweaking, especially the bent paper clips. Sometimes the piece that slips through the eyehole of the motor strikes the platform. Watch closely to see if this is a problem. Make sure the long end of the paper clip is wrapped snugly around the mallet so that it grabs the mallet firmly on each strike. You may also need to shorten the tall, vertical portion of the paper clip to reduce the distance the mallet head needs to travel to sound its note.

Remember, you hand-made these motor clips; they weren't stamped out by a machine, so there is room for error. Don't be discouraged. Work at it, try different bends, and be proud you could do this all with your hands!

CONCERT ON THE GREEN

Now all the hard work is done, and you'll be relieved to hear that the solar xylophone is not difficult to operate. On a sunny day, put on your favorite sunglasses, tuck a blanket under your shoulder, and carry the xylophone out to your lawn or local park. Before you even make it to the grass, the xylophone will probably start hammering away.

If it's triggering notes too quickly for its own good, adjust the trimpot on each circuit to a more fitting speed. You may have noticed that the mallets fired at a fairly pleasant rate during your tests inside, but now that you're under the midday sun, the xylophone has become a musical monster.

You can also put it in dappled tree shade for a randomizing effect. Solar cells work best with natural, full-spectrum sunlight, much better than with artificial sources like incandescent or compact-fluorescent light bulbs.

THEMES AND VARIATIONS

After you've built a full-sized solar xylophone, it's easy to purchase smaller versions and build modular, portable models. Arrange your portable solar xylophone in a space, and experiment with spatial sound. Weatherproof your xylophones, add hooks or magnets, and attach them to locations in your neighborhood. Add sound anywhere live music is least expected: a garden, outside your window, in sneaky and precarious urban hiding spots.

You can also play around with the orientation and arrangement of the solar cells. With a full-sized xylophone, you can build a truly unique instrument that emphasizes different notes depending on the time of day — an interesting take on the idea of a sun clock. To do this, arrange the solar cells along an arc-shaped bridge, and orient the xylophone so that the arc parallels the sun's path across the sky from east to west.

Want to build more solar instruments? Think about the rudimentary ways that sound is produced — by scratching, hitting, or plucking — and design a solar mechanism that does just that. Use solar cells to automate an already-popular instrument,

RESOURCES AND INSPIRATIONS
Solar Beam community site: solarbotics.net

Socrates Sculpture Park, in Long Island City, N.Y., filled with strange sculptures and environmentally powered sound pieces: socratessculpturepark.org

LEMURplex in Brooklyn, N.Y., another great New York institution dedicated to electronic music and interactive art: lemurbots.org/aboutLEMURplex.html

MonoBox is a small, inexpensive powered speaker that amplifies the output of your headphone music player. It's little but it's loud! All the circuit parts are available from RadioShack, while the speaker and cabinet are left to your preference.

You'll learn how to assemble and solder an audio power amplifier using an integrated circuit (IC) chip, and how to choose a speaker and install it in a cabinet with the amplifier.

The core of MonoBox is a compact and efficient audio amplifier based on the LM386 power amp chip. It will run on 200mA of current using power supplies from 6V–15V DC. This gives you the flexibility to power it from a wall adapter, a 9V battery, or a car accessory outlet.

You're probably thinking, "Sure, but it's so small. Does it rock?" Fair question. The prototype has been exhaustively tested and it does indeed rock. Maximum volume output is 90dB, and with the added bass boost, your socks will be rocked clean off!

✏ **TIME:** 6 HOURS   *✏* **COST:** $$

MonoBox Powered Speaker

Gregory Hayes

Build an amp and speaker for your music player. It's little but it's loud!

Written by
Ross Hershberger

MATERIALS

For links to all RadioShack parts on this list, visit makezine.com/projects/monobox-powered-speaker-2.

» **Power jack, DC, size N coaxial, panel-mount**
» **Printed circuit board, general purpose, for ICs**
» **Audio jack, stereo, ⅛", panel-mount**
» **Audio amplifier IC chip, LM386, low voltage, 8-pin DIP package**
» **IC socket, 8-pin**
» **Resistors: 10Ω, ¼W, 5% (R3); 10kΩ, ⅛ W (R2); 100Ω, ⅛ W (R1)**
» **Capacitors, 50V, 10%, PC mount: 0.01μF (for C1), 0.047μF (C2), and 0.022μF (also C1)**
» **Capacitors, electrolytic, 35V 20%, radial lead: 100μF (C4) and 470μF (C3)**
» **AC adapter, 300mA, 6V–12V DC output, with type N center-positive plug** such as RadioShack Enercell 9V/300mA
» **Loudspeaker driver, full range, 6Ω–12Ω, <150Hz resonance, >87dB sensitivity**
» **Wire, solid core, 22–24 gauge, various colors**
» **Box or other enclosure, about ¼ cu. ft. volume** See Step 1, below.
» **Dacron pillow filling, about ¼ cu. ft.**
» **Fabric or screen, acoustically transparent**
» **Machine screws, with nuts and washers: #6-32×1½" (2), and a second size (4–6)** to mount the circuit board and the speaker, respectively

» **Standoffs, plastic, ¾" long** for the #6-32 screws. I made mine from ¼" plastic tubing.
» **Sheet cork, solid cardboard, or foamcore board (optional)** to fabricate a speaker gasket if needed
» **Audio cable, ⅛" stereo plugs**

Optional, for battery power:
» **Battery and battery holder, 9V**
» **Battery snap connector, 9V, heavy duty**
» **Power plug, DC, size N, coaxial**

TOOLS

» **Soldering iron, fine point, with solder**
» **Wire cutter/stripper**
» **Helping hands tool** aka third hand tool
» **Pliers, needlenose**
» **Magnifying glass**
» **Hobby knife and scissors**
» **Marking pen, fine-tipped**
» **Crayon and paper**
» **Ruler**
» **Rotary tool with cutoff wheel, or small saw** such as a Dremel, or a scroll saw
» **Sandpaper, 100 grit, half sheet**
» **Drill with bits: ⅛" or 4mm, 6mm, and 10mm**
» **File, round, tapered, small**
» **Screwdriver(s)** to match your screws
» **Hot glue gun or caulking gun (optional)** if needed to seal the speaker box

1. Select your cabinet.

There are many options for housing your MonoBox. RadioShack sells project boxes made of black ABS plastic, which is very easy to work. Good sizes for this project are 7"×5"×3" (model 270-1807) or 8"×6"×3" (270-1809). I've used nice old wooden cigar boxes; you could also use a lunch box, small toolbox, or plastic kitchen storage container.

Choose something between ⅛ and ¼ cubic foot (equal to 6"×6"×6" and 6"×6"×12", respectively). Make sure it is deep enough for your desired speaker, and has one surface suitable to mount your speaker on. A box that can be tightly sealed against air leaks will provide the best bass sound.

Wood and plastic are good cabinet materials, as they're easy to work. Metal is more challenging. Almost anything relatively rigid can be used. Construct a box from cardboard in any shape you like, and cover it with colored

duct tape!

2. Choose your speaker.

The driver (loudspeaker) will determine the sound quality to a great extent. Good-quality drivers are available cheap as manufacturing surplus. Look for a driver at least 3" in size that's described as "full range." This type will reproduce the entire frequency range of sound from one driver.

Good online sources of drivers include Parts Express, Madisound, and any websites that sell overstock or surplus parts. Look in the

TIP Test a speaker's bass by temporarily mounting it in a cardboard box, or in the middle of a panel of cardboard at least 20" square (**Figure A**). Connect it to any stereo. The panel, or "test baffle," will separate the front and rear bass waves, allowing you to hear what it will sound like in a finished cabinet.

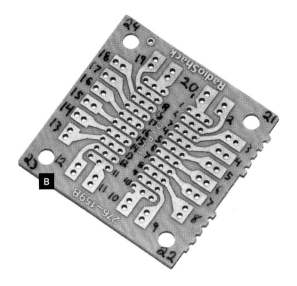

B

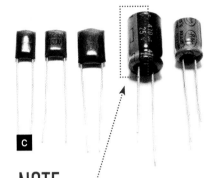

C

NOTE The electrolytic capacitors are polarized and must be installed in the proper orientation. The negative (–) lead is identified by a vertical band on the housing as shown in **Figure C,** above.

D

Gunther Kirsch

Specials, Closeouts, or Bargains sections for great deals. The driver I used in my prototype only cost $0.98 from Parts Express and it sounds great.

If specifications for the driver are stated, the impedance should be between 6Ω and 12Ω. Resonance frequency should be below 150Hz for good bass. High frequency range should extend to at least 8,000Hz (8kHz). And sensitivity of 87dB or higher will provide better volume output.

Drivers repurposed from table radios, computer speakers, etc. are often perfectly acceptable for DIY projects. Car speakers work, too.

3. Mark the circuit board.

Now you'll build the guts of the MonoBox: the amplifier circuit board. The board has 20 copper traces, each with 5 holes connected together. Eight of these traces will be used for the chip socket. Others will be used for component interconnections and off-board wires.

Mark the hole designations on your board before installing components, because it can be difficult to tell the holes apart once some of them are obstructed. On the soldering side of the board (with the copper traces), use a fine-tipped marker to label the traces, from 1 at the upper right clockwise to 20 at the upper left (**Figure B**).

Flip the board over to the component side (without the copper), and label the traces from 1 through 20 on this side too. Note that the handedness reverses when you flip the board.

Now label the 5 holes of each trace A through E, with A at the center of the board and E at the outer edge. Thus the inner hole on the trace at the upper left is 1A and the outer hole of the trace at the upper right is 20E. (You can see the component side labels in **Figure D,** above.)

4. Solder the socket and caps.

Insert the DIP-8 socket for the amp chip into holes 2A–5A and 19A–16A, orienting the notch in the socket body toward 2A and 19A (up). Flip the board over and solder the socket leads to the traces.

Insert capacitor C3's (470µF) negative lead into hole 11A and bend its positive lead over to hole 16B. Insert capacitor C4's (100µF) negative lead into hole 20C and bend its positive lead over to hole 17C (**Figure D**). Solder and clip the leads.

Insert capacitor C2 (0.047µF) into hole 13D and bend the other lead to hole 15D. (The film caps aren't polarized, so it doesn't matter which lead is which.) Solder and clip the leads.

Capacitor C1 (0.033µF) consists of 0.010µF and 0.022µF capacitors in parallel. Install them together by inserting one lead of each into holes 16C and 14C (**Figure E**). Solder and clip the leads.

E

NOTE If you have a film capacitor with a value of 0.030µF to 0.035µF, you can use that for C1 in place of the 2 in parallel.

5. Add resistors and jumpers.

Insert resistor R3 (10Ω) into holes 15E and 16E. Insert resistor R2 (10kΩ) into holes 1D and 2B. Insert resistor R1 (100Ω) into holes 4B and 5B. Solder and clip leads.

Insert a short narrow-gauge jumper wire, such as a cut-off lead, from hole 3C to 5C. For all other wires, solid-core insulated wire is recommended for ease of working.

Cut a 5cm jumper wire, strip 3mm at each end, and run it from hole 1E to 14E, passing around 19 and 20 as shown in **Figure F**.

Cut a 5cm jumper wire, strip 3mm at each end, and run it from hole 13E to 5C, passing around 8 and 9 as shown.

Cut a 4cm jumper wire, strip 3mm at each end, and run it from hole 20D to 5D, passing around 1 and 2 as shown.

6. Connect the off-board wires.

Wires to reach the power

F

TIP Where a resistor body is longer than the distance between its holes, stand it vertically on one hole and bend the other lead down toward the second hole.

G

input socket, shown red and black (**Figure G**), are soldered to holes 20E (black, −) and 17E (red, +).

Wires to reach the signal input socket, shown purple and gray, are soldered to holes 4E (gray, signal) and 5E (purple, ground). Wires for the speaker, shown blue and yellow, are soldered to holes 11E (blue, +) and 20A (yellow, −).

H

I

7. Check your work.

Carefully examine both sides of the board. On the component side, check the connection holes against the assembly instructions.

On the solder side, use a magnifier to look for missed solder joints, cold joints, or accidental solder shorts between traces (**Figure H**). This is a tiny board and problems are easily overlooked by the naked eye.

If you think you see a solder bridge between traces, run a knife point between them to scrape it away.

8. Add the amp chip.

Carefully plug the LM386 amp chip into the socket. Note the orientation of the dot on the back of the chip (**Figure I**).

The amplifier circuit is now complete.

9. Cut and drill the cabinet.

Trace the shape of the speaker on paper using a pen or crayon. Cut out your tracing and use it as a template to mark the box surface for cutting (**Figure J**).

Place the speaker on the box and mark its mounting holes. How you cut the box will depend on the material. For my wooden cabinet, I roughed out the speaker hole with a 50mm×1mm Dremel cutoff disk, then finished it to size with a 2" sanding drum followed by 100-grit sandpaper. For the plastic box shown here, we used a laser cutter, but a Dremel works too (**Figure K**).

In a rear corner, locate and cut a 10mm hole for the power socket and a 6mm hole for the audio signal input jack (**Figure L**, following page). My box was so thick I needed to cut a little relief inside so the power jack would reach through.

Position the circuit board on the inside cabinet surface and mark mounting hole locations through 2 of its corner holes. For #6 screws, drill these using a ⅛" bit.

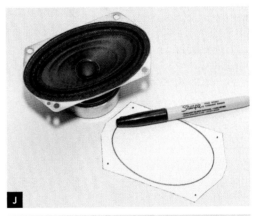

J

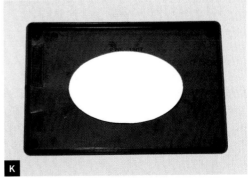

K

10. **Finish the box (optional).**

If you've got a box that wants paint or varnish, now's the time. Krylon Fusion spray paint is made for plastics; it's easy to apply to ABS and gives a great finish on project boxes (**Figure M**).

11. **Install the amp and jacks.**

Solder the audio ground wire (gray) to the outer tab of the audio input jack. Solder the audio signal wire (purple) to both left and right jack signal tabs.

The power jack mounts from the outside. Thread the black/red power wires through the power jack's nut and washer, then pass the wires out through the cabinet power jack mounting hole. Solder the black wire to the outer power jack tab and the red wire to the inner tab (**Figure N**).

The circuit board requires standoffs to give it about 1" of clearance from the cabinet. I used sections of ¼" plastic tubing. Pass two #6-32×1½" screws through from the outside, and slip the standoffs onto them inside. Slide the circuit board onto the screws and install the nuts (**Figure O**).

Put the audio jack through the 6mm hole and secure it with a nut on the outside. Pull the power jack through its hole and secure it with a nut on the inside.

L

TIPS Place the jacks in a corner so the cords will be low whether the box is placed vertically or horizontally.

Drill an undersized hole, test-fit the part or fastener, then use a small round file to adjust hole size and position. This gives you a chance to second-guess and adjust the hole rather than having to drill precisely and accurately in one try.

M

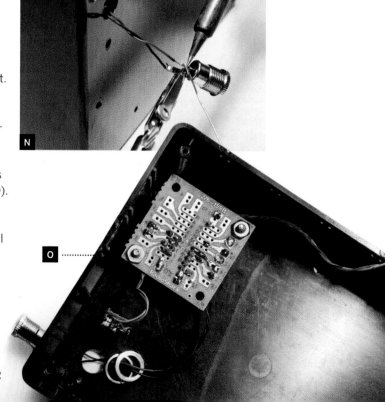

N

O ············

12. Make the speaker grille.

The speaker needs an acoustically transparent fabric grille to protect it. You can use a variety of materials including speaker grille cloth, cane material, metal screen, or anything else that will pass sound through.

My speaker needed a gasket to keep the cone from hitting the grille. I recommend sheet cork. You can use pasteboard, foamcore, thin wood, plastic, or other stiff foams, but soft foams and corrugated cardboard won't work.

Use the speaker frame as a template for the outer edge of the gasket, and the cabinet speaker hole as a template for the inner edge. Drill screw holes in the gasket as needed (**Figure P**).

13. Install the speaker.

I used #6-32 flat-head brass screws and finishing washers because they look good with my wooden box. Steel hardware looks nice with a plastic box. Again, use a ⅛" drill bit for #6-32 screws (**Figure Q**).

After mounting your speaker, solder its + terminal to the blue (+) wire from the amplifier and its − terminal to the yellow (−) wire.

For best bass performance, seal any gaps or air leaks in the box with hot glue or caulk, then fill the box with lightly packed Dacron pillow stuffing (**Figure R**). Fiberglass insulation works well too, but it's an irritant and should be handled carefully.

If you don't want to permanently seal your box, you can apply felt or adhesive foam to the lid rim to stop air leakage.

Done! Cool! Now we just have to arrange for power and signal connections, and your MonoBox will be ready to play.

14. Power up your MonoBox.

Your power source needs at least 300mA of current capacity. Voltage as low as 6V works for driving 4Ω speakers, but 12V is best for 16Ω speakers, so 9V is a good compromise.

RadioShack's Enercell 9V/300mA AC adapter is ideal. Install the type N power plug with the center (tip) positive. Now you can plug your MonoBox into wall power.

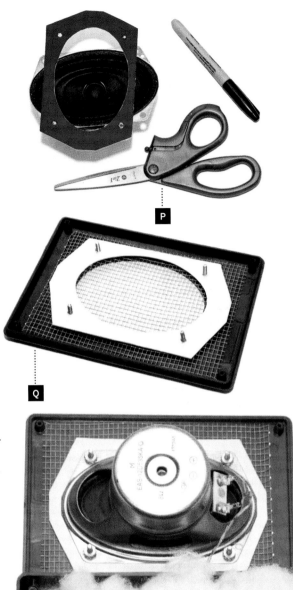

TIP After screwing the power plug's body together, you can fill it with hot glue to secure the wires in place.

T

NOTE The MonoBox has no power switch. To turn it off, unplug the power from the back. Leaving a battery plugged in will drain it overnight even if no music is playing.

To build a battery adapter, solder the red (+) lead of a 9V battery snap connector to the center contact of a size N coaxial DC power plug. Solder the black (−) lead to the outer contact. Attach the battery to the back of the cabinet with the 9V battery holder (**Figure T**).

To use a car or boat's 12V power system, make an adapter for the accessory "cigarette lighter" socket. Again, use a size N coax power plug and solder the negative lead to the outer contact, positive lead to inner contact.

Other DC power supplies can be used the same way, but make sure the voltage does not exceed 15V DC.

15. Hook up the music.

Audio signal input will depend on your music player. Purchase or make a signal cable suitable for your player with a ⅛" stereo plug on one end to plug into your MonoBox. Most smartphones and MP3 players accept a ⅛"

plug in their headphone jack.

Plug in the power source and the audio signal cable. Set your music player's volume to minimum, start the player, and increase the volume until you hear sound from MonoBox.

Crank it up! MonoBox has no volume knob, so just use your music player's volume control. Also use any tone controls or equalization on the music player to adjust the tone of the sound to your liking. ◪

➕ For video instructions, links to materials, schematics, and reader comments, visit makezine.com/projects/monobox-powered-speaker-2.

Ross Hershberger is the author of "Econowave Speakers" in MAKE Volume 20 and "Squelette, the Bare Bones Amplifier" in Volume 23. Formerly a mainframe systems analyst, a restorer of vintage tube audio components, and a tooling machinist, he is now a field service engineer for high-powered industrial lasers.

GOING FURTHER

What if you really want stereo? Make 2 MonoBoxes! In each, connect only one channel (left or right) of the input jack directly to the amplifier circuit board, and use a SPST switch to connect or disconnect the other channel, so each box has its own mono (or 1-channel) selector switch. Set both boxes to 1-channel, then use a stereo headphone splitter to connect both MonoBoxes to your music player.

Gregory Hayes

How Your Amp Works

C4
100µF
+9V to +12V

6
3
+
U1
LM386
5
2
−
8
1
4

2 O
1 O
MP3 player

R1
100Ω

R2
10kΩ

C1
0.033µF

C3
470µF
+

R3
10Ω

C2
0.047µF

SPEAKER
8Ω

» The amplifier circuit is designed to be fed by a headphone output, so the input impedance set by R1 is a relatively low 100Ω. This helps load down the source and eliminate noise. Use a 10kΩ resistor for R1 if you're driving your MonoBox from a line-level source like a home stereo CD player.

» Most small speakers need bass boost. R2 and C1 provide a high-pass feedback loop to boost bass

by reducing frequencies above 200Hz. If the sound is too bassy for your speaker, R2/C1 can be eliminated or disconnected.

» For the best bass from a 4Ω speaker, C3 can be increased to 1,000µF. For a 16Ω speaker, C3 can be reduced to 250µF without losing bass performance.

» R1 provides a load for the signal source and ground reference for the chip input. C4 decouples

and filters the power supply. R3/C2 is a Zobel network to ensure a low impedance load at high frequencies and to damp oscillations.

» Pin 5 is the audio output of the IC chip. This pin has a DC voltage of 4.5V added to the audio signal. Capacitor C3 blocks the DC voltage from reaching the speaker and passes only the audio signal.

Electronic Drum Kit

Velocity-sensitive impact sensors in PVC pipes interface to a micro drum machine.
By Tom Zimmerman

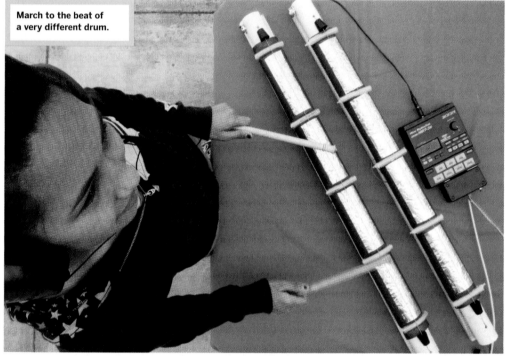

March to the beat of a very different drum.

Photograph by Sam Murphy

An electronic drum is basically a switch that triggers the playback of a digitally recorded drum. Here's how I built tubular drum controllers out of PVC pipe and connected them to a studio drum machine to create a professional-sounding electronic drum kit.

Each controller has a guitar string suspended above 4 strips of aluminum tape. When you strike the string with a drumstick, it touches the tape and closes a circuit to trigger the corresponding sound from the drum machine. Foam covering the pipe softens the blow and provides a nice bounce. Underneath each controller, a pressure-sensitive piezoelectric device lifted out of the drum machine detects the force of the hit, to determine the relative volume.

The brief contact between the struck string and the foil is too short for the drum machine to detect,

so a pulse-stretching circuit lengthens the signal, by charging a capacitor.

Two male-to-female serial cables let you unplug the controllers from the drum machine. I cut the cables in half and connected them to the controllers and the drum machine. To plug-and-play, you simply mate each connector to its former other half.

Tom Zimmerman is a member of the User Sciences & Experiences Research laboratory at IBM's Almaden Research Center. An MIT graduate, he was profiled in MAKE, Volume 04.

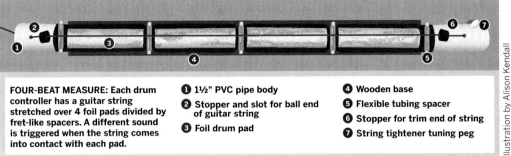

FOUR-BEAT MEASURE: Each drum controller has a guitar string stretched over 4 foil pads divided by fret-like spacers. A different sound is triggered when the string comes into contact with each pad.

① 1½" PVC pipe body
② Stopper and slot for ball end of guitar string
③ Foil drum pad
④ Wooden base
⑤ Flexible tubing spacer
⑥ Stopper for trim end of string
⑦ String tightener tuning peg

Illustration by Alison Kendall

MATERIALS

Clear vinyl tubing, ½" outer diameter (OD) × ⅜" inner diameter (ID) × 5' long **Home Depot SKU #702-229; many parts available in stores are *not* listed on homedepot.com**
White polyethylene tubing, ⅜" OD × ¼" ID × 5' **Home Depot SKU #301-762**
Heat-shrink tubing
Foil tape, 1.89"×5' **Home Depot SKU #915-245**
#4 size (1" diameter) rubber bottle stoppers (4) **Home Depot SKU #755-441**
Lag screws, ¼"×1½" (6) **Home Depot SKU #654-884**
Guitar strings (2) **the thicker the better**
Guitar tuning pegs (2) **aka machine heads**
#30AWG insulated wrapping wire in 2 colors **I used RadioShack #278-501 and #278-502.**
Serial extension cables, DB9 M/F (2) **RadioShack #26-117**
11" cable ties (12) **Home Depot SKU #295-858**
Foam pipe insulation, ⅜" thick x 1⅛" ID × 6' **Home Depot SKU #420-048**
1½" PVC pipes, 36" long (2) **Home Depot SKU #193-844**
1×3 wood boards, 36" long (2)
¼"×½" bolts and matching nuts (2)
Zoom MRT-3B Micro RhythmTrak drum machine **(See "Update" on page 237 for alternatives to the Zoom MRT-3B.)**
Dual general-purpose IC PC board **RadioShack #276-159**
CD4066 quad CMOS switch, DIP package (2)
14-pin socket (2)
Project enclosure **RadioShack #270-1802**
Momentary push-button switch **RadioShack #275-1556**
Resistors: 470kΩ (8), 1kΩ (2)
0.1µF capacitors (8)
Clear silicone sealer

TOOLS

Marker
Drill and bits: ¹⁄₁₆", ¼", ⅜", 1"
Small flat file
Soldering iron and solder
Wire stripper
Diagonal cutters
Screwdrivers, Phillips and slotted
Pencil eraser and sandpaper
Multimeter

1. Build the controller bodies.

Our Zoom MRT-3B drum machine has 7 trigger pads and 1 bank select switch. We'll make 2 controllers to drive it, each with 4 pads. For simplicity, I'll describe building 1 controller; just double each step to make both. The controllers are physically identical, but we'll wire them slightly differently later.

1a. Draw a reference line straight down the PVC pipe; a doorjamb makes a good guide. Cut five 3" pieces each of vinyl and polyethylene tubing, then insert the polyethylene pieces into the vinyl pieces and thread a cable tie through each (Figure A, next page). These spacers will flank each drum pad.

1b. Center a 30"×3" strip of foam over the pipe's reference line and secure it down with 5 spacers, spaced 7" apart. Make sure the foam lies flat. Stick four 6" strips of foil tape to the foam, centering them between the spacers and avoiding wrinkles.

1c. Orient the pipe left-to-right, the way you'll play it. Along the reference line at the left end, drill a ¼" hole 1" from the end, and file a ¹⁄₁₆"-wide slot going ½" to the right (Figure B).

1d. At the right end, drill a ¼" hole along the line 1½" from the end, and another ¼" hole, for the tuning peg, on the far side of the pipe, 90° around from the line and ¾" from the end. Drill ¹⁄₁₆" pilot holes and install the tuning peg with the screws that came with it (Figure C). Don't overtighten or you'll strip the threads. Drill a ⅛" hole near the upper left corner of each foil pad, outside the foam and next to its adjacent cable tie.

1e. Mount the pipe to its base by turning it over and attaching the wood with 3 lag screws in countersunk holes. Solder a 6" wire to a guitar string's brass ring and slide the rubber stoppers, large ends pointing

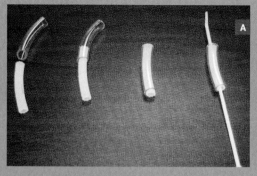

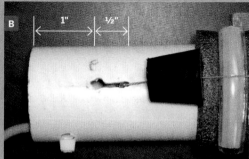

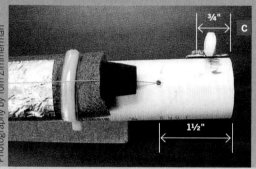

Photography by Tom Zimmerman

Fig. A: Nested vinyl and polyethylene tubing make fret-like spacers between drum pads. Fig. B: The cable end of the pipe, with a keyhole slot for the string barrel.

Fig. C: The right end of the pipe, with the tuning peg for tightening the string. Fig. D: Taping stripped wire to the foil drum pad.

toward each other, onto the string. Pass the wire through the slotted hole in the pipe, and anchor it by sliding the brass ring under the slot. Feed the other end of the string down through the hole at the opposite end and thread it onto the tuning peg inside. Slide 1 stopper to each end, and tighten the string so that it doesn't touch the foil.

2. Connect the pads.

2a. Feed a wire through each of the ⅛" foil pad holes and out the left end of the pipe. Strip 3" of insulation off the pad-side wires, lay the bare copper along the edge of its foil pad, and tape it down with a ¼"-wide strip of foil tape (Figure D).

NOTE: The adhesive on the bottom of the foil is non-conductive, so the copper wire must touch the top surface of the pad foil.

2b. Cut a serial cable in half. Take the female half and tie a knot 6" from the cut end. Use solder and heat-shrink tubing to connect the pad wires to the wires for serial pins 1, 2, 3, and 4, as specified in the schematic diagram, available

at makezine.com/projects/electronic-drum-kit. Pads are numbered from left to right. Use a multimeter to associate the wires in the cable with the corresponding pins on the connector.

3. Connect the piezo elements.

The Zoom MRT-3B drum machine has 2 pressure-sensitive piezoelectric elements that detect the force of pushes on its drum pad buttons. We'll remove them and put 1 underneath each pipe, so they'll perform the same function there.

3a. Pry the volume knob off the drum machine and unscrew and remove the back of the case. Remove the 4 screws inside that hold the battery case and the 2 screws on the MIDI connector. Write down where they go, and save them in a cup.

3b. The piezo elements are the disks behind the circuit board from the drum pad buttons (Figure E, bottom). Unsolder both and gently remove them. Solder and heat-shrink two 36" wires to each, and thread the wires through the side pad hole near the

Fig. E: The piezos inside the drum machine are the 2 disks behind the drum pad button contacts.

Fig. F: The piezo element between pipe and base, before and after being encased in silicone.

middle of each pipe. Slip the piezo element between the pipe and the wood, but don't force it, or it will crack. Encase the entire piezo element and its wires in silicone sealer and let it set overnight (Figure F).

3c. Following the schematic, connect the piezo's red wire to serial cable pin 7 and its black wire to pin 8. Also connect the string wire to pin 9. This completes the controller's serial cable connections. Drill two ¼" holes, one above the other, about ½" from the end of the pipe, and cable-tie the serial cable knot to the inside of the pipe (Figure G, next page).

4. Wire up the drum machine.

4a. Push the power switch into the case to dislodge the top circuit board, then unfold it to expose the board underneath. Remove the remaining screws, lift the boards from the case, and remove the white silicone pad membrane from the board along with the buttons, pads, and display. Gently sand the carbon coating off the top right corner dot of the switch pad contacts to reveal copper pads (Figure H).

CAUTION: Don't rub too hard or you'll scrape away the pads themselves.

4b. Solder 12" lengths of 30-gauge wire to the right contact of each drum pad, and pairs of wires to the Pad Bank and Function switches (Figure I). Replace the circuit boards in the case and thread the wires out the pad holes in front. Solder wire pairs to the power pads and to each of the piezo element pads on the back of the circuit boards (Figure E) and thread those out the front as well. Reassemble the drum machine in its case.

5. Build the pulse-stretching circuits.

Trim the mini PC board to fit in the project box. Clean its copper pads with a pencil eraser and solder a socket into the middle of each half. Follow the schematic to build the rest: connect the controller wires from the serial cable to the quad switch's control pins (pins 5, 6, 12, and 13), hanging a grounded 470kΩ resistor and 0.1µF capacitor off of each. Connect the quad switch's V+ (pin 14) to the controller's string (serial cable pin 9) through a 1kΩ resistor. Ground the specified quad switch pins. Finally, install both 4066 chips in the sockets.

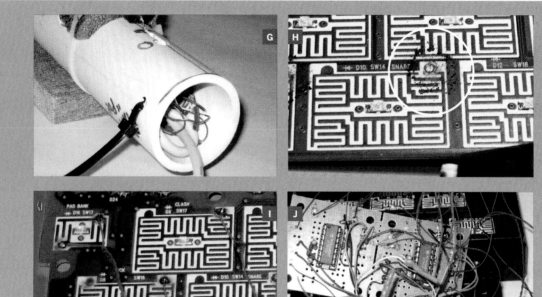

Fig. G: The drum controller's connection to the serial cable. **Fig. H:** Remove the coating from the upper right drum pad button contacts. **Fig. I:** The drum machine button pad contacts, wired. **Fig. J:** The pulse circuits wired to the serial cables and drum machine.

6. Connect the drum machine.

6a. Bolt the project box to the drum machine, pass the male serial cables through ¼" holes drilled in the sides, and strain-relieve them with cable ties. Mount the push-button switch on the side of the project box and solder it to the pair of wires from the drum machine's Function switch.

Follow the online schematic to connect the rest: solder the +V (pin 14) and Ground (pin 7) of one of the 4066 chips to the V+ Power and Ground of the drum machine circuit board (Figure E, previous page). Connect the pad button leads to the quad switches' signal pins. Connect the Pad Bank leads to pins 1 and 2 of controller #2 (Figure J).

6b. Plug the 2 drum controllers into the circuit, and you're ready to play. The circuit gets its power from the drum machine, so you should see its LEDs light up when you hit each pad. If they don't, swap controller cables to determine whether the problem is with the controller or the circuit. Since we wired directly to the pads, all the functions of the drum machine will still work. Pad 5 controls the Bank select. Strike it to select an alternate drum set.

Enjoy your electronic drum set. Bust out some beats and start a band!

For project schematics, and video clips of disassembling a Zoom MRT-3B and playing the Electronic Drum Kit, visit makezine.com/projects/electronic-drum-kit.

UPDATE

Since this article's original publication in *Make:* magazine, I have received comments asking about interfacing the drum circuit to other drum machines and keyboards. You are not limited to the Zoom drum machine mentioned in the article. You can hook up any musical or noise toy you find at a yard sale or flea market. Basically, any battery-powered device that has a mechanical switch can be connected to the 4066 CMOS electric switch, enabling the mechanical switch to be electronically pressed when the drum pad is hit. Just connect the two sides of the mechanical switch to any of the pairs of pins of the 4066 CMOS switch (pins 1 and 2, 3 and 4, 8 and 9, 10 and 11).

These days, my favorite sound source is the WAV Trigger board designed by Jamie Robertson and Jordan McConnell (sparkfun.com/products/13660). It has sixteen trigger inputs, so you don't even need the 4066 electric switch. Instead, wire the guitar string to ground, and connect each conductive drum pad directly to one of the sixteen trigger inputs on the WAV Trigger board (pins 2, 4, 6, 8, 10, 12, 14, 16 on jacks J6 and J7). You still need a 0.1 uF capacitor on each trigger line to stretch the pulse (otherwise you will miss some beats), but you don't need the 470k resistor. You will have to load some wav samples onto a microSD card, as the WAV Trigger board does not come with installed sounds.

Some sources for wav file samples include 99sounds .org/percussion-samples and sampleswap.org. Granted it is a bit of a hassle collecting wav samples and converting them into the specified format (44.1k, stereo, 16 bit, uncompressed), but free and low-cost tools like Audacity (audacityteam.org) and dBpoweramp (dbpoweramp.com) make editing and reformatting wav files easy.

Finally, there is the issue of robustness. Smashing a taut nickel-plated steel wire into a thin piece of aluminum foil mounted on pliable foam hundreds of time does lead to significant wear and tear. If you are a hard-hitting drummer, you will quickly tear the foil apart. Of course, you can replace the foil and be gentle in your drumming technique. Or, better yet, you can make the drum pads out of conductive fabric (sparkfun.com/products/10056) glued with silicon sealer to strips of cut-up mouse pads glued in turn to the PVC pipe. Sew the edge of each conductive fabric pad to wire using conductive thread (sparkfun.com/products/10867). The conductive fabric will take more abuse than the aluminum foil, the fabric-coated mouse pad provides a nice gluing surface for the conductive fabric, and the mouse pad will keep its original shape better than the foam pipe insulation.

If you go with the WAV Trigger board, you have sixteen inputs — enough to support four drum tubes. So invite your friends over and have a drumming circle!

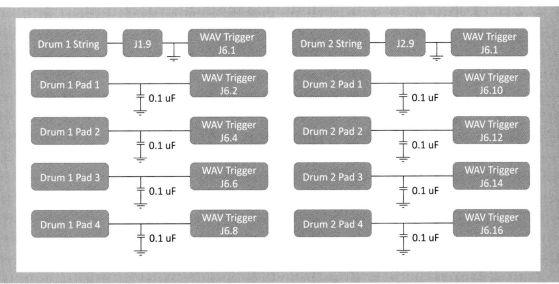

Photography AND Video PART 7 >>>>>

Larry Cotton enjoys photographing birds, but he found it difficult to do well. Being a retired engineer, however, Larry figured out a way to put the birds where he wants them—by making a remote-controlled bird feeder that slowly rotates his subjects into the best position. Add a long-distance shutter control for the camera, and the old quip, "Look at the birdie!" takes on a whole new meaning. Larry reveals his system to you here so you can try it yourself.

Have you ever wished you could get a good overhead shot of your house or property? Jim Newell needed to, and he figured out an approach to aerial photography that's a lot simpler than building and piloting a drone. He'll show you how to fashion a lightweight, programmable camera assembly that can take a ride with a helium balloon up to 300 feet and snap pictures of everything below.

Richard Kadrey is best known as the author of the Sandman Slim series and other urban fantasy novels and short stories. He's also an excellent photographer. In his article "Looking at the Low End," Richard explains how to take eerie, unearthly looking photos using infrared.

The popularity of 3D films has waxed and waned over the years, but they have never been more popular than they are now. Eric Kurland's commitment to 3D has never wavered. He's president of the LA 3-D Club, director of the LA 3-D Movie Festival, and CEO of 3D SPACE: The Center for Stereoscopic Photography, Art, Cinema, and Education. Eric is also a 3D consultant and stereographer for independent and studio productions. He'll show you how to create your own 3D video camera setup and create a portable system for viewing your 3D movies.

Maker Media's Tyler Winegarner is an outstanding video producer and cinematographer. If you have ever longed for the kind of lighting unit the professionals use but can't afford it, Tyler will show you how to make your own professional-quality unit for less than a hundred dollars—out of a brownie pan!

Make: contributing writer and all-around maker Sean Michael Ragan presents another money-saving DIY project. Enhance your in-home viewing experience by making a retro-reflective, glass-bead projection screen that's even more luminous than a titanium white screen for your home theater.

Speaking of screens, artist and UCLA film professor Bill Barminski is more interested in the green kind. If you have ever wanted to add special effects to your home movies, green screens are the key—or, perhaps more accurately, the *chromakey.* Bill will show you how to set up your own green screen backdrop, light and place figures in front of it, and find inexpensive software to start making your own FX.

SPIN THE BIRDIE

By Larry Cotton

Photograph by Sam Murphy

BIRD SHOT

Birds make lousy subjects for digital photographs. They're fearful, fidgety, and, well, flighty. But you can improve your odds of getting awesome avian photos by moving your camera closer to the birds — and you farther away. And while you're at it, why not get them to pose for you?

Since converting from film to digital photography more than ten years ago (I bought an Epson PhotoPC with a half-megapixel resolution for $500 in 1996), I've sought ways to take advantage of the "tons o' shots to get a winner" phenomenon. Recently I discovered that bird photography definitely falls into that category.

But digitally shooting birds taxes a camera's resolution limit big-time: your 6 or 10MP digicam suddenly becomes another PhotoPC when you throw away valuable pixels by cropping. So, ya gotta get closer — but getting closer makes the birds more skittish and shy. It's a digital catch-22.

We *can* have our cake and eat it too, though, thanks to a long shutter-release cable and a gadget that typically controls model vehicles: an R/C radio.

Larry Cotton is a retired power-tool engineer, musician, part-time math teacher, coffee roaster, and bird harasser living in eastern North Carolina.

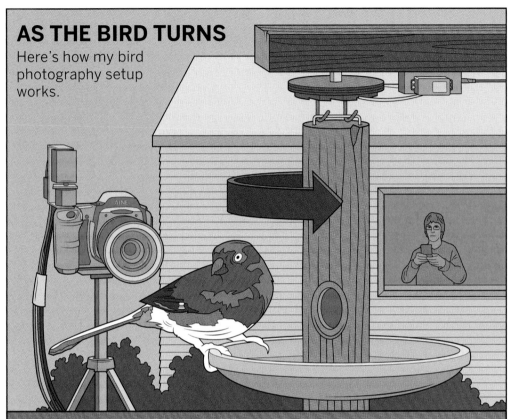

AS THE BIRD TURNS

Here's how my bird
photography setup
works.

To pose the birds, and to prolong camera battery life, I use an old R/C radio and its 2 servos: one actuates the camera's power switch, and the other slowly turns the bird feeder!

To control my camera's shutter release button, I use a long 3-conductor cable with 2 switches — the easiest, cheapest way to put your camera near the subject and shoot from a distance.

A shutter release button does 2 things: pressed halfway, it initiates metering and focusing; fully pressed, it fires the shutter. Some cameras have a jack for a shutter release cable, connected in parallel with the button; 2 switches at the other end are sequentially operable. The jack on my Pentax *ist DS camera accepts a standard ³⁄₃₂" (2.5mm) stereo plug, as do other DSLRs including the Canon EOS Digital Rebel, Samsung GX-1L, and Pentax K10D.

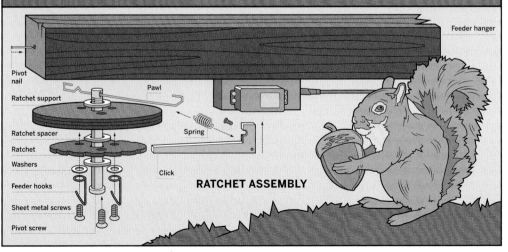

RATCHET ASSEMBLY

Feeder hanger

Pivot nail

Ratchet support

Pawl

Ratchet spacer

Spring

Ratchet

Washers

Click

Feeder hooks

Sheet metal screws

Pivot screw

Illustration by Timmy Kucynda

SET UP.

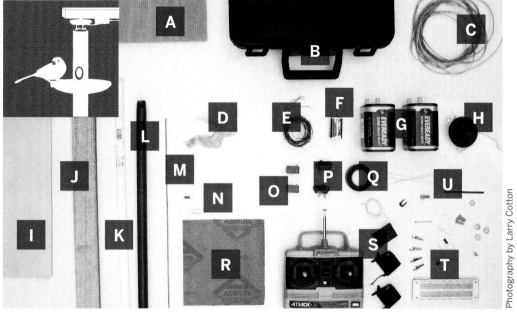

Photography by Larry Cotton

MATERIALS

[A] ½" plywood, 4"×4"

[B] Case **for carrying it all. An old power-tool case works well.**

[C] 3-conductor cable, at least 50' **such as an old telephone cable; 6- or 8-conductor is even better.**

[D] Sheet metal ¹⁄₆₄", several square inches **from home supply store, or from a can or metal sign**

[E] 24-gauge wire, several feet **insulated or not**

[F] 9V alkaline battery

[G] 6V lantern batteries (2) **to power the R/C radio**

[H] Small speaker **such as All Electronics SK-214 or SK-285**

[I] Laminated plastic 3"×3"×5" thick **such as Formica**

[J] Scrap wood

[K] Aluminum bar 1"×¾", about 14"

[L] PVC riser ½" internal diameter **I used Home Depot 86106 (10 pack)**

[M] Brass brazing rod ³⁄₃₂" diameter **for feeder hooks. A coat hanger also will work.**

[N] Large paper clip

[O] Pushbutton switches (2) **I used All Electronics SMS-229, but any will do.**

[P] Single pole single throw switches (2) **Most any SPST switch works.**

[Q] Electrical tape

[R] Acrylic sheet ⅛" thick **from home improvement store or glazier, or other ⅛" material, for ratchet spacers**

[S] Tower LXGRM7 R/C radio with 2 servos and receiver **from Tower**

Hobbies (towerhobbies. com). **Bigger servos have more torque. Hardware varies according to configuration.**

[T] BREADBOARD ASSEMBLY
• Photocell **also called a CDS photoresistor**
• IC breadboard **I used RadioShack 276-175. Perf board is OK, too.**
• 1K resistor, ½-watt
• 0.47µF capacitor **All Electronics RMC-474**
• 0.1µF 35V capacitor
• 555 timer IC **I used RadioShack 276-1718. CMOS also works.**
• Small alligator clips (3–4)

[U] FASTENERS
• ¼" wire staple
• 4"–6" wire ties **for wiring harness**
•#16×1" wire nails (7)
•#16×1½" wire nails (4)
• ¼"-20×½" machine screw
•#10×¾" sheet metal screws (3)
•#10 flat washers (3)
•#8-32×3" machine screw

•#6×½" sheet metal screws (#5)
•#4×½" wood screws (2)
•#6-32×¼" machine screw and nut
• ¾" weak tension spring **or small rubber band**

[NOT SHOWN]
Stepladder

C-clamps (1 or 2)

Stereo plug 2.5mm (³⁄₃₂") **from RadioShack or from yourcablestore.com. See the previous page for cameras that will take it.**

Tripod

Camera **DSLR preferred, but point-and-shoot OK**

Solder

9V battery clip

Plastic O-ring **or ¼M flat plastic faucet washer**

Hanging bird feeder and bird seed

Music wire, .032" diameter

MAKE IT.

BRING YOUR SUBJECT AROUND

START >>   Time: **A Weekend** Complexity: **Medium**

1. MAKE THE SHUTTER-RELEASE SYSTEM

1a. Construct the cable. You can buy a ³⁄₃₂" (2.5mm) stereo plug from RadioShack. But if you're not into delicate soldering, spend a few bucks on an adapter cable (for example, yourcablestore.com part #HP 3M-2M 6) and cut off the ⅛" plug. Connect at least 50' of cable to the ³⁄₃₂" plug. An old telephone cable with at least 3 conductors works great.

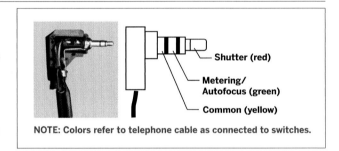

Shutter (red)

Metering/ Autofocus (green)

Common (yellow)

NOTE: Colors refer to telephone cable as connected to switches.

1b. Mount switches on a switch block. The switch block is a noncritical block of wood; it just holds your switches. Any pushbutton (momentary-on) switches will work, but 2 micro switches with different actuating pressures can be combined into a sequential switch.

Nail (yes, nail) them to the switch block. With 2 nails, mount the one with the greater actuating pressure below the one with the lesser actuating pressure, which pivots around its own single mounting nail (don't pound it all the way down). Tap a small nail through the block from the back, to act as a stop for the pivoting switch. Wire as shown here.

1c. Your cable is now testable! Plug the stereo plug into your camera's shutter-release cable jack. Power up the camera. Press the first (or upper) switch, then press a little harder to actuate the second switch.

NOTE: Go to makezine.com/projects/build-a-rotating-feeder-for-perfect-bird-pictures for tips on experimenting with your new shutter-release system.

2. BUILD THE CAMERA POWER SWITCH

2a. Make the switch coupling. My camera's main power switch rotates about 30 degrees and has a small lever-like protrusion. If yours is configured like that, Dremel-shape a small section of plastic plumbing riser (pipe) to match the camera switch on one end and the servo horn (the part that moves) on the other. Then screw the horn and your homemade coupling together with the #4×½" wood screws.

2b. Make the camera bracket. Mine is ugly and tortuously bent to hold the servo, receiver, and LED brackets. Make it from ⅛"×¾" aluminum extrusion. Be careful that the bracket doesn't block the optical viewfinder or the LCD; you'll need access to one of these for aiming and/or focusing the camera.

2c. Make the servo and receiver brackets. Made from thin sheet metal, these brackets attach the servo and receiver to the camera bracket. Use the assembly screws for mounting the servo and the #6-32 screw to mount the receiver.

2d. Mount the camera bracket loosely to the camera (for testing) by trapping its lower end between the camera and your tripod with a longer ¼"-20 machine screw, replacing the tripod's screw. Make sure the screw doesn't bottom out in the camera's tripod-mounting hole.

2e. Add the servo and receiver. Ensure the switch end of your coupling just snugly engages the camera's switch. Excessive pressure requires excessive torque (thus current) to turn the switch on and off.

2f. Test the servo coupling. With the ¼"-20 screw loosened and the coupling disengaged from the switch, connect and turn everything on. Press the appropriate joystick to observe how many degrees and in which direction the coupling moves. You can reposition the horn on the servo, tweak the trimmer on the transmitter, and/or limit the amount and direction of joystick travel by gluing a couple of pieces of thin plastic, such as Formica, near the joystick.

Photography by Sam Murphy

MAKE AN INDICATOR FLAG: Tighten the ¼"-20 screw and add a small but visible-from-a-distance "Camera On" flag to the servo's horn. Bend the music wire into an "L" shape and add colored tape or paper to the tip.

3. SPIN THE BIRDIE

3a. Make the feeder ratchet parts. The second R/C servo poses the birds by slowly ratcheting the feeder around. Make the ratchet parts by using the templates at makezine.com/projects/build-a-rotating-feeder-for-perfect-bird-pictures (the rest are pre-made, from the Materials list earlier in the article).

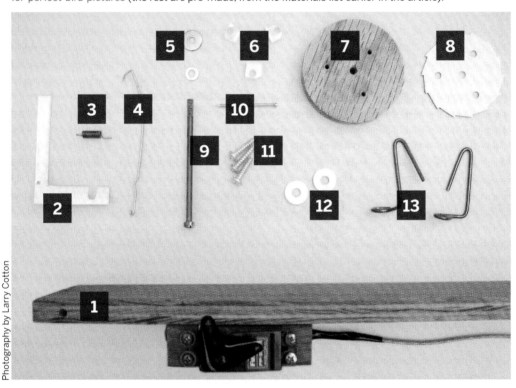

Photography by Larry Cotton

RATCHET ASSEMBLY
1. Feeder hanger Use a length of 3"×½" plywood to extend from your feeder support (pole, tree, etc.) to the feeder. Pre-drill appropriate holes in the other end to attach the hanger to your support.
2. Click Use thin ¹⁄₆₄" metal.
3. Spring
4. Pawl Use a paper clip.
5. Pivot screw washers

6. Ratchet spacers (3) made from ⅛" acrylic. Most any shape works as long as the hole is ¹⁄₁₆" from the curved edge. These must clear the pawl.
7. Ratchet support Use ½" plywood.
8. Ratchet Make it from ¹⁄₁₆" Formica using a band saw or a jigsaw mounted upside down; use a file to shape the teeth. The photo shows 14 teeth;

it's not critical as long as they're fairly evenly spaced and the throw on your servo is adequate to advance the ratchet 1 full tooth.
9. Ratchet pivot screw This pivot method using the 3" machine screw is almost foolproof. If you use a wood screw, make sure its threads are at least 1½" long.
10. Pivot screw nail
11. Sheet metal screws #10×¾"

12. #10 washers
13. Feeder hooks Since feeders vary, you'll customize these. Make them from ³⁄₃₂" brazing rod or a coat hanger. Bend an eye into one end of each hook for mounting to the ratchet. Use at least 2 hooks to smoothly transfer torque to the feeder.

NOTE: For parts templates, go to makezine.com/projects/build-a-rotating-feeder-for-perfect-bird-pictures.

3b. Assemble the ratchet. Clamp the feeder hanger upside down in a vise. Using Figures 12 and 13 (online at makezine.com/projects/build-a-rotating-feeder-for-perfect-bird-pictures) as a guide, put together the ratchet assembly — except for the pawl and click.

3c. Mount the servo to the feeder hanger with small wood blocks. (Servo configurations may vary.)

3d. Add the pawl and click. Connect a short, weak tension spring or rubber band between them to keep them engaged with the ratchet.

3e. Connect and turn on the transmitter and receiver. Move the joystick, which actuates the feeder servo. Each pawl-pull should cause the click to engage an opposite-side tooth to prevent backward rotation. Each full cycle rotates the ratchet 24°–26°. You may have to adjust the action by slightly bending the pawl and/or click.

NOTE: To connect the ratchet assembly, extend the servo's wires so that they're longer than the distance from the feeder to the camera rig.

4. MAKE YOUR CAMERA FEEDBACK BEEPER (OPTIONAL)

Only one challenge remains: a way to signal "Picture taken!" without having the flash go off in birdie's face.

On the back of my camera, there's a small LED that indicates memory-card access. A photocell (photoresistor) mounted face to face with this LED can change resistance in a simple 555 circuit (*see MAKE, Volume 10, page 62, "The Biggest Little Chip"*), creating or changing a tone in a speaker.

If your camera has this LED, make and mount another small bracket of thin sheet metal to bridge between the camera bracket and the LED. Drill a hole to fit the photocell, while holding it in good contact with the LED. Wrap black electrical tape around the assembly (don't cover the face of the photocell!) and put a small O-ring around the photocell to shield it from extraneous light.

Build the beeper per the schematic shown online, using prototype or perf board. Mount it near the R/C transmitter, then run 2 wires from the photocell, alongside the remote shutter-release cable, to the beeper.

NOTE: For the schematic on the feedback beeper, go to makezine.com/projects/build-a-rotating-feeder-for-perfect-bird-pictures.

5. MOUNT REMOTE CONTROLS AND WIRE IT UP

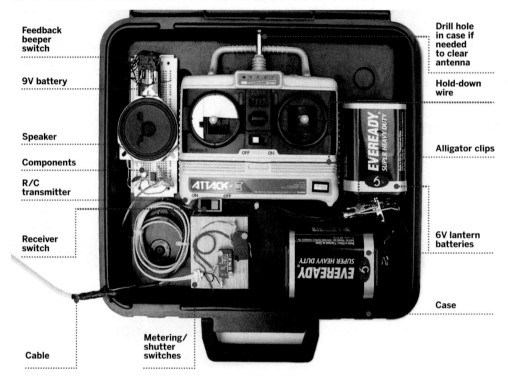

Feedback
beeper
switch

9V battery

Speaker

Components

R/C
transmitter

Receiver
switch

Cable

Metering/
shutter
switches

Drill hole
in case if
needed
to clear
antenna

Hold-down
wire

Alligator clips

6V lantern
batteries

Case

Since you're running wires anyway, run another pair to power the receiver.

I mounted all the remote stuff in an old plastic power-tool case using hot-melt glue and wire pulled through holes in the case to hold everything in place. You can also glue foam blocks to the inside of the top.

Mount the meter/shutter switch block near the camera-on joystick, for one-hand operation. You'll need the other hand to move the feeder-turning joystick.

Three more switches turn everything on: 1 in the transmitter, 1 for the receiver, and 1 for the feedback beeper. The 2 you add can be any SPST switches.

One 9V and two 6V (lantern-size) batteries power everything. To replace the (usually) 8 AA batteries in your transmitter, solder wires to the first (+) and last (-) battery terminals, run them out of the housing, and connect them to the lantern batteries, in series, with alligator clips. Wire the whole rig according to the diagram online at makezine.com/projects/build-a-rotating-feeder-for-perfect-bird-pictures.

FINISH ⊠

NOW GO USE IT »

USE IT.
GET A BIRD'S-EYE VIEW

NOTE: If you're using the camera feedback beeper, you'll hear a beep (or a pitch change) from the speaker when you take a picture.

Set the camera up and move away from it, as in the shutter-cable testing procedure.

Turn on the transmitter, receiver, and feedback beeper (if using). Move and hold the joystick to turn the camera on. (Watch the flag on the camera for confirmation.) While holding the joystick on, actuate the metering and shutter switches.

Now practice rotating the feeder by repeatedly moving its joystick s-l-o-w-l-y.

Take another shot or two, then retrieve your memory card and download the picture(s) to your computer to check focus, exposure, composition, etc. (Unless you're extremely lucky, you'll probably just have beautiful shots of your bird feeder.)

Finally, reinsert a formatted memory card, get comfortably away from the feeder, and hold the plastic case in your lap. Don't turn on the camera until a bird lands on the feeder.

Using both joysticks and the meter/shutter switches, take lots of pictures while you slowly rotate the feeder to the best camera angle. You may also turn the feeder between bird visits; they may prefer certain feeding positions to others.

Do the birds like to be rotated? Eventually, most of them adapt. Some are more skittish than others; some actually seem to enjoy it. The female red-wing blackbirds stop eating and look up at the sky. Blue jays bolt. And cardinals just smile.

Photography by Larry Cotton

HELIUM BALLOON IMAGING "SATELLITE"

By Jim Newell

Photograph by Sam Murphy

INFLATION WATCH

Snap aerial photos from 300' up by suspending a hacked drugstore camera from 3 tethered helium balloons.

The first time I saw a satellite photo of my house on Google Earth, I expressed shock at the "Big Brother" implications of an all-seeing, commercial eye-in-the-sky. But meanwhile, I was also secretly disappointed with the picture quality and clarity because (Orwellian angst aside) I needed better overhead images for my own use — to help me lay out a new driveway and complete a bird's-eye-view CAD drawing of our lot.

So I decided to design and fabricate a simple helium balloon "satellite" camera platform, tethered to the ground for ease of control and retrieval, and dedicated to a single purpose: to capture aerial images of my house and surroundings.

Here's how I completed this project using inexpensive and readily available components — helium balloons on a nylon kite string, a drugstore camera perched on a platform made out of an old CD, and a PICAXE microcontroller housed in an empty pill bottle.

Jim Newell has degrees in mechanical engineering, physics, and business administration, and has worked in the aerospace industry for the past 28 years. With interests including electronics and home automation, he's also an avid guitar player and singer/songwriter.

Balloon Cam: How It Works

A tiny $3 microcontroller chip tells your aerial camera when to snap.

A cheap, lightweight **A** digital camera, modified for electronic shutter triggering, snaps photos. The camera's lens aims down through the center hole of a spare **B** CD (or DVD), which serves as the camera's support platform. The CD is suspended from the bottom of a **C** pill bottle that holds the trigger board and battery. The **D** trigger board uses a tiny PICAXE-08M **E** microprocessor that's programmed (in BASIC) to send repeating pulses from one of its output pins. The pulses are applied to a **F** reed relay, which opens and closes the connection between the camera's shutter control contacts, repeatedly triggering the shutter. Power for the board comes from a **G** 9-volt battery via a **H** voltage regulator that drops the circuit's voltage to 5V. The compiled microprocessor code is uploaded to the PICAXE through a **I** programming header connected to a computer.

Calculating the Number of Balloons Needed

Total weight = weight of payload + weight of tether at max height = ½lb + ¼lb = ¾lb

Lift from helium at sea level (approx.) = 0.067lbs per cubic foot (ft^3)

Volume of helium to lift ¾ pound = 0.75/0.067 = 11.2ft^3

Volume of a sphere = $4/3 \times \pi r^3$

Balloon radius needed (single balloon) = $(¾ \times 11.2 / \pi)^{1/3} = 1.4'$

To account for the added weight and helium backpressure from the balloon itself, ensure sufficient lift for multiple flights, and add redundancy, I tripled this minimum figure and used three 3'-diameter balloons.

Illustration by Tim Lillis

SET UP.

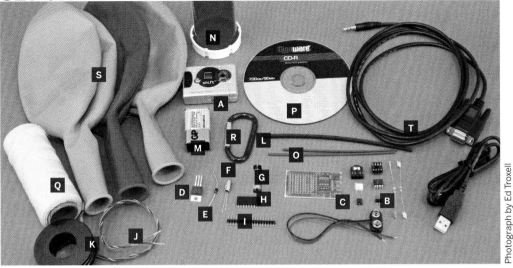

Photograph by Ed Troxell

MATERIALS

[A] Digital camera, mini "keychain" $10–$40 from chain drugstores; brands tested were Shift3, Aries

[B] PICAXE-08M2 microcontroller SparkFun Electronics item #COM-10803 (sparkfun.com), $3

[C] PICAXE 8-Pin Proto Kit SparkFun #DEV-08321, $4

[D] Voltage regulator, 5V, TO-220 package, Digi-Key #LM2940T-5.0-ND (digikey.com)

[E] Diode, 1N4001 Adafruit #755 (adafruit.com), $1

[F] Electrolytic capacitor, 22µF RadioShack #272-1026, $2

[G] Relay switch, 5V reed aka reed relay. RadioShack #275-0232, compact 5V DC/1A SPST; the cylindrical package will work, but the box-shaped one is too large to fit.

[H] Connector header, female, 9×1, 0.1" spacing Digi-Key #S7042-ND, $1

[I] Breakaway headers, male SparkFun #PRT-00116, $2

[J] Electrical wire, 18–20 gauge, insulated

[K] Solder

[L] Heat-shrink tubing or electrical tape

[M] Battery, 9-volt

[N] Pill bottle with child-proof cap large enough to hold microcontroller and battery

[O] Solid wire, 12 gauge

[P] Compact disc or DVD

[Q] Kite string, 150lb test, I used Dacron Archline from Conwin (conwinonline.com), 200yds for $15.

[R] Small carabiner

[S] Balloons, 3' diameter (3) Try BalloonsFast (balloonsfast.com), $4 each

[NOT SHOWN]

Helium, 42 cubic feet You can have balloons filled at a balloon supplier or a supermarket, but the supermarket balloon person will probably not be happy with you using so much helium.

Duct tape

Double-sided foam tape

TOOLS

[T] PICAXE USB programming cable PICAXE part #AXE027, SparkFun #PGM-08312, $26. Or, with a Windows or Linux PC, you can save money by buying the $15 PICAXE-08M Starter Pack, SparkFun #DEV-08323, which includes the PICAXE 8-Pin Proto Kit (see Materials list) and a serial programming cable rather than USB.

[NOT SHOWN]

Computer with internet connection

Soldering equipment

Wire strippers

Wire cutters

Phillips screwdriver, small to unscrew camera case

Flathead screwdriver, small to pry open camera body

Tweezers

Multimeter or ohmmeter

Drill and drill bits: ⅛", ¼"

Pliers

Scissors

Helping hands with magnifier (optional)

Tabletop vise (optional)

MAKE IT.

BUILD YOUR BALLOON-SAT

START >> Time: **1–2 Days** Complexity: **Moderate**

1. MAKE THE TRIGGER BOARD

For all connections, refer to the project schematic seen here (downloadable at makezine.com/projects/balloon-imaging-satellite). To stabilize voltage regulator operation, I added an additional 22μF electrolytic capacitor across the power leads, in parallel with the Proto Kit's included 100nF cap.

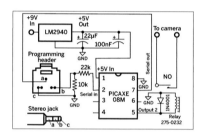

1a. Assemble the PICAXE Proto Kit, soldering the following components at the locations indicated on the printed circuit board (PCB): 8-pin IC socket, stereo download socket, 3-pin header, 10kΩ resistor, 22kΩ resistor, and 100nF capacitor. These components are small, so a helping hand with magnifier or a tabletop vise will come in handy. Don't connect the battery clip yet.

Once all parts are in place, carefully place the microprocessor chip in its socket, with pin 1 (indicated by the notch) pointing away from the prototyping area (or you can place it later; see Step 1e). Also, move the jumper on the 3-pin header to the PROG side to enable it for programming.

1b. Thread the leads of the battery clip through the 2 holes in the Proto Board PCB, and solder the black wire into place on the bottom of the board.

1c. Insert the voltage regulator through 3 holes near the center of the PCB. Run the red wire from the battery clip across the top of the PCB and connect it to the regulator's input pin (indicated by a dot). Using a wire jumper, connect the middle (ground) pin of the regulator to the black wire from the battery clip. Use another wire jumper to connect the LM2940 output pin to the PCB, at the location marked "RED," where the red wire from the battery clip would normally attach.

1d. Insert the 22µF capacitor through the 2 PCB holes indicated by (+) and (−), and solder it in place. Be sure to watch polarity; the stripe on the cap goes on the (−) side.

1e. Insert the relay into the PCB at the forward edge so that the single switch terminal dangles off the side of the board and the other 3 pins run through holes. The bottom left pin should run through the hole that's 3 up from the bottom of the board and 2 from the left.

1f. Insert the 1N4001 diode into the through-holes that connect to the relay coil terminals so that its body drapes over the top of the relay. This diode protects the PICAXE from back electromotive force when the relay is de-energized. Using jumper wires, connect one of the relay coil terminals to the PCB ground pin and the other to PICAXE output 2 (pin 5 on the chip).

1g. Solder about 1' of 18- to 20-gauge wire to each of the relay switch terminals. Use wire cutters to cut a 2-pin length from the male breakaway headers, and solder the 2 pins to the wires' other ends. Cover the solder joints with heat-shrink tubing or electrical tape to prevent shorts and to add strength.

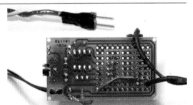

Photography by Jim Newell

2. PROGRAM THE TRIGGERING

2a. Download and install Revolution Education's free AXEpad software from rev-ed.co.uk/picaxe.

2b. Download the BASIC file *Camera_Timer.bas* from makezine.com/projects/balloon-imaging-satellite, then open it up in AXEpad.

This simple, 14-line routine waits 20 seconds from the time of initial power-up to give time to replace the pill bottle cap, takes one picture to confirm that it's running, waits another 20 seconds to let the balloon rise, then begins snapping pictures every 2 seconds. You can modify this to suit your needs.

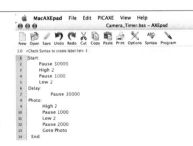

2c. To load this code into your Proto Board, connect the PICAXE programming cable between your computer and the board's programming jack, then click the Program button in AXEpad, in the upper right.

3. MODIFY THE CAMERA

The specifics of this step will depend on the camera you use, but it's a simple mod, and readers with basic electronics skills should have no problem. The camera shown here is Shift3 brand and was purchased for $11 at a Rite Aid pharmacy.

3a. Remove the stick-on label from the front of the camera (or the side, for the Aries camera) to reveal a screw that holds the case together.

3b. Remove the screw with a small Phillips screwdriver.

3c. Gently pry the camera shell open using a flathead screwdriver.

3d. Remove the 2 screws holding down the circuit board, and also unscrew the keychain clip, which we don't need (with the Aries camera, remove 3 screws to detach the board).

3e. Turn the board over so that the lens is visible. Handling the board by the edges only, and without touching any parts, use tweezers to remove the black potting material from around the shutter switch, exposing solder terminals at its base.

3f. Solder wire leads to the 2 newly exposed switch terminals.

3g. To make room for the shutter switch wires to exit the case, use pliers to cut a hole in the plastic on the side opposite the lens (with the Aries camera, instead of cutting the case you can remove the pop-up viewfinder lens assembly and route the wires out of its hole).

3h. Replace the board and screw it back into the case, routing the shutter switch wires out, and reassemble the case.

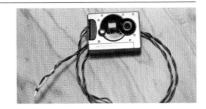

3i. Cut a pair of adjacent female connector headers, and solder one of the wires to each. This will connect to the male header pair from the trigger board.

4. ASSEMBLE THE SATELLITE STRUCTURE

4a. Drill 4 equidistant ⅛" holes around the bottom of a pill bottle large enough to hold the microcontroller board and battery (about 2" in diameter and 4" tall). Drill another hole through the center.

4b. Insert 6" lengths of stiff 12-gauge solid wire into the 4 perimeter holes and extend them downward from the bottom of the bottle. Inside the bottle, bend the tops of the wires so they stay in place when you pull the wires from below.

4c. Mark and drill 4 uniformly spaced ⅛" holes around the periphery of a spare CD or DVD.

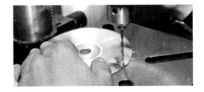

4d. Thread the 4 wire standoffs from the bottom of the pill bottle through the 4 holes on the CD, bend them to lay flat underneath the surface of the CD, and secure them in place with duct tape, or by twisting them up and around.

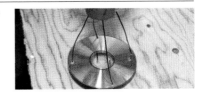

4e. Grab the spool of kite string, and route the free end up through the center hole in the CD and through the center hole in the bottom of the pill bottle. Tie it off to the 12-gauge wires inside the bottle.

5. ATTACH THE CAMERA AND BALLOONS

5a. Mount the modified camera to the top surface of the CD with double-sided foam tape so that the lens of the camera looks down through the hole in the center of the CD.

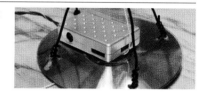

5b. Insert the PICAXE board and 9-volt battery into the pill bottle, but don't connect the battery yet (powering up the board will start the program running).

5c. Drill a ¼" hole in the center of the pill bottle cap and 4 more small holes around the periphery, uniformly spaced and as close to the outside diameter as possible.

5d. Route 4 pieces of kite string, 8"–12" each, through the perimeter holes and tie them all together securely inside the cap. Then tie together the other ends of the 4 strings, extended from the top side of the cap, to a small carabiner.

5e. Inflate the balloons with helium, tie each one off with a knot, and tie on a 1'–2' piece of kite string. Tie the balloons together to form a tight group, and tie them all to the carabiner. When finished, you'll have the balloon group attached to the cap of the pill bottle, and because the balloons attach to the satellite only through this cap, it must be of the childproof variety to make sure it stays on securely.

5f. Finally, attach the bottle cap, routing the camera connector wire through the center hole. You're ready to fly!

FINISH ▣

NOW GO USE IT 》

USE IT.

NUDE SUNBATHERS BEWARE!

BALLOON-SAT OPERATION

Once the balloon-satellite is fully assembled you're ready to launch. Here's how:

1. Plug the headers together from the trigger board and the camera, and turn the camera on.

2. Unscrew the top of the pill bottle, being careful not to let it fly away. Inside the bottle, connect the 9-volt battery to the Proto Board battery cable, and quickly screw the top back on.

3. Let the satellite go — up, up, and away!

If you look carefully at my photos, you can see the kite string along with a knot I had to tie due to some poor planning. I hope that readers can plan their string routing a little better than I did and keep the images knot-free.

ENHANCEMENTS

You can extend the trigger board for greater functionality by adding a second relay to switch the camera on and off. That way the keychain camera could turn on once it reaches altitude and start capturing aerial home movies in video mode. Not even Google Earth can compete with that!

Photography by Sam Murphy (balloons). James Newell (house)

Looking at the Low End

Infrared photography reveals a world invisible to the naked eye.
By Richard Kadrey

Photography by Richard Kadrey

For the human eye, the lowest visible wavelengths are red light measuring about 700 nanometers (nm). Below that, infrared radiation runs from about 750nm down to 1mm. When photographed in this part of the spectrum, leaves and grass glow with energy, as if the entire natural world is lined with fiber optics. Skin is luminous and perfect, like alabaster. Infrared photography gives you an inhuman view of the world, and it's a beautiful one.

In the beginning, infrared photography was nothing you needed to know about. It was a high-tech procedure reserved for laboratories and mapping satellites. Even when artists got their hands on the stuff, it required special film that had to be kept in an ice chest until it was used, and special processing that required access to a darkroom with the right chemicals, and all the expenses those items entailed.

Digital photography has made infrared accessible to everyone. That's great news to those using IR for the first time, because this is when you're liable to make the most mistakes. Better yet, you don't need an expensive camera to take great shots. In fact, cheaper and so-called "dinosaur" digital cameras can be the best ones for IR shooting. The reason is simple: most high-end cameras come with a built-in infrared-blocking filter (sometimes called "hot glass") that sits right in front of the camera's sensor chip. Cheap cameras don't always have this IR filter, and they're easy to hack if they do. But remember when picking your cheap camera to make sure it has a Preview mode. This will allow you to see your infrared shot and make adjustments on the fly.

1. GET A DIGITAL CAMERA THAT CAN SEE INFRARED

You may already own a camera that can see infra-red. To find out, point a TV remote control toward your camera (in low room lighting), press any button on the remote, and watch it on your camera's LCD screen. If you see the tiny lens on the end of the remote glowing white, your camera is sensitive beyond the human visible spectrum.

I shoot infrared with a Sony Cybershot DSC-F707. When it was released in 2001 it was considered a fairly high-end "prosumer" camera. Most photo geeks now think of this 5-megapixel machine as a kind of steam-powered relic, something that would have wowed Jules Verne, but that's just amateur-hour snobbery spouted by people who think that the only thing that counts is who has the bigger megapixel count. Even shooting in regular mode, it's not hard to make good 8×10 prints with the F707. In fact, 95% of the prints in my first gallery show were

shot with the F707. And it packs plenty of power for infrared work.

One reason the Sony F707 is such a great off-the-shelf infrared shooter is that it comes with a NightShot feature, which is a built-in infrared system. Unfortunately, this works only in the dark.

Sony crippled the ability to shoot IR in daylight because, supposedly, some fabrics are transparent to infrared, effectively turning all of Sony's cameras from that period into a voyeur's favorite new toy. But any hack a corporation comes up with can be broken by patient geeks on a mission. By putting the F707 in NightShot mode and adding infrared and neutral density filters, we trick the camera into thinking that daytime is nighttime.

Infrared photo wiz Chris Maher shoots with other dinosaur cameras, such as the Nikon 950 and 990. I loved the 950 and 990, generally loathed by "real" photographers when they came out. Their bodies swiveled independently of the lens, so using

Preview mode, you could shoot over and around people without anyone knowing what you were doing.

Another favorite of infrared enthusiasts is the old Olympus 3030. Instead of trying to unload it for $10 in a garage sale, you can use it to see the world in a whole new light — literally.

2. GET SOME INFRARED FILTERS

By hacking your digital camera, you're tricking it into thinking that it's still shooting in the visible spectrum. To do this, all you need are a few simple filters. When you put the filters on your lens, you can forget about all your regular camera settings. This is why you need Preview mode.

The rim of the F707's lens, like most lenses, is threaded so that you can screw on standard filters, such as polarizers for outdoor shooting. If your lens doesn't have threads, you can simply hold a filter over the lens with one hand and shoot with the other.

Infrared filters come in all the standard lens sizes. The F707 takes 58mm filters. The first IR filter I used was the Hoya R-72, which allows only infrared rays longer than 720nm to pass through. Other infrared filters block different parts of the spectrum, giving your shots different looks, bringing out different details in the sky, foliage, and foreground objects.

The best way to find out what works for you, and works in different situations, is to experiment. Since you can pick up an F707 on eBay for $225–$250, you may have some money left for filter shopping. Filters can run from $35–$200, depending on the size of your lens.

3. GET SOME NEUTRAL DENSITY FILTERS

Along with your infrared filter, you may need 1 or 2 neutral density filters to cut the amount of visible light entering your camera, without changing its color. Think of it as a piece of welding glass you could use to look at a solar eclipse. My first neutral density was an ND-400.

Without an ND filter your infrared shots can be much too bright. On a sunny day, the highlights will be completely blown out. Even with a single neutral density, you can end up with too much light. Since

THE LOOK OF INFRARED: Infrared radiation reflects off solid surfaces at different frequencies, depending on the surface's composition. Leaves and branches can turn snowy white, and water an opaque black.

the filters come threaded, you can add a second ND, along with the infrared filter.

4. LEARN MORE

One of the best sites on infrared shooting is infrareddreams.com, with an excellent IR primer at infrareddreams.com/how_to_shoot_ir.htm.

You can find infrared filters at most decent camera shops, as well as good online photo sites such as bhphotovideo.com. You can even find them in Amazon's Camera area.

Have fun shooting. Experiment. Take chances. And be prepared to see a world that you have never seen before.

Richard Kadrey has written about technology and culture for magazines such as *Wired*. He's also a fiction writer; his latest novel is *Killing Pretty: A Sandman Slim Novel*.

Go Green!

Special video effects are available to anyone with a cheap camcorder and $25 of software. Greenscreen is the most powerful of these, and is surprisingly easy to use. By Bill Barminski

Photography by Bill Barminski

Would you like to make a video of yourself standing on the moon? There are two ways to do it. You can build a rocket and fly there — expensive, not to mention dangerous. Or you can use a greenscreen to make it look as if you are there. Yes, a greenscreen. I hope I won't be shattering too many illusions when I tell you that this is how they did a lot of that cool stuff in *Star Wars*. They placed an actor in front of a greenscreen and filmed the scene while he pretended to fight a giant space squid. A technique called *chroma keying* was then used to remove the green color, allowing a new piece of video to be placed behind the actor.

This is called a *composite shot*, and the process is called *keying*. In the past you needed high-end software costing hundreds if not thousands of dollars, but today you can do it for $25 plus some cheap paint and lights.

But just let me issue a word of caution: greenscreening can be tricky. There are many variables that can affect the outcome. Even professional filmmakers run into unexpected problems from time to time.

1. MAKE A BACKDROP

First you're going to need a screen, which can consist of colored fabric or a painted wall. Lime green is most commonly used, because it is so freaking ugly that the exact-same color is unlikely to appear on anyone or anything else in the shot, and thus it can be earmarked for replacement. (This means your subject can't wear a lime green tie.)

You can buy special greenscreen fabric and paint, but they're expensive. I've used very cheap green fabric from the local fabric store with decent results. I've even used a lime green blanket I found at a thrift store for $4. Look for something sheer that resists wrinkles, which will show up and make it harder to pull your key; iron the wrinkles out if you need to.

If you have a wall you can paint, so much the better, since there will be no wrinkles. Go to any paint store and pick out the worst lime green color you can find. Be sure it has a flat finish, not glossy. The exact shade is unimportant, since our software will find it for us when the time comes to replace it with our desired background image.

2. LIGHTING AND PLACEMENT

The biggest problems with greenscreen shots stem from poor lighting and placement of your subject. You want to illuminate your greenscreen with a flat, even light, so that it has no shadows or highlights. Don't use spotlights for this.

The placement of your subject in relation to the greenscreen is also crucial. The subject needs to be as far from the green as possible, to avoid picking up reflected green light. This is tricky because the reflected green is hard to see. Of course the farther away you put your subject, the bigger your greenscreen must be. If you're doing this for the first time, frame your subject from the waist up. Don't try a wide shot of the whole person.

The cheapest lighting source is the sun. If you can shoot outdoors, that's great, provided you find a place that gets even lighting with no shadows on the background. A gray, overcast day is actually best for shooting since it produces an even, flat light.

If you shoot indoors, you'll need 2 sets of lights, one for the greenscreen and one for the subject. Don't try to use the same set of lights for both.

To illuminate the greenscreen you can use cheap fluorescent tubes. They give a smooth, even light.

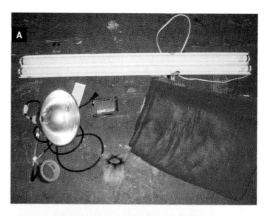

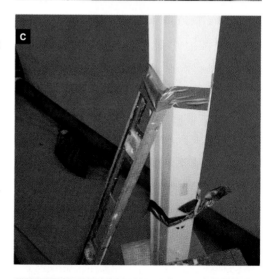

Fig. A: Fluorescent fixture and tubes ($12), scoop light ($10), camera (duh), duct tape, fake beard (optional), green blanket (about $4). Fig. B: Here's a greenscreen that's been thumbtacked to the wall. Be sure to stretch it tight to avoid wrinkles. Fig. C: This is why you need duct tape, unless you want to spend 200 bucks for light stands. PREVIOUS PAGE: The author in front of the "fire."

GREEN SCREEN TIPS. Fig. D: The subject has been filmed against a greenscreen. A mistake made here is the use of glass. As you can see, the green shows through. Avoid mistakes like this.

Fig. E: Subject after the green has been keyed out and new video inserted. The problem with the glass is evident. Other issues show up too, such as the green bounce light hitting the side of the chair. Color correction can help to remove or limit the green.

For about $12 you can buy a 4-foot fixture containing 2 tubes. Depending on the size of your greenscreen you may need 3 or 4 of these fixtures. I use sticks and lots of duct tape to anchor them. You can then light your subject with a couple of workshop clip lamps using bulbs rated from 100 to 500 watts. Remember that they must not cast shadows on your screen.

3. KEYING YOUR VIDEO

Now, let's say you've shot your footage and you're ready to key out the green. On a Mac, you can use iMovie with a plugin called Stupendous Software Masks & Compositing, which costs $25. If you have Windows, you can find free software such as ZS4 (zs4.net/downloads), or economical all-purpose editing software (with greenscreen feature included) such as Video Edit Magic, available for free in a trial version or for $69 fully featured, from trusted sites such as tucows.com.

3a. In iMovie, first import the video that you shot against a green background, and place it in the timeline. Then import the footage that will replace the background, and place it next to your video.

Select the first clip, go to the Effects category, and choose Green Screen, Smooth. This effect has 3 controls: Outside Fill, Inside Fill, and Choke. Play around with these settings. The little preview window will show a black and white sample. The 2 fill settings determine how crisp the outline will be. You basically want your subject to appear all white and the green area to appear solid black; avoid shades of gray.

The choke allows you to bite into the cutout to remove jagged edges. Once you think you have the settings right, click the Apply button. It will take several minutes or longer to render your shot. Once it's done, you can watch your clip. You may need to go back and change the settings a few times to get the best results.

3b. In Video Edit Magic, place your background video in the Video 1 timeline. Place your foreground video (with greenscreen) in the Video 2 timeline. Click the Video Transitions tab in the Collection window and drag Chroma Key Color to the Transition timeline. In the window that pops up, you can click your green background to sample it, and drag the Similarity slider to adjust the tolerance.

You should see the green vanishing to reveal your new background. Once you have it the way you want it, stretch the transition to the desired time span, then render and save.

Other software will take you through steps very similar to those described above. For an instructional video dramatizing the greenscreen process, check out makezine.com/go/green and browse the many other greenscreen tutorials available on YouTube.

There's really no need for dull video backgrounds when you can key your own!

Bill Barminski is an artist, videographer, and lecturer in the Film Department at UCLA.

BROWNIE PAN
LED LIGHT PANEL

Roll your own for a fraction of the cost of pro units.

Written and photographed by Tyler Winegarner

Time Required:
4-6 Hours
Cost:
$70-$100
Hack LED strips with a better dimmer and save hundreds.

TYLER WINEGARNER
is a filmmaker and photographer based in San Francisco. When he's not busy shooting, he's down in his shop grinding, soldering, or hacking at his next project. Follow him at @the_real_tylerw.

OPINIONS ARE MIXED ON LED LIGHTING UNITS, BUT ANYONE WHO HAS EVER DONE LOCATION WORK CAN'T DENY THEIR UTILITY — they're lightweight, they have very low power draw, and they generate very little heat. They're an excellent tool to have in a one-man band style of shooting. They're also expensive. But now, with the proliferation of LED lighting kits for home use, you can build a very good equivalent to $500 off-the-shelf products for under $100. It even looks good — and that might be handy, depending on who your clients are.

At the core of this project is the adhesive-mounted strand of LED lights. These are usually sold in kits with an external power supply and an inline dimmer. Unfortunately this dimmer operates at a relatively slow cycle — it looks steady to your eyes, but in camera you'll see the flicker. So we'll be using the guts of a better, external dimmer to get the results we need.

There's a lot of soldering in this build, but none of it is very tough, so it's a good project to help you build your skills.

1. CUT THE PANELS
Cut the plexiglass and corrugated plastic to size. Drill a ¼" mounting hole in each corner of the corrugated panel, ½" inward from the edges.

2. PREPARE THE HOUSING
Using the holes drilled in the corrugated panel as your guide, mark and drill four corresponding ¼" holes on the back of the baking tin, starting with the 1/16" bit for a pilot hole.

In the bottom of the long edge of the housing, drill 3 additional holes: one ¼" hole in the center for the mounting hardware, and about 2" away, ⁷/₁₆" and ⁵/₁₆" holes for the DC jack and the dimmer knob. Keeping the jack close to the mounting hardware will make cable management easier when using the light.

Use your grinding wheel to roughen the metal (and remove any teflon coating for better adhesion) around the inside of the ¼" center hole.

If your pan is teflon coated, also remove about a 1"×2½" patch of teflon roughly 2" above the holes for the DC jack and dimmer knob.

3. MOUNT THE INTERNAL THREADING
Use the grinding wheel to score one side of a ¼-20 nut. Use epoxy to bond the nut to the inside of the housing, centered on the ¼" center hole. Let the epoxy set.

4. CUT AND SOLDER THE LED SEGMENTS
You can only cut the LED strip at the marked areas, which appear every 3 lights. Starting with the end that has the loose wires, cut fourteen 10" segments.

Cut fourteen 2"–3" segments of speaker wire, and split and strip each end.

Now use the wire segments to solder all the LED segments back together. I find the best technique is to melt a small dot of solder onto each terminal of the LED segment, and then heat up that solder while poking the wire end into it. Connect all the LED segments end to end, making sure you don't cross up the positive and negative terminals.

Use a voltmeter to verify continuity.

5. MOUNT THE DIMMER AND JACK
Disassemble the PWM dimmer and remove the PCB and potentiometer. Use double-sided foam tape to mount the PCB to the inside of the housing, on the patch you prepared.

Use a 3"–5" length of speaker wire to connect the DC jack to the input

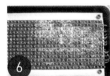

terminals on the dimmer PCB, following the manufacturer's instructions for both. Connect the positive and negative wires from the first LED segment into the output terminals of the dimmer PCB.

Mount the DC jack and dimmer potentiometer into their holes. Connect your DC jack to power and check your circuit to make sure everything lights up and dims when you twist the potentiometer.

6. MOUNT THE LED PANEL
Mount the corrugated plastic panel to the housing by fitting the Allen screws and washers into the coupling nuts.

Peel the adhesive backing from your first LED segment and mount it to the corrugated panel just above the washers. Mount each consecutive segment in the same way, zig-zagging your way up the panel.

Check your circuit again, and resolder as needed.

7. MOUNT THE COVER PANEL
Use the original holes at the ends of the baking pan to mark the plexiglass. Drill ¼" holes at your marks, then mount the plexiglass cover to the front of the housing using the remaining Allen screws, washers, and nuts.

If you have access to a bandsaw, you can cut the plexiglass to match the shape of the baking pan.

Use the grinding wheel to carefully clean up any rough edges. You're done!

USE IT
You can use the threaded hole on the ballhead to mount to any ¼" tripod stud, or use the shoe adapter included with the ballhead to mount to any shoe mount.

Because the LEDs generate little heat, you can use household items like baking parchment to diffuse the light. Binder clips can be used to attach diffusion material or gels to the outside of the housing.

There's a lot of variation you can do with this project, from panel size to color temp to lots of other configurations. Happy shooting! ◐

Materials
- » **Baking tin, 11"×7"** with holes in the rim, such as Wilton #2105-960
- » **Clear plexiglass, 12½"×8"**
- » **Corrugated plastic, white, 6"×10"** aka Coroplast
- » **LED strip, 5 meters long, self-adhesive** I'm using a 600-LED, daylight balanced, non-weatherproof model, Amazon #B005ST2I90.
- » **PWM Dimming Controller, for LED lights** Amazon #B007V1B0W8
- » **Power supply, 12V 3A DC, with size M plug** Amazon #B00BPCL0MY
- » **Coaxial DC power jack, size M, panel mount** RadioShack #274-1563
- » **Speaker wire, 2-conductor, 22 gauge,** braided, 36"–48" length
- » **Tripod ball head** with ¼" threaded hole on the bottom and ¼" stud at the top
- » **Nuts, ¼-20 (3)**
- » **Nylon washers, ¼" ID (6)**
- » **Cap head Allen screws, ¼-20, ½" long (6)**
- » **Double-sided foam tape**
- » **Coupling nuts, ¼-20, 1" long (4)**
- » **Machine screws, ¼-20, ½" long (4)**

Tools
- » **Electric drill with drill bits: ¹/₁₆", ¼", ⁵/₁₆", and ⁷/₁₆"**
- » **Grinding wheel for drill** or Dremel with grinding wheel
- » **Soldering iron**
- » **Wire cutter / stripper**
- » **Masking or painter's tape**
- » **Utility knife**
- » **Voltmeter / continuity tester**
- » **Bandsaw (optional)**

For full instructions and photos, go to makezine.com/brownie-pan-led-light-panel

GLASS BEAD
Projection
Screen

⚡ **TIME:** 8 HOURS OVER 3 DAYS   ⚡ **COST:** $50–$100

Need a bright surface for your projector? Get high gain at low cost using house paint and sandblasting beads.

Written and photographed by *Sean Michael Ragan*

Titanium dioxide is the most common white pigment in paint, sunscreen, and even food products. It's cheap, safe, and almost unsurpassed in whiteness. It's also the baseline for calculating an optical property called *screen gain*, which is the amount of light reflected from a projection surface divided by the amount of light reflected from a titanium dioxide reference surface. Since titanium dioxide is the pigment used in most white paint, a smooth wall painted flat white has a screen gain very close to 1.

But you can do better. This method applies a high-gain optical projection surface using common, cheap materials – flat white latex paint and glass sandblasting beads. I started out trying to directly mix them (which doesn't work) and happened on this "sprinkling" method by accident. It gives a much brighter screen surface than paint alone.

1. Determine your screen size.

Set up the projector as you will use it. Turn it on. Measure the height and width of the image. Plan the size of your screen accordingly. In my case, a single 4'×8' panel made a convenient size.

2. Build the screen.

Cut 1×2 frame members with a miter box and saw. Here's the cut list for my 4'×8' screen.

- » 2 sides 96" long on outside edge, mitered ends
- » 2 sides 48" long on outside edge, mitered ends
- » 2 braces 45" long, square ends

Tack the 1×2s in place on the hardboard with hot glue, then secure with ¾" wood screws every 10" or so. Install the screws from the front side of the unfinished screen, countersink them, and fill the depressions with wood putty.

3. **Lay out the dropcloth.**

A plastic painter's dropcloth will protect your floor, but it's also useful for collecting loose glass beads after you apply the screen surface, so use a fresh one without holes. Spread it in a clean area with a smooth floor, and tape the edges down. Set your screen down in the middle.

4. **Apply the basecoat.**

First paint the edges of your screen with a brush, then apply a smooth, even coat of paint to the surface with a roller. It's easiest to just pour the paint directly from the can onto the screen — Jackson Pollock-style — then smooth it with the roller. Let the basecoat dry 24 hours.

5. **Ready the glass beads.**

Pour out your supply of beads into the tub. A cheap plastic dustpan makes a convenient applicator for sprinkling. With a bit of practice, it's easy to get an even sheet of falling beads.

6. **Apply the topcoat.**

After 24 hours, paint the screen edges again, then pour about ½qt of paint onto the screen, distributing it more-or-less evenly. Roller the paint smooth. You want a quick, even, heavy coat.

7. **Sprinkle the glass beads.**

While the topcoat is still wet, sprinkle beads generously over the entire surface. You'll recover any excess later, so go ahead and apply them all, being careful not to miss any spots. Once the paint is dry, the surface is hard to touch up.

8. **Brush off and recover excess beads.**

Remove the excess beads using a soft brush.

Give the screen back a few thumps to dislodge any

remaining loose beads, then stand it upright for a final brushing. Peel off any flash around the edges with your fingers.

Gather up the tarp from the edges into a "sack," and lift it into your tub. Release one edge and slowly work the tarp out from underneath the mass of beads. I recovered 16 of the 25lbs of beads I applied.

9. Install wall hardware.

I put in a row of 4 self-drilling wall anchors behind the top edge of the screen, and drove in their screws, leaving about ½" sticking out from each. Then I just hung my screen on the screw heads. It's easy to adjust horizontally, but not as secure as I'd like. A French cleat would probably be the best solution.

10. Hang the screen.

Put on clean gloves before handling the finished screen to avoid getting oil on it. Lift it into position and hang it in place.

11. Use it!

Turn on your projector and refocus and adjust the image as necessary.

Conclusion

The final surface — glass beads embedded in latex house paint — is surprisingly tough. I was concerned that flexing the screen would cause the beads to flake off, but the latex paint is still flexible after 2 years. I almost think you could apply it to a thin surface that actually rolls up.

Another pleasant surprise was that, at viewing distances, the surface treatment is remarkably tolerant of small imperfections, and does not require a very smooth texture. I believe it could even be applied directly to a textured wall, thus eliminating the need for a separate screen altogether. ◪

CAUTION: Depending on the size and weight of your screen, you may want to get help lifting it. Be careful!

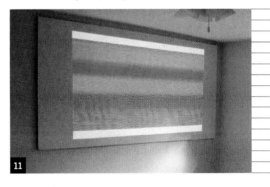

▲ Bare wall   ▲ Just paint   ▲ Paint + Beads

Sean Michael Ragan is technical editor of MAKE magazine. His work has appeared in *ReadyMade*, *c't – Magazin für Computertechnik*, and *The Wall Street Journal*.

HOMEBREW
DIGITAL
3D MOVIES

Build your own
stereo video camera
and 3D viewer.
By Eric Kurland

Photograph by Amy Crilly

**The author shoots a movie
of the photographer with
his 3D dual-camera rig.**

I have two eyes. And because of that simple fact, I also have *stereopsis*, the ability to perceive depth. When I was about 7 years old, I gazed into a View-Master toy and saw an amazing three-dimensional picture, and I was hooked. Today, I create 3D videos, using various homebrew camera rigs and displays. I'll introduce you to a few of my devices, but first, a quick history lesson.

In 1838, British scientist and inventor Sir Charles Wheatstone theorized that seeing with two eyes together is what allows us to see in 3D. Wheatstone deduced that each eye observes a slightly different view of the world, and our brain fuses these two perspectives together, interpreting the parallax differences as depth. He called his discovery stereoscopic vision (literally meaning "to see solid") and built an optical device, the stereoscope, that allowed three-dimensional viewing of pairs of drawings.

With the invention of photography, and later cinema, real-life images could be captured with two lenses and viewed in 3D. The popularity of stereoscopy has persisted over the years. In the 1890s, arcades offered 3D peep shows as entertainment, and the handheld stereoscope was a common item in home parlors — the TV of the Victorian era. The 1950s and 1980s both saw 3D movie "booms" come and go, due to the technical limitations of the times. And currently, in the age of digital video, stereoscopic 3D is seeing a major rebirth.

My own foray into 3D video began a few years ago, after I attended the monthly meeting of the Stereo Club of Southern California. Many of the photographers at the meeting had pairs of digital still cameras mounted side by side for shooting 3D photos, and it occurred to me that I could build a similar hand-held rig for use with small camcorders.

Shooting 3D

Starting with a pair of Sony Handycams, I set out to build a stereoscopic rig. My plan was to make the distance between the lenses, called the *interaxial*, equal to my *interocular*, or the distance between my eyes. This would give a natural-looking 3D depth to my footage, and would allow me to view 3D while shooting, just by looking through both camera's viewfinders. Putting the lenses so close required removing the hand strap from the left camera.

I attached the cameras to a metal bar using quick-release mounts for easy removal, in order to access the tape and battery compartments. I fashioned a bracket from some spare parts to hold both cameras securely at the top and keep the lenses aligned. Inspired by director Mike Figgis' steering wheel-like camera stabilizer (the "Fig Rig"), I bolted together

a pair of photographic flash bars with handgrips, salvaged from a flea market dollar bin, and created a "handlebar" stabilizer. This allows me full mobility with the rig, and puts the center of rotation between the two cameras.

To control recording, I use a device called the 3D LANC Master. Developed by Dr. Damir Vrancic of Slovenia, the 3DLM connects the cameras via the LANC ports and provides simultaneous control of most camera functions. It also keeps the video recording in sync by continuously polling the timing frequency of one camera, and adjusting the frequency of the other up or down to prevent drift. This is very important when shooting 3D, as any time disparity between the camcorders will result in nonmatching left and right views. Schematics and software for the 3DLM are open source under a GPL and are available for free.

Viewing Live 3D

With my camera setup complete, my next task was to build a portable stereoscopic video monitor, so others could watch live 3D during shooting. In movie theaters, stereopsis is achieved by projecting left and right images through two oppositely oriented polarizing filters onto a reflective screen. By viewing through 3D glasses made from matching polarizers, each eye sees only the corresponding projection. I decided to use the same principle for my monitor.

I started with two small LCD monitors capable of showing NTSC video, the kind that are strapped to the back of car headrests. The video output from each camera is input to one of these monitors. Because LCDs have a polarizing layer, these displays appear black to one eye and visible to the other when viewed through polarized 3D glasses. I found that the monitors had a clear plastic protective sheet glued over each LCD. These had to be carefully peeled up and removed, as they were depolarizing the light from the screens.

On one display I needed to flip the picture horizontally like a mirror image, so I opened the case and wired pin 62 of the PVI-1004C LCD controller chip to ground. I attached the LCD displays to each other at a 90° angle, their screens facing inward, and mounted a piece of half-mirrored glass between

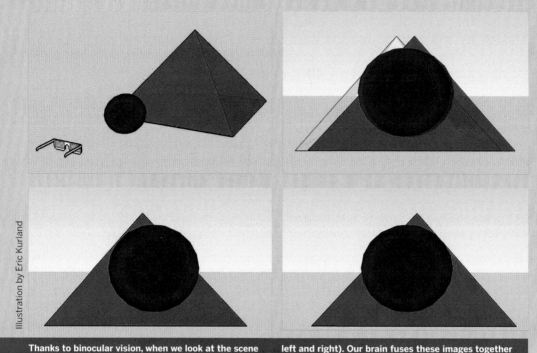

Illustration by Eric Kurland

them. This glass superimposes the reflection of one screen on top of the other. When viewed through polarized glasses, the reflected image and its polarization are reversed, each eye sees only one screen, and we have live 3D video of whatever is being shot.

Editing 3D

Shooting with two cameras creates two individual video files, which are digitized into the computer for editing. First, I use the freeware application Stereo-Movie Maker, developed in Japan by Masuji Suto, to correct misalignments in my footage, which can cause eyestrain.

In StereoMovie Maker, I am able to load both the left and right videos and visually transform, scale, and rotate them while viewing in 3D with anaglyph glasses. Anaglyph is the method in which the two pictures are combined into a single image with one eye in red and the other in cyan. Primarily used in printed stereoscopy, anaglyph also provides a means of viewing depth on any computer screen using inexpensive red-cyan glasses.

Once satisfied with the alignment, I save my videos as a single file, formatted side-by-side in a split screen, and twice as wide as a normal video

picture. I prefer this format, as it ensures that the two views always remain in sync throughout the editing process. The footage can be cut together in any standard video editing program. My system is PC-based, so I use Adobe Premiere, but the same techniques would apply to a Mac Final Cut Pro system. One thing to take into account when editing 3D is that drastic depth changes between consecutive shots can cause eyestrain.

To watch my completed movies in 3D, I use Peter Wimmer's excellent Stereoscopic Player program, a full-featured media player for stereo video files that converts on-the-fly to the many different viewing formats required by stereoscopic displays and projectors. Both Stereoscopic Player and StereoMovie Maker are Windows-only applications, but they will run on Intel Macs running Windows.

Showing 3D

In order to show 3D video to audiences, I have a dual-projector setup, just like the 3D theaters of the 1950s, using two projectors, polarizers, and a silver screen. The only real difference is that my projectors are small DLP digital models, and my "film" is a file played back by computer. This arrangement works well for large audiences, but I also wanted some method of carrying my 3D movies with me to show at a moment's notice — a portable stereoscopic media player.

The Sony PSP looked like it would be the answer. The PSP can play MPEG-4 files from a flash memory card and it has a nice, wide screen — wide enough to hold side-by-side-formatted left and right images.

In fact, the PSP is just about the same size as a standard vintage stereo card. I decided it would be cool, and somewhat steampunk, to mount a PSP onto a circa-1904 stereoscope.

As luck would have it, the PSP fit almost perfectly between the two card-holder wire clips. I didn't want to physically alter the viewer, as it's an antique, and I wanted the PSP to be removable, so I cut two loops of thin velcro strapping, just long enough to go around the PSP and hold it firmly to the slide bar.

To get my videos onto the PSP, I converted them to x264 compressed MPEG-4 files at the PSP's screen resolution of 480×272 pixels, and copied them to a PSP Memory Stick. Sure enough, side-by-side video files played on the PSP and viewed through the stereoscope's eyepiece are seen as a single three-dimensional movie.

The whole setup works perfectly. I can easily carry around a bunch of my homemade 3D movies on a Memory Stick in my pocket, and quickly show them to people through the "PSPscope" — a perfect marriage of 19th- and 21st-century technologies.

1. Build the 3D dual-camera rig.

1a. On one of the video cameras, remove the hand strap. This will be the left-eye camera.

1b. Attach the quick-release mounts to the bottom of the camcorders. Make sure that the mounts are perfectly straight, and tighten them down with a screwdriver or a coin. Position the cameras as close to each other as possible on the twin-camera bar, and tighten the thumbscrews until both cameras are rigidly secure (Figure A, next page).

1c. Optional: If you have the hot-shoe microphone extenders, place them on the cameras and tighten the thumbscrews. Attach the extenders to the 6" flat steel bracket by placing bolts through the bracket and tightening a wing nut onto each bolt (Figures B and C). The mic extenders should be parallel. Make sure that the camera lenses are aligned.

1d. Optional: Create a camera stabilizer by placing the flash bars end to end and attaching them to the

MATERIALS

FOR THE 3D DUAL-CAMERA RIG:

Matching pair of compact video cameras Any pair of camcorders will work, but the 3D LANC Master will maintain sync only on certain older Sony MiniDV cameras. Mine are Sony DCR-PC100s.
Twin-camera mounting bar from a specialty photography store
Quick-release camera mounts (2)
3D LANC Master camcorder controller Build your own from the schematics at dsc.ijs.si/3dlancmaster or order one pre-assembled from inddd.com.
Screwdriver
Hot-shoe microphone extenders (2) (optional) These were an accessory with the camcorder microphones.
6" flat metal bracket (optional)
Adhesive velcro squares (optional)
Camera flash bars with handgrips (2) (optional) Any 2 should do; mine are a mismatched pair from a flea market.
Dog training clicker (optional)

FOR THE LIVE 3D MONITOR:

Audiovox EX50 5" LCD video monitors (2) You should be able to use other LCD panels, but you'll need to check the polarization.
5"×7" half-mirrored glass aka flat glass beam splitter. You can use a cut piece of window glass but the results may not be as good.
Short length of sheathed wire such as telephone wire
Self-adhesive velcro squares
½" velcro strapping
10"×6" piece of plywood
1½"×8½"×3½" metal L-brackets (2) I had these in a scrap parts bin; you'll need to find a similar substitute or perhaps use wood blocks instead.
½" wood screws (6)
Linear polarized 3D glasses There are 2 types of polarized 3D glasses, linear and circular. Only the linear kind will work with these LCD panels.
Precision screwdriver set Phillips and flathead
Fine-tipped soldering iron and solder
Magnifying glass

FOR THE PSP STEREOSCOPE VIEWER:

Sony PSP
Holmes-style stereoscope vintage or new
½" velcro strapping
Scissors
Pliers
Adhesive rubber foot or nut and bolt (optional)
Electric drill (optional)

bottom of the twin-camera bar with the handgrips pointed upward (Figure D).

1e. Remove the cameras from the bar. Insert blank tapes and charged batteries. Place the cameras back on the bar. Use velcro to attach the 3D LANC Master to the rig (Figure E).

1f. Connect the 3D LANC Master cable to the cameras' LANC ports (Figure F). Power up the Handycams, use the 3D LANC Master's controls to reset the cameras and get them in sync, and you're all set to record.

TIP: If you can't get a 3D LANC Master, you can still shoot 3D video with your twin camcorders. Use a dog training clicker to make a sync "pop" on your soundtrack. In editing, you can use the clicker peaks on the audio tracks to align the videos. Just be aware that on long takes your scenes may drift out of sync.

2. Make the live 3D video monitor.
2a. Open the case of one of the LCD monitors. First, use a small flat-head screwdriver to pop the

rubber feet out of their holes, and remove the 4 Phillips-head screws from the case; 2 of these are under the hinge, so rotate the hinge until you see the rubber feet. Make sure to note which screws go where, as they are different sizes (Figure G).

2b. Carefully separate the 2 halves of the case. Remove the 2 Phillips screws holding down the printed circuit board (PCB) below the screen (Figure H). Very carefully flip over the PCB. Locate the LCD timing controller chip, a 64-pin IC marked "PVI-1004C." Using a magnifier, find pin 62 and follow its trace to a small board-mounted resistor. Carefully solder a short piece of wire from this resistor to the ground point on the PCB, next to the ribbon connector (Figure I).

2c. Close up the case, replacing the screws and rubber feet.

2d. Using the thinnest flat-head precision screwdriver, pry up a corner of the protective plastic sheet covering the LCD screen (Figure J). Slowly peel the plastic from the case. Repeat with the second LCD.

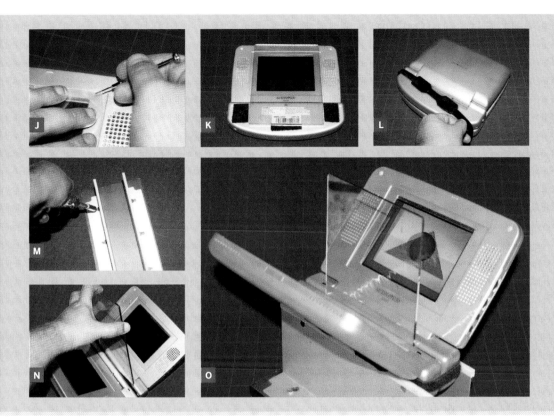

2e. Place the half-mirrored glass against the hinged base of the first monitor so that it extends over the display. Peel and stick several velcro squares onto the base around the glass (Figure K). Repeat with the second monitor, using the opposite halves of the velcro squares.

2f. Remove the glass and align the second monitor so that it faces the first. Press the 2 bases together, fastening the velcro.

2g. Cut a length of velcro strap long enough to encircle the base with a few inches of overlap. Wrap the strap around the base tightly and fasten it to itself (Figure L).

2h. Make a platform to hold the monitors. Using wood screws, fasten the L-brackets to the plywood so that they create a 2¼" opening (Figure M). Fit the monitors into the opening so that they stand upright, then bend each one back on its hinge 45°, so that the 2 form a 90° angle.

2i. Slide the glass between the bases of the LCDs (Figure N). It should clear the velcro squares and fit very snugly.

2j. Attach the power supplies, connect the video inputs of the LCDs to the video outputs of the cameras, and power everything up. Look down through the glass at one LCD, and you should see the other display reflected in the glass (Figure O). Tweak the angle of the monitors until the images are perfectly superimposed. Adjust the brightness levels of the panels until they match.

2k. Put on some 3D glasses and you can watch live stereoscopic video.

3. Make a PSP stereoscope.

3a. Make sure your PSP has the latest system firmware that supports playing full-screen MPEG-4 files. To check for this, go to the PSP's system menu and select Network Update.

3b. Cut two 10½" lengths of velcro strap. Loop each piece, and overlap the ends 2½", fastening securely (Figure Q).

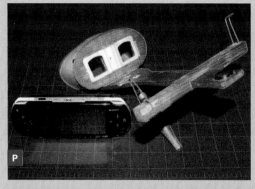

P

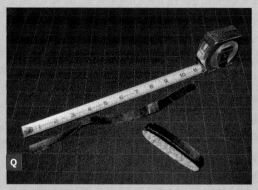

Q

R

S

3c. Place the PSP on the sliding bar of the stereo-scope, between the wire card-holder clips. If the PSP doesn't fit, use pliers to bend the wires.

3d. Holding the PSP firmly in place, slide a velcro strap over one end of both the PSP and the sliding bar (Figure R). The fit should be tight, but still allow you to move the strap on and off easily. If necessary, adjust the overlap to tighten or loosen the strap. Fit the second strap on the opposite end.

3e. Position the straps right at the edges of the screen without obscuring it (Figure S).

3f. Optional: Because of the weight of the PSP, the sliding bar has a tendency to slide off the back of the stereoscope. To prevent this, stick a rubber foot on the main bar, near the end.

3g. Save your 3D videos as PSP-compliant MPEG-4 files, using x264 compression, at a resolution of 480×272 pixels. Place your files into the root directory of a Memory Stick Duo. Fire up the PSP, load a side-by-side stereoscopic video, and enjoy a 3D movie.

RESOURCES
Stereo Club of Southern California: la3dclub.com

3D LANC Master free schematics and software: dsc.ijs.si/3dlancmaster

StereoMovie Maker free download: stereo.jpn.org/eng/stvmkr

Stereoscopic Player: 3dtv.at

Eric Kurland is an award-winning independent filmmaker, President of the LA 3-D Club, Director of the LA 3-D Movie Festival, and CEO of 3-D SPACE: The Center for Stereoscopic Photography, Art, Cinema, and Education. Operating out of a Secret Underground Lair in Los Angeles, he specializes in 3-DIY (do-it-yourself 3-D) and consults on every stage of 3-D production and postproduction, from development through exhibition. His 3-D clients have included National Geographic, Nintendo, and NASA's Jet Propulsion Laboratory. He has worked as 3-D director on three music videos for the band OK Go, including the Grammy nominated *All Is Not Lost*, and he was the lead stereographer on the Oscar-nominated 20th Century Fox theatrical short *Maggie Simpson in The Longest Daycare*. In 2014, he founded the nonprofit organization, 3-D SPACE, which will operate a 3-D museum and educational center in Los Angeles. He sometimes wears a gorilla suit and space helmet.

Fun
AND Games
PART 8 >>>>

Daryl Hrdlicka teaches and performs historical reenactments at events and pageants that hearken back to the Civil War and the Victorian era. In his project for Make:, he delves even deeper into the past. At the Jeffers Petroglyphs historic site in Minnesota, he teaches young visitors how to use the atlatl—a hunting tool that's even older than the bow and arrow. Daryl will teach you how to make an atlatl and darts, as well as how to use them.

David Simpson deals with projectiles, too—in this case, model rockets. A private pilot and a flight instructor, he wanted to measure the G-forces withstood and the altitudes reached by his rockets. David created a handmade aerospace device that you can make for around $5 and fit into the payload chamber of your model.

Frank Yost always loved remote control cars, but the artist in him yearned for something more classic looking. When he discovered the tether racing cars, or "spindizzies," of the 1930s, he decided to adapt that look to his own r/c cars. With some sheet metal and a few workshop tools, you can follow along with Frank's instructions to create your own classic cars.

You'll find more classic styling in Tom Martin's wooden mini-yacht. Tom is more likely to be found sailing among the clouds with the gliders he crafts for his company, Aerosente. But here he'll show you how to build an elegantly simple wooden boat to sail across pools and ponds.

Edwin Wise likes to scare you. He's a special effects technician and make-up artist who has worked on several movies and plays. During Halloween, he's usually busy setting up scare effects and performing as an actor at haunted houses. As Edwin points out, you can come up with all sorts of ways to scare people, but nothing works better than a good, old-fashioned startle. His Boom Stick delivers a sudden, loud noise that will have your victims jumping out of their socks.

John Mouton shows you how to have more fun than you've ever had before with remote control cars. Sure, it's always fun to watch them race around the room. But it's even more fun to get "inside" and sit in the driver's seat, which you can do with an onboard camera and a pair of VR goggles. John will show you how it's done.

The French existentialist philosophers would probably have loved Brett Coulthard's "Most Useless Machine." Does it signify the absurdity of existence? Or is it just absurdly entertaining? Whatever conclusions you draw, you'll have fun building it and watching the expressions on your friends' faces when they see what happens. The Most Useless Machine even made an appearance on *The Colbert Report*, where host Stephen Colbert was so delighted he kept it!

RETRO R/C RACER

By Frank E. Yost

Photograph by Sam Murphy

RIVETING TALE

Using scrap sheet metal and pop rivets, you can construct a model 1930s British Midget racer that combines vintage "tether car" styling with modern R/C capabilities.

Radio control (R/C) toys are fun, but their plastic bodies are so obviously mass-produced. To create something more interesting, I wanted to find a metal toy racer 10 inches or longer that I could transplant some R/C insides into. I soon discovered that the best candidates were all precious collectibles.

I decided to build a metal body myself and soon stumbled upon an old hobby called "tether cars," a type of model racer that predates R/C (see sidebar, page 291). After ordering blueprints from a tether car enthusiast group in England, I learned how the old toys were made, and then ported the style to my R/C project. With a few modern adaptations, such as pop rivets instead of solder, and coil spring shocks, I built a couple of vintage-style R/C cars that would make any antique dealer jealous. Here's my latest.

Frank E. Yost is an amateur artist who lives in Andover, Minn. His interests include drawing, woodcarving, bronze casting, welding garden art, and building tether cars. In 2002 he won an award for his comic book *Cookie and Butch*.

BLOCK, SHOCK, AND BUCKLE

R/C hobby manufacturers produce countless prefab parts.
Some are great for combining with parts you make yourself.

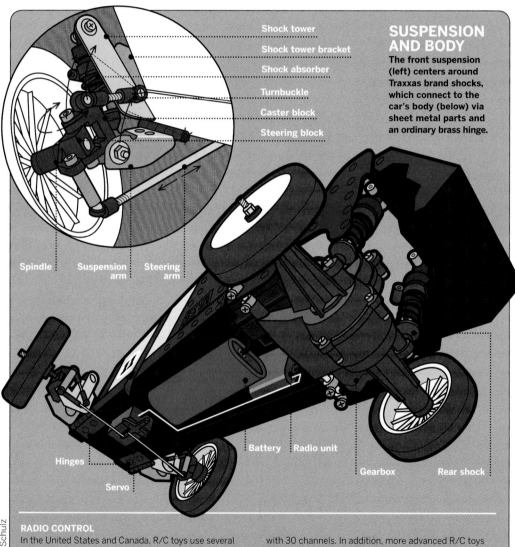

Shock tower
Shock tower bracket
Shock absorber
Turnbuckle
Caster block
Steering block

SUSPENSION AND BODY

The front suspension (left) centers around Traxxas brand shocks, which connect to the car's body (below) via sheet metal parts and an ordinary brass hinge.

Spindle | Suspension arm | Steering arm

Hinges
Servo

Battery | Radio unit

Gearbox | Rear shock

Illustrations by Nik Schulz

RADIO CONTROL

In the United States and Canada, R/C toys use several radio bands, most commonly the 27MHz band, which has 6 channels spread between 26.995 and 27.255MHz. When users play near each other, they avoid accidental signal hijackings by flying colored flags from their toys' antennas to indicate the channel they're using. R/C aircraft can also use a 72MHz band that has 50 channels, and surface vehicles such as cars and boats can use a 75MHz band with 30 channels. In addition, more advanced R/C toys requiring an FCC amateur license operate at the 50MHz band (10 channels) and 53MHz band (8 channels).

R/C toys transmit control information over these carrier frequencies in various ways, including AM, FM, and PCM encoding schemes. With cars, the 2 values typically transmitted are speed and direction, which control the car's main motor and steering servomotor, respectively.

SHEET METAL MODELING: A MINI-PRIMER

Sheet metal is a major component of junk everywhere, and with pop rivets and simple skills, you can turn this durable material into a model racer — or almost anything else.

TYPES OF SHEET METAL

Regular sheet metal made from steel, is very forgiving: bend it wrong, and you can just hammer it flat again and start over. Thickness varies; I use between 18 and 24 gauge. Most sheet metal is coated to prevent rust. Tin-plate steel, also called simply "tin," has a uniform shine. Galvanized steel, coated with zinc, sports a distinctive crystallized pattern.

Stainless steel tends to be harder than regular steel, so it's more difficult to work. If your tinsnips aren't strong enough to get through it, try cutting with a rotary tool (Dremel).

Aluminum is lightweight and soft, easy to work, but easy to dent and scratch. Aluminum also weakens when bent in the same spot too many times. But it has a beauty unto itself, and was once even considered a precious metal.

OPERATIONS

Bending For most bends, you should use a sheet metal brake. Cheaper brakes can let the metal slip a bit, which puts the fold in the wrong place, but you can discourage this by first applying masking tape along the fold line and clamping the metal down. To bend small pieces of stiffer metal such as stainless, clamp them in a vise between wood blocks and hammer them over.

Cutting Make paper templates for all the shapes you're cutting. Then either stick them directly onto the metal with double-sided tape, or glue them to cereal-box cardboard, cut out, and trace around them on the metal with a fine-point marker. To help anchor larger templates, cut holes in the middle and tape over them. Cut shapes out with tinsnips or a Dremel cutting wheel, but note that a Dremel can sometimes make tape melt and slide around. File or sand the edges smooth after cutting.

Drilling Center-punch holes before you drill, and use a high-quality drill bit with a sharp tip. I've bought some surprisingly bad drill bits, which wind up dancing around on the metal and biting in at the wrong place. Drill at ⅛" for standard pop rivets.

Riveting Insert a rivet through your hole. For extra strength, back it with a washer on the other side. Then squeeze with the pop rivet tool. To remove a rivet, drill it out using the same ⅛" bit you made the hole with. To test-fit a rivet, you can hold it in place temporarily with a nail.

PROTOTYPING WITH CARDBOARD

Anything you want to make with sheet metal you can prototype first with cardboard. Just cut the pieces out, then fold and tape them together. This is an easy, fast, and cheap way of checking out a design before building.

SET UP.

MATERIALS

Steel sheet metal, tin or galvanized, about 20 gauge I used the case from an old Compaq tower.

Thin steel sheet metal, about 26 gauge, 2'×2'

Thin aluminum I used an 18"×13" cookie sheet.

Thick aluminum I used an old street sign.

Stainless steel I used the reinforcing brackets from an old desk, around 18 gauge.

Perforated sheet metal I used the ventilation panel from a very old computer.

1/8" metal rod, about 12" long I salvaged this from an old typewriter.

Wire coat hanger

Old inner tube

Double-sided, masking, and electrical tape

Scrap cardboard for templates

Scrap wood or plywood not balsa

Thin wire or twist-ties

Mini toggle switch and wire (optional)

Double-sided foam tape (optional)

1/2"×3/4" aluminum angle 1/16" thick, 3' length

1/2"×1/2" aluminum angle 1/16" thick, 4' length

1/8" aluminum pop rivets and pop rivet washers (100)

1½"×1¼" brass hinges (2)

5/32" round copper tubing, 6"

Assorted small machine screws and bolts

FROM A HOBBY STORE OR R/C ENTHUSIAST

Any donor R/C car will work. I used mostly parts from a Bandit, from a box of used parts I bought from a neighbor for $25. If your parts differ from the following, you may have to adjust dimensions and construction details to fit.

Traxxas PARTS: Steering blocks and wheel spindles (2) part #1837 Pro-series caster blocks, 25° (2) part #1932

Big bore shocks, short (2) part #2658

Wheels, front (4) part #2475

Tires, 2.1" spiked, front (4) part #1771

Suspension pin set, hard chrome part #1939

Tom Cat/Spirit gear-box assembly various parts find one on eBay, assembled

Hot Bodies purple threaded touring shocks (2) part #HB24010

Tamiya TT-01 turnbuckles for R/C cars (2)

2-channel radio control (R/C) system and compatible electronic speed control (ESC) I used a Futaba T2PH with a Tazer 15T ESC, bought on eBay for $20.

¾" turnbuckles for R/C cars (2)

Pack of miscellaneous brass bushings

Pack of miscellaneous lock collars aka brass wheel collars or Dura-Collars

Lacquer paint and primer

TOOLS

Pop rivet gun, sheet metal bending brake, tinsnips, rotary tool with cutting and grinding wheels such as a Dremel, hacksaw, disk and belt sander, drill and 1/8" drill bit, metal files and sandpaper, hammer, pliers (regular and needlenose), wire cutters, center punch, ruler, C-clamps, scissors, drafting triangle and tri-square, screwdriver for R/C parts, vise, extra fine-point marker, white glue, a penny, a couple of nails, protective eyewear, fine paintbrushes, solder-ing equipment (optional) if installing switch

MAKE IT.

BUILD YOUR
VINTAGE R/C CAR

START ⋰ Time: **A Month of Evenings** Complexity: **Medium**

1. PRINT OUT THE BLUEPRINTS

Download the project blueprints at makezine.com/projects/retro-rc-racer and print out at full size.
The images are oversized (16"×24"), so with a letter-size printer you'll need to print multi-page and tape
the pieces together.

2. MAKE THE FRAME

2a. Cut section 1A out of the blueprints, glue to some scrap plywood,
and let dry overnight. This is what
we'll be building on.

2b. Use a hacksaw to cut two 9"
lengths of the ½"×¾" aluminum
angle. Use masking tape to hold the
2 pieces in their places on top of
Section 1A.

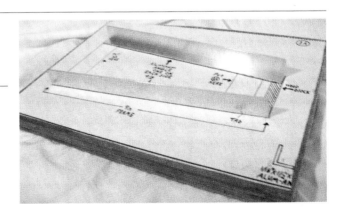

2c. Cut 3A out of the blueprints, trace the template onto tin or
galvanized sheet metal, and carefully cut out the part using tinsnips. Shape by folding along the dotted lines using a brake and
pliers. Trace, cut, and shape part 2A. Note that the longest bend
goes the opposite way from the other 3. Sand a small piece of
wood to ¹¹⁄₁₆" thickness and cut to the shape of template 4A.
Slide wood block into 2A.

2d. Place the 2A/4A combo into the front of the frame and 3A
into the back. Drill ⅛" holes in parts 2A and 3A where indicated by
the + symbols, and pop-rivet the frame together. Trace around a
penny at the front end of each aluminum piece, then use a belt
sander to round out the front of the frame.

Photography by Frank E. Yost

2e. Cut and drill brass hinges following pattern B7. The sides of the hinges to cut are the ones with 2 loops around the pin, not the ones with 3. Slide the uncut sides of the hinges under the sides of the frame about ½" from the front. Tape and clamp everything back down without blocking the holes on 2A and 3A, then re-drill those holes, going through the frame and the hinges. Unclamp everything, and then pop-rivet the hinges to the frame, using washers underneath for added strength.

3. MAKE THE FRONT SUSPENSION

This is the most difficult part of the project — everything must be adjusted just right. Old tether cars had leaf springs, but I wanted a buggy-style independent spring suspension.

3a. Cut out sheet metal parts B3 and B4, the suspension arms that connect the hinges to the shocks and caster blocks. Drill ⅛" holes where indicated, and shape using pliers. Pop-rivet them to the hinges. Cut the front shock towers (B1 and B2) out of something strong. These parts take a lot of stress, so don't use thin steel or aluminum. I used stainless, cutting the parts with a rotary tool and then filing them down. Use a vise and a hammer to bend the shock towers following the patterns. Bend them to only about 80° and make sure the 2 angles are the same. These towers connect the frame to the tops of the shock absorbers.

3b. Drill towers B1 and B2 with the ⅛" holes indicated by arrows on the blueprint. Pop-rivet to the car frame, over the hinges. Gauging with a drafting triangle, pivot each tower to lean slightly forward so that its top edge lies ⅜" from vertical. Clamp them in position, then drill and rivet the 2 holes next to the existing rivet to secure the towers to the frame.

3c. Cut and bend shock tower brackets B5 and B6 out of the same metal as B1 and B2. Clamp them into place on the front of the frame so that the flat sides sit parallel to the back of 2A and the side tabs sit flush against the tower. Drill and rivet the bracket tabs to the towers and the frame. Don't drill all the way through the wood; just go far enough to accommodate the rivet, which serves as a nail here.

3d. Use machine screws and bolts to attach each front shock between the top of its tower and the front of its suspension arm. Bolt the bottom of each caster block into its arm. Assemble the car's turnbuckles, substituting shorter ¾" rods for the originals, then connect one end to the top of the caster block and bolt the other to the back of the shock tower at about the same horizontal level. Connect the steering blocks to the caster blocks using suspension pins.

4. MAKE THE BODY

4a. If you're using a Traxxas Tom Cat/Spirit gearbox, trace the cab pattern 1C onto thin sheet metal, cut out using tinsnips, and flatten using a hammer and scrap plywood. With a smaller gearbox you can use pattern 2C.

4b. Smooth the edges of the metal using a file and sandpaper, then bend the cab following the pattern. Use a brake for the large bends and pliers for the smaller ones. Make sure the body is square. The blueprints also include an optional side cab, part 3C, which you can add on for looks.

4c. Trace pattern D1, for the hood, onto aluminum, then cut out slowly using a cutting wheel (don't use tinsnips). File edges. Use a brake and pliers to bend the hood along the lines indicated on D1.

NOTE: Beginners might cut piece D1 out of tin, which is more forgiving if you need to tweak the bends.

4d. Cut out template G1, the radiator grille, and slide it into the front of the hood. Trim the template to fit, then trace onto perforated sheet metal. Cut out the sheet metal grille and check the fit carefully before riveting. Drill and rivet the grille onto the hood.

4e. Cut paper template D1 into parts D2 and D3, as indicated. Tape the scuttle template, D2, onto the hood, and then trace onto the metal. Use a cutting wheel to carefully cut the hood along the line you just traced, then file down any marks.

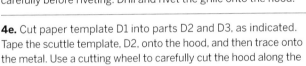

4f. Slide the cab, scuttle, and hood into the frame. C-clamp the scuttle to the frame, remove the hood, and then drill and pop-rivet the cab and scuttle into place. Cut, smooth, and bend hood lip F1. Fit under the scuttle, then clamp it down, drill, and rivet into place.

NOTE: Be sure that F1 points slightly downward so that the scuttle will be at the same angle as the hood.

5. MAKE THE REAR SHOCK TOWERS AND TRUNK

5a. Cut two 2" lengths of the ½"×½" aluminum angle. Drill according to patterns 4C and 5C, then rivet side-by-side to the inside back of the cab, centered 1½" apart and vertical. These are the rear shock towers.

5b. Cut 2 more pieces of the ½" aluminum to 3½" in length, and drill as parts E3 and its adjacent E2, the trunk brackets. Rivet to either side of the back of the cab, so they stick out vertically with an outside distance of 3½" apart.

5c. Out of thin steel, cut out 2 trunk side pieces following template E1 and 1 trunk cover piece, E2. Drill, bend using a brake, and rivet together. Fit the assembled trunk over the trunk brackets, drill through from the sides, and rivet into place.

NOTE: We'll get to installing the gearbox and rear wheels later.

5d. For the antenna bracket, fold a 1¾"×¾" piece of sheet metal around some ⁵/₃₂" copper tubing using some pliers. Cut 1" of the tubing to remain in the bracket, then center-punch and drill holes on each side of the folded tab. Drill and pop-rivet the bracket next to the trunk. The plastic radio antenna will slide perfectly into the tubing.

6. FINISH THE FRONT END

6a. Trace windshield template H1 onto thick aluminum and cut out with a cutting wheel. Sand and file all the edges smooth. File the inside edges by hand. Trace and cut windshield bracket H2 onto thin steel. Bend the tabs on H2 using pliers, and check the bends for proper fit by placing it on top of the scuttle. Cut 2 side curtains H3 out of thicker steel, bend in a vise, cut the inside using a cutting wheel, and file smooth outside and inside.

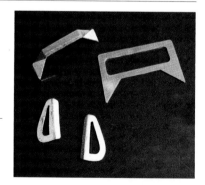

6b. Fit bracket H2 onto the scuttle using small clamps, then clamp windshield H1 onto the bracket. Unclamp the pieces from the scuttle, then drill and rivet together. Clamp, drill, and rivet the side curtains H3 to the windshield, then position this assembly back on the scuttle and drill through the bracket and scuttle. Bolt into place using small nuts and bolts.

NOTE: Use an old rag in the vise to prevent the metal from getting marred.

6c. Optional: I wanted a switch on the dash, so I made a bracket out of stainless, following template I1, and used a large bit to drill the hole for the toggle. I bolted the switch to the scuttle, and hooked it up by cutting the receiver unit's power wires and soldering them to the switch's terminals.

6d. As in a vintage car, the hood is held down with a strap. Make it by cutting an 11"×¾" strip from an old bicycle inner tube. Then cover each end with electrical tape and cut a slit for attaching it to the hooks under the car. On another model car I used leather instead of rubber, making it more authentic.

6e. Make the hood strap hooks out of a wire coat hanger. Cut 2" pieces, then use needlenose pliers to make hooks on one side and loops on the other. Pop-rivet the loops under the frame.

7. INSTALL THE GEARBOX, STEERING, AND RADIO UNITS

7a. Cut off the ball joint pivot that protrudes from the front of the gearbox assembly. Thread a ⅛" metal rod through the horizontal sleeve on the gearbox that sits just under the old ball joint.

7b. Attach the rear shocks to the back of the gearbox, then drill and bolt the other ends to the rear shock towers. Connect the front of the gearbox by drilling holes in the frame for the metal rod to thread through, then trim the excess length and secure it with lock collars at each end. Put brass bushings onto the steering shafts, then install the front wheels and tires.

7c. Cut frame bracket 5A out of thick aluminum and rivet it flat into the frame. Bolt the servo unit to 3A and 5A so that the steering rods run underneath the frame. To boost undercarriage clearance, I raised the steering unit up on ½"-thick wooden spacers. Install the receiver and speed control boxes behind the servo, on opposite sides of the frame, under the scuttle. I cut and riveted small brackets to hold them, but you can also just use double-sided foam tape.

7d. Use scrap metal and rivets or bolts to connect the rods up to the steering block arms. Or, if the rods are long enough, you can bend them up to reach. I used some junk brass rod with rivets.

7e. Connect the battery unit and install it in front of the servo. I drilled holes in 3A and 5A, looped wire tightly between the 2 on either side, and used more wire to suspend the battery between the 2 loops.

8. TEST, PAINT, AND REASSEMBLE

8a. Adjust the turnbuckles so that the front wheels are straight. Confirm that everything is connected and test the car out with the radio.

8b. If everything works properly, dis-assemble the body pieces and paint with lacquer, starting with a coat of primer. I used Plasti-Kote Burgundy Metal Flake for most of the body and black for the inside. In honor of the issue of MAKE where this article appeared, I painted on a number 11.

8c. Let lacquer dry, and reassemble the car. You're done!

FINISH ⊠

NOW GO USE IT »

MAINTENANCE AND ADJUSTMENTS

Keep all moving parts working freely, with no rubbing of any kind. If a part is rubbing the wrong way, adjust it with pliers or file it down. Also, make sure the front shocks don't press down too hard on the hinges, or they won't rebound fast enough.

Make sure the nuts and bolts that hold the caster blocks in the front suspension arms aren't too tight. The block should be able to pivot; if not, add some washers and oil to the joint.

Photograph by Sam Murphy

REMOTE POSSIBILITIES

Photograph from *Model Craftsman* magazine, 1940-1941, courtesy of tethercar.net

TETHER CARS: A SHORT HISTORY

Tether cars, also known as spindizzies, are the miniature racing car hobby predecessor to the R/C era. Cars race one at a time around a circular track, tethered to a steel pipe that sticks up in the center, and accelerate with each lap. The hobby started in California in the 1930s, and back then the 10cc gas-powered cars would often exceed 50mph. Many cars looked like the midget racers (sprint cars) that were popular at the time.

Tether cars have no brakes. To stop them, the car's owner lowers a stick, flag, or broom over the car's path as it passes underneath. This knocks down what looks like an antenna sticking up in back, but is actually a cut-off switch for the fuel or electrical supply.

During World War II, tether car racing was put on hold in the United States, but it began flourishing in England. The British government grounded toy gas-powered airplanes and kites, fearing that spotters might mistake them for Nazi aircraft (I also wonder if they wanted to hide their near-ground radar capabilities). Then in 1942, D.A. Russell and A. Galeota published a design in the magazine *Aeromodeller* for a simple spindizzy powered by a 2.55cc gas engine that was originally built for model airplanes. It was an instant success, and tether car clubs soon formed all over England.

Tether cars still have a following today, with enthusiasts divided into two camps: collectors and vintage-style builders who love the aesthetics of model making, and competitive racers more interested in competition and speed records. Traditional tether cars look like real cars, but the ones that compete today at dedicated tracks in the U.S., Europe, and Australia have more abstract designs, optimized for speed. Some of these rockets-on-wheels can top 200mph. That's fast! (But personally, I'm more interested in the artistic side.)

Plans for vintage tether cars are available in books and online (see Resources), but they're tough to build today because some of the transmission and drivetrain parts haven't been manufactured

for more than 60 years. You can still construct the bodies, though, and I think that steering and accelerating via radio control is more interesting than running around in a circle anyway. A vintage-style tether car body with modern R/C combines the best of both worlds.

RESOURCES

Spindizzies: Gas-Powered Model Racers by Eric Zausner, EZ Spindizzy Collection, 1998.

Vintage Miniature Racing Cars by Robert Ames, Graphic Arts Center Publishing Company, 1992

Reprints of old British and U.S. tether car books: tethercar.com

American Miniature Racing Car Association: amrca.com

Retro Racing Club, Louth, Lincolnshire, U.K.: +44 (01507) 450325

➕ More resources at makezine.com/ projects/retro-rc-racer.

PARADOX BOX

Last year I saw a video of the "Leave Me Alone Box" built by Michael Seedman. Flip its switch on, and an arm reaches out of a door to turn the switch back off. That's what it does, that's *all* it does, and it will not stop until its circuit is dead.

I had to have one of my own, so I made one. Seedman's design uses a microcontroller to run two servomotors: one to open the lid, and another to push the switch. This makes for an impressive performance, but seemed too complicated, and actually, his circuit remains powered even when the box is idle.

For existential purity, I wanted a super-simple machine that *really* turned itself off. So I came up with a single-motor design controlled by a 555 timer chip, with a curved arm that both lifts the lid and flips off the switch. I called it the "Most Useless Machine" and posted it on Instructables along with a YouTube video of the box in action. The project soon went viral, attracting millions of viewers, thousands of comments, and many builds and design variations. Whew!

Along the way, Instructables member Compukidmike came up with an even simpler version that dispenses with the 555 circuitry entirely by using a gearmotor and two switches. The resulting project, presented here, is the ultimate in technology for its own sake, a minimal assemblage of parts that, through its one meaningless act of defiance, speaks volumes.

Brett Coulthard (frivolousengineering.com) has a short attention span, which explains his varied interests. He lives and pushes buttons near Moose Jaw, Saskatchewan.

Turn-ons and Turn-offs

The Most Useless Machine has a toggle switch, a motor, and a micro lever switch. The toggle runs the motor forward or backward, depending on which way it's flipped. Running forward, the motor swings a wooden arm that flips the toggle switch the other way. The motor then runs backward, and as the wooden arm returns to its original resting position, it triggers the lever switch to cut the power off.

SUICIDE-BOT 3000

BONUS: Download this custom design and stencil by Rob Nance for your useless machine at makezine.com/ projects/the- most-useless- machine

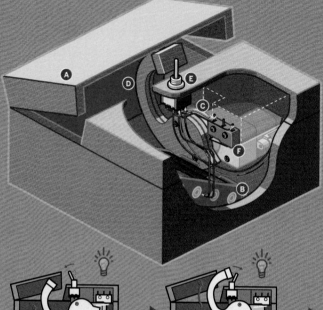

1.   2.   3.

The plain-looking box **A** has a hinged lid and a toggle switch on the outside.

Inside the box, a 4×AA battery pack **B** powers the motor.

The motor **C** is a gearmotor whose built-in gearing gives its small DC motor enough torque to move a wooden arm.

When the motor runs forward, the curved wooden arm **D** swings up, pushes open the lid, and knocks into the toggle, switching it the other way.

The toggle switch **E** is a double-pole double-throw (DPDT) switch that's wired to feed power to run the motor forward or backward, depending on how it's switched.

The micro lever switch **F**, located under the wooden arm, is a single-pole double-throw (SPDT) switch. It's wired from its normally closed (NC) pin to cut the motor-backward power when the wooden arm pushes down on its lever.

CONTROL SEQUENCE:
1. When the machine is idle, the circuitry is fully powered down. The toggle switch is in its "off" position. Inside, the lever switch is held down in its "off" position by the wooden arm.

2. When a person flips the toggle switch "on" (forward), it powers the servo to rotate the arm into the toggle switch and flip it back "off" (reverse).

3. Once the toggle switch is flipped "off," the servo moves in reverse until the arm rests on the lever switch, which is the real "off" switch. This shuts the motor off again.

ORIGINS
The Most Useless Machine has a proud heritage. The first machine that simply switched itself off was built by information theory pioneer Claude Shannon in 1952, based on an idea by artificial intelligence pioneer Marvin Minsky. The device sat on Shannon's desk at Bell Labs, and in 1958, sci-fi author Arthur C. Clarke dubbed it "The Ultimate Machine" in *Harper's Magazine* and in his nonfiction book *Voice Across the Sea*. Since then, knockoffs and variations on the theme have ranged from mass-produced novelties to works of art.

Illustration by Rob Nance

SET UP.

MATERIALS

[A] Heavy paper, card stock, or scrap cardboard

[B] ¼" plywood scrap about the same size as one side of the wooden box [L]

[C] Wood glue or cyano-acrylate gel glue or other good permanent glue for wood and plastic

[D] AA alkaline batteries (4)

[E] 4xAA battery holder Digi-Key #BH24AAW-ND (digikey.com) or RadioShack

[F] Insulated solid-core wire, 24-gauge, different colors Scavenge from dead telephone cable, Ethernet cables, thermostat wire, intercom cable, and anything with tiny colored wires. You can't have too many different colors of wire.

[G] Insulated, stranded hookup wire, 22-gauge or thereabouts

[H] Gearmotor part #GM2 from Solarbotics (solarbotics.com). You can also use a standard R/C servomotor modified to ignore signal input and allow for continuous rotation, if it doesn't already. See make-zine.com/projects/the-most-useless-machine for sources and instructions. A GM2 is less expensive, but if you have an extra servomotor already, the mods are easy.

[I] Gearmotor mount and mounting bracket Solarbotics #GMW and #GMB28, if you're using the GM2 gearmotor.

[J] DPDT toggle switch Digi-Key #EG2407-ND, RadioShack #275-636, or salvage this and the micro switch from common electronics.

[K] SPDT lever micro switch Digi-Key #EG4544-ND or RadioShack #275-016

[L] Small wooden box with lid large enough to fit the battery pack, motor, and arm in resting position (down). The one I used was purchased at a Dollar Giant store. If your lid isn't hinged, you'll also need some small hinges.

TOOLS

Table saw or handsaw that can cut wood straight

Miter box (optional) but helpful for handsaw straight cuts

Jigsaw, coping saw, or scroll saw or other saw that can cut curved wooden pieces

Pencil and eraser

Ruler

Scissors

File and sandpaper

Drill and drill bits: ¹⁄₁₆", ¼"

Small screwdriver for hinge and motor mounting screws

Soldering iron

Wire strippers

Side cutter

Hobby knife

Glue gun and hot glue

MAKE IT.

BUILD YOUR
USELESS MACHINE

START ⟫   **Time: An Afternoon Complexity: Easy**

1. PREPARE THE BOX

One half of the lid hinges up, while the other half carries all of the machine's workings. The workings all mount onto the same piece so that they'll stay aligned.

1a. Remove any latches and hinges on the box's lid.

1b. Cut the lid approximately in half through the middle cross-wise, undercutting at a slight angle so that the hinged half won't bind when opening. (You can see this undercut in Step 2c or the overview illustration on page 294.) Before you cut, make sure the machine half has at least enough space to fit both the motor and the micro switch lined up lengthwise.

2. DETERMINE THE LAYOUT

2a. Use the pencil to draw scale paper templates of the motor, toggle switch, micro switch, and the machine half of the lid, all in side-view, and cut them out. Also cut a template for the motor's mounting wheel or horn, and mark the axle on both the motor and mount templates.

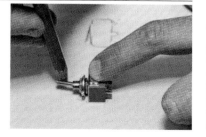

Photography by Ed Troxell

2b. Use the component templates to figure out the shape of the arm and how everything should attach to the lid. Download sample templates at makezine.com/projects/the-most-useless-machine. You want the back of the arm (or a mounting horn) to push against the lever switch when it's retracted. Then the arm should rotate and clear the lid while its "hand" swings over and pushes the toggle switch.

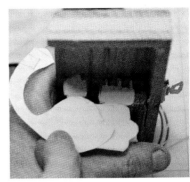

Cut templates for the arm and for a standoff bracket that holds the motor and the lever switch. Refine their shapes and sizes and move the pieces of paper around until you're sure that they all work together, while still leaving room for wire connections.

2c. Mark the positions on the templates where they meet: the toggle and bracket's position on the lid, and the motor and switch's position on the bracket.

2d. Trace the template shapes onto ¼" plywood and cut the pieces out. File and sand the edges smooth.

2e. Mount the arm to the motor's mounting wheel or servo horn. Drill ¹⁄₁₆" pilot holes in the arm and mount it with small screws (usually included).

2f. Measure and mark a centerline across the machine half of the lid, perpendicular to its cut end. Then mark the toggle's distance along this line, following its position marked on the templates.

3. BUILD THE CIRCUIT

3a. Now it's time to fire up the soldering iron. If your motor already has leads attached, solder them to each of the 2 middle legs of the toggle switch. Otherwise, cut, strip, and solder wire leads between each motor terminal and the middle toggle pins; 4" long is plenty for all connections in this circuit, and you may want to shorten them later for neatness.

3b. Solder a short jumper wire diagonally between 2 opposite corner legs of the toggle switch, then solder separate leads to the remaining 2 legs at the other corners.

3c. Solder the 2 free leads from the toggle to the lever switch. Connect one to the common tab (marked C), closest to the lever's pivot. Connect the other lead to the normally closed (NC) leg, farthest from the pivot. Don't connect anything to the normally open (NO) tab in the middle.

3d. Solder the battery pack's leads to the 2 legs at either end of the toggle. The circuit is complete! Test it by loading batteries into the pack. The motor should run, the toggle should reverse its direction, and the lever switch should shut it off in one direction. If it all checks out, remove the batteries, leave the toggle thrown in the direction that the lever interrupts, and mark or note this direction on the motor.

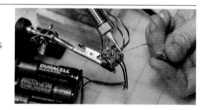

4. ASSEMBLE AND ADJUST

4a. Drill a ¼" hole in the machine side of the lid, at the toggle position you marked in Step 2f.

4b. Fit the toggle switch up through the hole, positioned with the toggle thrown in the direction opposite the lid cut. Don't glue it in yet.

4c. Hold the wooden bracket, motor, and lever switch in place against the lid so that they align properly with each other and with the toggle. Mark their positions with the pencil.

If the motor turns in the opposite direction from what you antici-pated while determining the layout (if the interruptible direction turns the arm out, not in) you should reverse the connections to the motor, or else position the motor the other way and arrange the pieces in mirror-image on the opposite side of the box.

4d. Temporarily hold all the pieces in place with a bit of hot glue. Put the lid on the box with the other half off, load the batteries, and check to see that everything works perfectly (which is unlikely).

4e. Tweak the components' placement and the shape of the arm as needed, ungluing and regluing with hot glue, until everything does what it should. You may need to file down part of the arm so it clears the bottom of the box, or fine-tune the position of the all-important micro lever switch.

When everything checks out, mark the final locations. Then mount the motor to the bracket with its included screws and attach the other components in place with permanent glue.

4f. For the other half of the lid, replace (or install) the hinges on the narrow end opposite the machine half, drilling ¹⁄₁₆" pilot holes.

4g. At this point, you should have a fully functioning Useless Machine. Don't wear it out!

FINISH

NOW GO USE IT »

USE IT.

LEAVE ME ALONE ALREADY

CUSTOMIZED USELESSNESS

The Most Useless Machine, like Claude Shannon's original, is a desktop or tabletop conversation piece. I went with a minimal aesthetic that leaves it most open to interpretation, but you can dress it up by labeling the switch positions, using a recognizable object like a doll's arm for the arm, or otherwise decorating it.

On a much larger scale, Swiss artist Hanns-Martin Wagner built a version that used an old wooden trunk as a box, a weathered prosthetic arm, and an air compressor for power (see below).

I was amazed at the response to my original Instructable. Everyone wants one of these boxes, and wants to share details of their own build! Its social appeal was also shown this past spring, when the Birmingham, U.K., hackerspace FizzPop (www.fizzpop.org.uk) hosted a Useless Machine-

making workshop led by Nikki Pugh (at right).

➕ For part templates, videos of the Most Useless Machine in action, alternate versions, how-to videos, and other resources, visit makezine.com/projects/the-most-useless-machine.

Photography by Nikki Pugh (top); and Hanns-Martin Wagner/www.sinnwerkstatt.ch (bottom)

OUTDOORS

THE ATLATL

Make the ancient tool that hurls 6-foot spears at up to 100mph. By Daryl Hrdlicka

Before the bow and arrow there was the atlatl*, or spear-thrower, an ancient weapon that could throw a spear or dart with enough force to penetrate a mammoth's hide. It was used in North America for about 10,000 years, and used by native Australians and Aleuts as recently as 50 years ago.

It's easy to make your own atlatl, and throwing with it is fun and very satisfying. Here's how to make one in the style of the Kuikuru (kwee-KOO-roo) of the Amazon Basin, who still use the spear-thrower today. I'll also explain how to make darts for it, and how to throw. But never forget that the atlatl is a *weapon*. It is *dangerous*. A dart will go through a side of beef. So I'll go through some precautions as well.

*Most people say "at-LAT-l," or "AHT-laht-l" but pronunciations vary. Find one you like, get your friends to pronounce it the same way, and you'll be right.

Atlatl Basics

Atlatls range in form from the simple to the very ornate, but they all have the same 3 components: the hook, the grip, and the shaft. The grip is where you hold the atlatl, the hook engages the back of your projectile and propels it, and the shaft connects the two and acts as a lever to multiply the speed of your arm.

A typical length for an atlatl is 18"–24", although some have been found as short as 6" (in California) and as long as 48" (in Australia). Length is mostly a matter of personal preference, but it needs to fit the length of your arm and of the dart you're throwing.

The simplest atlatl is the first kind ever used — the basic branch atlatl. To make one of these, just find a tree branch that measures about ¾" in diameter and has a smaller branch angling out of it. Cut it

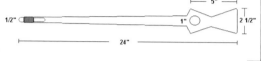

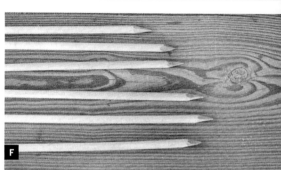

Fig. A: Template for traditional Kuikuru spear-thrower. Fig. B: Outline traced onto pine board, and ready to cut. Fig. C: Atlatl shaft cut and sanded smooth. Fig. D: Drilling the hole for the peg or hook.

Fig. E: Peg installed. If you're using pine (as here), it's a good idea to reinforce the peg by lashing it with some cordage. Fig. F: Dart points sharpened with a hobby knife.

MATERIALS

1×4 lumber, 24" long **A standard piece is actually ¾"×3½". That's fine.**
Wooden peg or dowel, 5/16"×1¼"
Cordage for wrapping (optional) **If you're using pine, you'll wrap it. Natural or artificial sinew works great, or you could even use kite string.**
Wooden dowels, 3/8"×48" **one for each dart**
Duck feathers, 8" long **two for each dart; available at most craft stores**

TOOLS

Saber saw **aka reciprocating saw**
Drill with 5/16" bit, 1" wood boring bit, and ¼" bit
Utility knife
Sandpaper and sanding block
Wood glue
Hot glue and gun **for the dart**
Strapping or electrical tape **for the dart**

Making a Kuikuru Spear-Thrower

1. Size and trace the template above (Figure A) onto your piece of wood, and cut it out (Figure B). Use the wood boring bit for the finger hole. I normally make a 1" hole, because I have fairly large fingers and I share my atlatls a lot. People with smaller hands can use an atlatl with a 1" hole, but ideally it should fit close around their index finger; my wife's atlatl has a ¾" hole. I'd say start with 1", and possibly adjust this for your next one.

2. Use the utility knife to round off the edges, then sand it all smooth (Figure C).

3. Drill a hole in the end for the peg; this will act as the hook. Go in at about a 45° angle (Figure D).

4. Put some wood glue in the hole and insert the peg (Figure E). You're done!

5. If you're using a softwood, reinforce the peg joint by wrapping it with sinew or cordage. I normally use pine because it's cheap and easy to shape, but it tends to break near the peg, so wrapping it helps.

just below the smaller branch, clip off the other end about 18" farther up, and then clip off the smaller branch. Now you have a functioning atlatl. To make it a little easier to handle and control, you can add a finger loop. Just attach a 10"×¾" strip of soft leather about 7" from the narrow end, looping it around on the side opposite the branch stub.

Fig. G: Feathers hot-glued to the tail end of the dart. These slow the tail end down, keeping it in back during flight. Fig. H: First dart finished; now make some more! Fig. I: Larger feathers glued and taped in back so they won't fray as fast. Fig. J: At the Hedoka Knap-In Primitive Skills event, an atlatl thrower steps into his throw. Fig. K: The dart is sent on its way. Note the downward flex in the shaft.

You don't need reinforcement with a hardwood such as oak or maple, but those also require more skill and the proper tools to work effectively — two things I don't have. For your first one, I recommend pine.

Making a Dart

1. Sharpen one end of the dowel (Figure F). I just use a utility knife. You can also add stone, bone, or steel points to darts, but you should probably gain more experience before making your darts lethal.

2. Drill a dimple in the other end. This is the "nock" that the peg fits into. I usually drill with a ¼" bit and just touch the end, going in about ⅛" or so.

3. Hot-glue the feathers on, one on each side, with the quills forward (Figure G). Then wrap tape over the glue to help keep it on, or wrap it with artificial sinew if you want it to look better (Figure H).

4. With large feathers, glue and tape down the trailing ends so that they don't fray as fast (Figure I).

Your dart is done! While you're at it, you might as well make another 5 or so. Otherwise, you'll get very tired chasing it after each throw, and you'll get less practice. Since you're using dowels, your darts will be closely matched, which will help you practice.

Two factors that govern your dart's flight are its flex (also called its *spine*), and the feathers (its *fletching*). Flex is the amount of pressure it takes to make the shaft bend. An atlatl generates 6–10lbs of pressure (depending on your throw), so your darts need a spine of 6–10lbs. Less than that and you won't be able to throw it; more, and it won't fly right.

To measure the spine of a piece of wood or bamboo, press it lengthwise onto a bathroom scale. When the shaft begins to bend, look at the number. That's it. Common ⅜"×48" dowels typically flex in our desired range, as do ½"×72" dowels, which are better for target practice. It's satisfying to make a dart out of a natural piece of wood and hone it to the proper flex, but it's also quite a bit of work. So to start with, I'd say use dowels.

Unlike with an arrow, the feathers on a dart don't act as vanes. They add wind resistance, which slows the rear end so the sharp end stays in front. You can use other materials besides feathers, such as birch bark, cornhusks, cloth, and duct tape, but you can't beat feathers for the look.

For our 48" dowel, a pair of 8" feathers should work fine. With less fletching, the dart will travel farther but won't be as accurate. More fletching means the dart will be more accurate, but won't go as far.

Using the Atlatl

Now let's get out there and throw! The 3 basic steps are the grip, the stance, and the throw itself.

The Grip Slide your index finger through the hole from the side opposite the peg, and grip the handle with your other fingers. Put the point of the dart on the ground, then fit the atlatl peg into the nock and hold the dart with your thumb and index finger. Squeeze them, almost like you're holding a pencil, but keep them on the sides of the dart, not over the back. The dart will come out of your hand at the proper moment, if you just let it.

The Stance Point your left foot at the target (if you're right-handed) and angle your right foot away from it, about a shoulder's-width back. You should feel comfortable and balanced. Turn your body sideways, in line with your left foot, and turn your head to look at the target. Point at the target with your left arm to help with accuracy and balance.

The Throw First, aim the dart by bringing your grip hand up by your ear and sighting along the shaft to your target. Next, bring your arm straight back as far as you comfortably can, but don't twist your wrist on the way back, which will point the dart off to the side. Unless you're a powerful thrower, tip your hand back so that the point rises up a few inches. This will give your throw an arc, making it travel farther. Pause to collect yourself and focus.

And now, the throwing motion itself: using an atlatl is like throwing a fastball — you need to put your whole body into it. Lean back, balancing on your back foot. Then step forward and shift your weight onto your other foot. Slide your arm forward, keeping the dart pointed at the target, and when it's almost fully extended, snap your wrist forward *hard*. It should all be one fluid movement, and the atlatl should end up pointing at the target. For an example, watch the video clips of atlatl throwing on Bob Perkins' website, atlatl.com.

Practice without a dart until you get used to it.

And don't worry about releasing the dart; it should come free on its own at the proper moment. Don't try to throw it hard — this will just mess you up. Just concentrate on throwing smoothly, and your speed and power will develop. Everything will click at some point, and it will be a thing of beauty.

Target Practice

Throwing the atlatl purely for distance is fun, but after a while you get tired of chasing down all your darts. Besides, you'll want to see what it would be like to hunt with one. You can use paper archery targets on hay bales, and 15yds is usually a good starting distance. If you switch to a heavier dart, you'll want to double up the bales.

Standard bull's-eye targets are fine for accuracy competitions, but I personally don't like them. The atlatl is for *hunting*, so I prefer animal silhouettes; 3D targets, your basic foam animals, are my personal favorite. You really feel like the "mighty hunter," and the first time your dart flies straight and true into the target, well, it's indescribable. You need to experience it.

In a pinch, almost anything will make a decent target. A friend and I once used some styrofoam coolers. We ended up hunting those "sheep" for about 3 hours, until it got so dark we couldn't even see them anymore! We were tied at the time (it's always about competition, you know), so we had to keep going, listening to see whether we'd hit them or not. If I remember right, he won. Barely.

ATLATL SAFETY — WHERE TO THROW

» When you throw an atlatl, make sure you have an open area that's at least 30yds long, with nothing breakable behind it.

» There should *never* be anyone in front of you when you throw.

» In spite of your aim, the dart *can and will* go out of control once in a while. It may go off to one side or go farther than you intended. *Make allowances for this.*

Daryl Hrdlicka (thudscave.com/npaa) lives in southwestern Minnesota with his wife and four kids. They homeschool, so making cool stuff is *always* on the menu. Daryl teaches the atlatl at the Jeffers Petroglyphs Historic Site.

G-METER AND ALTIMETER

 Double-duty aerospace instrument on a shoestring budget. By David Simpson

Here's an aerospace instrument you can build for $5 that will measure the crushing forces that a model rocket withstands and the rarified strata it attains. It isn't exactly six-sigma technology in terms of accuracy, but it's darn fun.

The device, which you install in the rocket's payload compartment, uses 2 small bands of heat-shrink tubing that slide over a dowel to record the maximum G-force and altitude attained. As the rocket accelerates, the G-force band is pushed down by washers on a spring, and as the rocket rises, the altitude band is pushed down the rod by the expansion of a pressure chamber made from a pill bottle and a rubber-balloon membrane.

The force of landing doesn't disturb the positions of the bands, which are heat-shrunk snugly over the rod and stay in place thanks to their relatively high coefficient of friction and low mass.

The "secret sauce" for both readings is the calibration step, where you mark positions on the dowel with their corresponding G-force and altitude levels. To calibrate the G-force meter, we stack increasing weights onto the spring and gauge its compression. The altimeter we calibrate using a kitchen vacuum food sealer and a commercial altimeter or barometer.

Build the Altimeter

Make the flexible membrane by cutting a 2"-diameter circle from a rubber balloon. Stretch it flat over the open end of the pill bottle, and secure it by winding button thread around several times, near the top. Tie off the thread and coat it with a thin layer of wood glue or epoxy, then trim away the excess rubber.

Cut a disk the same diameter as the pill bottle out of ¼" balsa or aircraft plywood. Drill a ⅟₁₆" pilot hole in the center of the disk and glue it to the bottom of

Photography by David Simpson and Linda Kennyhertz

MATERIALS

Cylindrical plastic pill bottle about 2½" long **to fit in a rocket payload compartment**
Rubber balloon **large enough to cut a 2" round piece from**
Button thread or heavy-duty sewing thread
³⁄₁₆"-diameter wood dowel, 16" or longer
Model aircraft plywood, ³⁄₁₆" thick, 2" square
Balsa wood, ⅛" thick, 2" square
Balsa wood, ¼" thick, 2" square
Scrap wood board, at least 4" square **for G-meter calibration stand**
Small cylindrical spring, ¼" diameter
Washers: ³⁄₁₆" ID (1), ⅞" OD (2) **You'll need more of each for calibration, which you can do in the hardware store aisle.**
Heat-shrink tubing, ¼" diameter
Wood screw, about 1½" long
Wood glue or epoxy
CA (cyanoacrylate) adhesive gel
High-tack, double-sided foam tape
Paper **Graph paper is best.**

TOOLS

Model rocket with transparent plastic payload section at least 5" long **I used the Estes HiJax (EST 2105), which has a 6" clear payload compartment. Or you could make your own from a spare body tube and nose cone mated to a matching "booster."**
Drill and drill bits: ¹⁄₁₆", ³⁄₁₆"
Saw **for cutting wood dowel**
Tool for cutting 1" rounds out of thin plywood **You can use a hole saw, but I rough-cut with a saw and used a Dremel tool with the 401 mandrel bit to turn the pieces against a sandpaper surface.**
Altimeter or barometer **a dedicated device, or you can use a hiker's wristwatch or a weather station, such as one from RadioShack**
Vacuum chamber or a vacuum food sealer and vacuum bag
2" PVC pipe, 12" long
Colored pencils or fine markers

the bottle. This base will later accept a wood screw to hold the chamber fast to the tube coupler at the bottom of the payload section (Figures B and G).

Build the G-Meter

Cut a disk of ³⁄₁₆" aircraft plywood just slightly shy of the inner diameter of the rocket's payload tube, and drill a ³⁄₁₆" hole in the center.

I did this in a fun way: first I traced the circumference of the body tube onto the plywood, rough-cut and drilled a ¹⁄₁₆" hole in the approximate center by eye. Then I threaded the disk onto a Dremel 401 mandrel bit (a shank with a screw-like head at the end) and used the Dremel as a sort of mini reverse-lathe, turning the disk down to diameter against a piece of sandpaper tacked to a board (Figure C). The mandrel bit is designed to hold polishing bits, but hey, here's a new use! After sanding down the disk, I redrilled its center hole out to ³⁄₁₆".

Glue a 3½" length of ³⁄₁₆" dowel into the disk, and mark a scale on the dowel from the base to the end, in ¹⁄₁₆" increments, alternating between contrasting colors to make it easier to read (Figure D).

Slip a ¼" length of heat-shrink tubing over the dowel, and shrink it down until it stays in place even if you shake it like a thermometer. This is the G-force band. Slide the spring over the dowel, followed by the smaller washer and then the 2 larger washers. Now shrink another ¼" length of heat-shrink — this is the altitude band.

Prepare the Adjacent Rocket Sections

Cut 2 more disks out of ⅛" balsa, to fit the rocket's inside diameter (Figure E). Drill one with a ⅛" hole in the center and use CA gel adhesive to attach it flat into the tube coupler piece that will take the bottom of the rocket's payload tube, about ½" down from the top edge. This disk will anchor the altimeter chamber.

Drill a ³⁄₁₆" hole in the center of the other disk and use CA adhesive to glue it into the nose cone, flush with its bottom edge.

The G-meter dowel will run through this disk and slide farther in when higher altitudes cause the pressure chamber membrane to balloon upward. The altitude heat-shrink band will meanwhile stay in place against the disk.

Assembly: No Bouncing Around!

Use a short wood screw to secure the payload tube coupler to the altimeter base (Figure F). Attach the G-meter base to the altimeter membrane with a ⅛"–³⁄₁₆" square of high-tack, double-sided foam tape. Use a thin film of epoxy to attach the G-meter base to the spring, the spring to the washers, and the washers to each other (Figure G).

Drill a ¹⁄₁₆" hole through the side of the payload section to allow external air pressure to reach the altimeter.

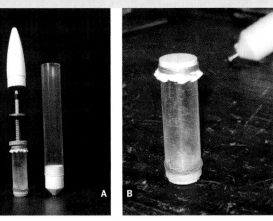

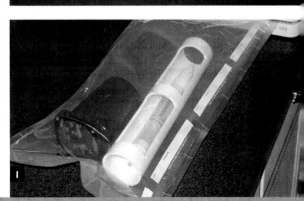

Fig. A: The combo G-meter/altimeter before installation in the payload compartment. Fig. B: The completed altimeter chamber before assembly. Fig. C: Turning wood disks to size with a rotary tool.

Fig. D: Mark the scale on the G-Meter dowel with contrasting colors for easy reading. Fig. E: Wooden disks for the tube coupler below the payload and the nose cone above.

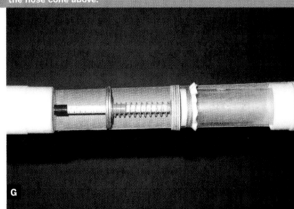

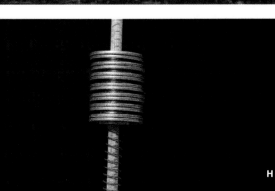
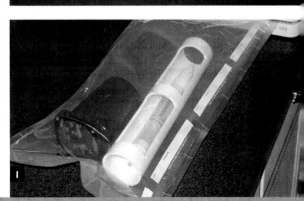

Fig. F: The tube coupler screws to the wooden base of the altimeter chamber. Fig. G: The G-meter/altimeter combo installed without the tube, to show position.

Fig. H: Calibrate the G-meter by stacking washers in multiples of carry load. Fig. I: Calibrate the altimeter alongside a working barometer or altimeter in a vacuum.

Calibrate the G-Meter

Calibrating this combo meter is more complicated than constructing it, but it's both cool and educational. To calibrate the G-meter, you first build a calibration rig. Glue a 12" length of ³⁄₁₆" dowel vertically in a wooden base and mark it the same way you did the G-meter dowel.

Slide the G-meter spring over the dowel and add the payload quantity of 1 small washer and 2 large washers. Check the reading on the scale. That's 1g, our baseline. Make a table that translates millimeter readings on the dowel to corresponding G-forces. Record the 1g reading, then add 1 more small washer and 2 more large ones, and record the position on the scale for 2g. Continue doing this until you've got readings for 40g (Figure H). Unless you need a lot of washers, you might opt to conduct this procedure in the aisle of the hardware store.

Calibrating the Altimeter

Calibrating the altimeter is even cooler, but more involved. You'll need access to a vacuum chamber; if you can't gain access to a scientific or industrial vacuum chamber, borrow (or buy) a vacuum food sealer. You'll also need an altimeter or barometer; I experimented with a RadioShack mini digital weather station, which included a barometer, but it was a challenge getting it to take instantaneous readings. I was able to do it by interrupting and re-energizing the circuit at the moment I wanted the reading. But I wound up using a good old airplane altimeter, which was simpler.

If you're using a vacuum food sealer, put the G-meter/altimeter in a cage made by cutting two 5" windows in a 12" piece of 2" PVC pipe. The cage prevents the bag from collapsing around the instrument. Put the reference altimeter or barometer and our instrument in the vacuum bag or chamber (Figure I).

Blip the vacuum button to reduce the pressure in small increments, and record the readings from the dowel and the commercial barometer or altimeter. Make an association table between the two, like with the G-meter. If you're using a barometer, you can convert pressure to altitude using the calculator at csgnetwork.com/pressurealtcalc.html.

Prepare for Liftoff

Make sure both heat-shrink bands are in their starting positions before each flight: the G-meter band at the top of the spring, just under the washers, and the altimeter band at the top of the dowel, so that it will be flush with the nose cone disk when you slide it on. If the nose cone is a little loose, secure it with masking tape. As with other model rockets, friction-fit the payload atop the booster, connect the elastic shock cord, and attach the parachute — or else you'll be digging your combo meter out of a hole!

Recovery, Reading, Reset, and Repeat

After "3, 2, 1, liftoff!" and *whoosh!*, recover your vehicle and peek inside. You may need to loosen the bottom wood screw in order to read the bands, then use your tables to translate their positions into max-g and max-altitude. You've got your max-g and max-altitude recorded on the combo!

A variation: Instead of marking millimeters on the dowels, inscribe the target units — gs and feet or meters — directly during calibration. Just take care not to bump anything while making the marks.

You'll find that different rockets, engines, and gross weights all affect the readings. You may need to add or remove washers and recalibrate to change the overall range of the G-meter; more powerful engines need fewer washers to register a reading while weaker engines need more.

Safe rocketeering!

David Simpson is Founder of Innovators Inc. (innovators-inc.net) whose mission is to make learning about science and technology fun and cool for kids through doing. He's developed maker programs for the Girl Scouts and Boy Scouts as well as the Liberty Science Center and the International Spy Museum. Dave has been a mentor with FIRST Robotics and the Civil Air Patrol (CAP) and was twice named CAP's New Jersey Aerospace Education Officer of the Year.

Make: TIPS !

The Key Hole

My first job was as a bicycle mechanic at a shop where we also cut keys. One fine day, a fellow came in to have his house key duplicated, and he asked me to drill a nice big hole in the key so it would be easy to find in the dark. It wasn't long after that when I drilled my own key! My current key ring has, among others, three identically shaped Schlage keys, so I cut a little off the wings of another to make it easy to identify. Try this trick — I'll bet you never go back.

—*Frank Ford*, frets.com

LIVING ROOM BAJA BUGGIES

By John Mouton

Photograph by Robyn Twomey

BUG'S-EYE VIEW
With wireless cameras on board, these radio-controlled racers give you virtual reality telepresence.

Do you like radio-controlled (R/C) cars? Do you like the desert, but hate the heat? Well, sit down and kick back as you engage in the excitement of Living Room Baja Buggy Racing. This off-road competition combines the fun of homemade R/C cars with the air-conditioned convenience of a fake, indoor desert landscape — without the big dollar price. There are no rules, no expensive automotive racing equipment, and a total disregard for public safety (because these cars are only 6" long and 4" tall).

You don't even have to look at your car as you negotiate the terrain — an onboard camera and a virtual reality headset can be installed for your lackadaisical safety!

So chase the dog, race against the kids, or just put the electrons to the PC board and hold on for a bug's-eye view of the whole racecourse. It's the best thing to hit urban living since the Man Cave.

FULL DISCLOSURE:
The author is an employee of Microchip Technology, which manufactures some of the components used in this project.

John Mouton is an applications engineer with Microchip Technology's Security, Microcontroller, and Technology Development Division (SMTD).

R/C COMMAND, MICRO CONTROL, AND MINI VIRTUAL REALITY

In this project, we'll build an R/C Baja buggy that you can drive around the house using a pair of virtual reality display goggles that give you a view of the world as if you were literally in the miniature driver's seat. All of this is controlled by the user (you) via a hand-held radio transmitter and a wireless camera that displays what the buggy sees through the VR goggles.

I started by designing a simple motor control circuit, a voltage regulator circuit, and a PIC12F683 MCU. This gave me a small drive circuit that would move the buggy forward and in reverse, electrically brake and stop the buggy, and provide variable speed control. I sent my schematic diagram and Gerber files to a vendor and got back a printed circuit board, ready to plug my components into.

After the circuit was built, I bought parts on the internet to build the buggy. When I put the buggy into action I found that it performed well — in fact, better than I expected.

I added a wireless camera and virtual reality goggles. Now I drive the buggy all over the house without leaving the comfort of my couch. You don't have to add the wireless camera and virtual reality goggles to have fun with this project. In fact, the buggy will be less expensive and lighter without them. Nevertheless, this article provides step-by-step instructions on how to build an R/C Baja buggy with the whole enchilada.

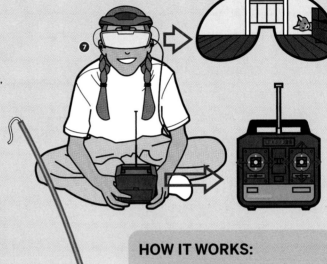

HOW IT WORKS:

❶ The Baja buggy is built on a Tamiya Buggy Car Chassis Set.

❷ The motor control circuit is built on a custom printed circuit board.

❸ The drive motor (red arrows) makes the buggy go forward and backward.

❹ A 9V battery and a lithium battery add weight to the back for improved traction.

❺ One servomotor (blue arrows) pans the webcam mounted on top.

❻ Another servomotor (orange arrows) steers the front wheels.

❼ The wireless webcam signal is sent to the VGA virtual reality goggles.

Illustration by Nik Schulz

SET UP.

 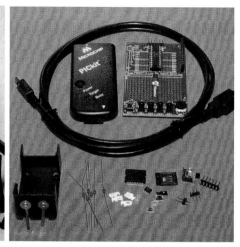

Photography by John Mouton (left) and Ed Troxell

MATERIALS

Wireless camera and tuner/ receiver **You need a wireless 9V DC, 300mA A/V security camera, and a 9V DC, 500mA tuner/receiver with 0.9G frequency control. Try** tigerdirect.com or frys.com.

VGA goggles **Any wireless-capable generic or name brand; try** amazon.com or tigerdirect.com.

Tamiya Buggy Car Chassis Set **#70112** tamiyausa.com

Tamiya Atomic-Tuned DC Motor **#15215**

E-Sky EK2-1003 **4-channel radio system which includes transmitter, receiver, and 1 servomotor (mini), or similar radio system**

E-Sky EK2-0500 servomotor **for the camera** Futaba FM75.470 MHz CH.64 crystal **For multiple**

buggies, use different crystals in the 75MHz–76MHz range to avoid cross-talk.

Thunder Power lithium polymer (Li-Poly) 1,320mAh/7.4V, 2-cell rechargeable battery **for the motor**

9V battery **for the camera**

Apache Smart Charger 2020 **for charging the Li-Poly batteries**

ElectriFly 2-pin male connectors **#GPMM3106, for all battery, camera, and charging connections**

Servo cable with 1 male S connector towerhobbies. com **#LXPWB5**

Male S connector **for servo cable, Tower #LXPWC3**

Ferrite ring **for radio-control applications, check** digikey.com.

Battery holder digikey.com **#1294K-ND**

#4³⁄₈" Phillips head screw, #4 flat washers (2), and 4-40 steel hex nuts (2) for tie rod hardware

NOTE: For the R/C transmitter, receiver, servos, and crystal, try hobbylobby.com and jameco.com. For the batteries and charger, try rctoys.com and electrifly.com.

TOOLS

PICkit 2 Starter Kit **for programming** microchipdirect.com

Electric drill, cordless or AC

Drill bits **to fit the rivet diameter**

Center punch

Soldering iron and solder

Double-sided tape

Sandpaper

Computer

Phillips screwdrivers **large and small**

Side cutters, modeling knife, and scissors

Pop rivet tool, hand-held

Heat-shrink tubing

Heat gun or hair dryer

Wire strippers

Needlenose pliers

Fake desert landscape

DOWNLOADS

Download the Baja buggy motor control program for your PIC microcontroller, *carprogorig.hex*, as well as the Gerber files. Then download a complete list of the circuit components. makezine.com/projects/ living-room-baja-buggies

NOTE: The whole project ran me $400, but if I didn't install the VR goggles and camera, it would total only $200: about $145 for radio-control components, $30 for the PCB and circuit components, and $20 for the buggy chassis set with motor. (I did order 10 PCBs for a unit price of $12. Ordering 1 PCB will run about $50.)

MAKE IT.

BUILD YOUR OWN VIRTUAL REALITY BAJA BUGGY

START ⋙   **Time: 4–5 Hours Complexity: Medium**

1. BUILD THE BUGGY CHASSIS

1a. Assemble the buggy chassis set as directed in part 3 of the instructions enclosed in the Tamiya Buggy Car Chassis Set box.

1b. Drill a hole in the back of the motor/gear bracket housing. You'll use it to rivet on the 9V battery holder. Use the same drill bit size as the size of your rivet.

NOTE: Center-punch the hole before you drill, and use a high-quality drill bit with a sharp tip. This will prevent the drill bit from skipping across the metal and biting in the wrong place.

1c. Modify the tie rod. Bend and trim a 16-gauge wire to about 1" high with arms that fit against the tie rod provided in the buggy car chassis kit, as shown. Then solder the 2 together. This will give the servo a way to move the tie rod left, right, up, and down with the suspension, thus steering the buggy.

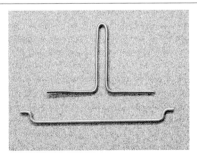

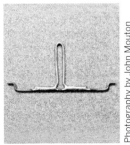

Photography by John Mouton

1d. Solder one 0.1-microfarad (µf) capacitor and a 24-gauge wire from each lead of the atomic-tuned DC motor to the motor case. Next, solder a third 0.1µf capacitor between the leads. This will prevent electrical motor noise from interfering with the receiver circuit.

1e. Install the wheels as directed in part 6 of the buggy car instructions. However, don't install the roll bar — it will just be in the way later when you install the camera servo and the motor-control circuit board.

2. BUILD THE MOTOR-CONTROL CIRCUIT

2a. Get a printed circuit board. I emailed the schematic and Gerber files provided at makezine.com/projects/living-room-baja-buggies to Circuit Express (circuitexpress.com). I chose their least expensive package, 10 boards for $120, and they built the boards and mailed them to me within 24 hours. Other vendors are less expensive for smaller quantities. Or you can buy a PCB build kit and make it yourself.

2b. Install the circuit components. All components used in the motor-control circuit can be purchased from Digi-Key (digikey.com) or any other electronic components supplier. For the battery, camera, and charger connections, I used ElectriFly 2-pin male connectors to ensure safe, secure connections. I chose the PIC12F683 MCU because it's small, has a pulse-width modulation (PWM) module onboard, and very simple to work with. Here's what the completed circuit board should look like after all the components are soldered. Use the schematic to guide you. A list of components and part numbers is available at makezine.com/projects/living-room-baja-buggies.

3. INSTALL CIRCUIT BOARD, BATTERY HOLDER, AND SERVOMOTORS

3a. Before you solder the motor leads, twist them together. This will keep electrical motor noise from interfering with the receiver. Now solder them to the circuit board at the locations marked M1 and M2.

3b. Use a small piece of double-sided tape to attach the circuit board to the wooden chassis, all the way back so that it touches the motor/gear bracket housing.

3c. To install the 9V battery holder, first sand off about ¹⁄₁₆" of its bottom end. This ensures that the battery will fit flush inside, and that the holder itself will fit evenly and spaciously between the rear wheels.

3d. Drill a hole in the center of the battery holder the same diameter as the hole you drilled in Step 1b. Then rivet the battery holder to the motor/gear bracket housing.

NOTE: To rivet, insert a rivet through the hole, and then squeeze with the pop rivet tool until the shaft pops off. To remove a rivet, drill it out using the same bit you used to drill the original hole.

3e. Extend the battery holder's positive and negative leads by soldering 24-gauge wires to them, long enough to comfortably reach J4 on the motor-control circuit board. Solder a 2-pin male connector to the ends of these wires, which will allow you to disconnect the 9V camera battery from the circuit board as needed.

3f. Using a small piece of double-sided tape, attach 1 servomotor to the motor-control circuit board as close to the motor/gear bracket housing as possible. This servo will be used to pan the wireless camera that will be mounted on top of it.

The servos come with an assortment of lever arms that can be attached to them, depending upon the application. Use the circular lever arm on the camera servo.

3g. Next, attach the other servo to the chassis just in front of the circuit board (the location is shown in Step 3e). This second servo will be used to steer the buggy.

3h. For the steering servo, use the straight lever arm. Drill a small hole in the lever arm the diameter of a #4 screw. Use the #4 screw, 2 washers, and two #4 nuts to loosely attach your modified tie rod to the servo lever arm. Now, as you drive the car over rough terrain, the servo will still be able to steer the front wheel even as the front suspension moves up and down.

4. INSTALL THE R/C RECEIVER AND WIRELESS CAMERA

4a. You can use just about any R/C transmitter/receiver set for this project. If you don't already have a set, shop around for a good deal. I found a great deal on a 4-channel radio system kit that included the transmitter, receiver, and 1 servo for $60.

The reason I chose a 4-channel dual stick transmitter (rather than a 2-channel pistol-type transmitter) is because I needed the second left-to-right stick to pan the camera.

4b. Using double-sided tape, attach the receiver to the top of the steering servo. To reduce the receiver's bulkiness, remove its cover. This allows you to easily install frequency crystals and plug components into the receiver.

4c. Plug the camera servo into channel 4, and the steering servo into channel 1. Plug connection JP1 of the motor-control circuit board into channel 3 of the receiver; this connection will supply power to the receiver and allow you to control the speed of the motor via the PWM signal generated from the PIC12F683 MCU. Make the PWM signal cable by crimping the male S connector's pins to the bare-wire end of the male servo cable using a pair of pliers. Now, in order to reduce electronic noise interference, loop the PWM cable through a ferrite ring before plugging it into channel 3 of the receiver. This ring will increase the cable's inductance, thereby filtering out high-frequency electronic noise.

⚠ **CAUTION:** When using R/C transmitters and receivers for cars, keep them in the R/C car frequency range of right around 75MHz. R/C airplanes operate right around the 72MHz frequency range.

4d. Before installing the wireless camera to the camera servo, shorten the camera's cable and solder another 2-pin male connector so it will plug into the motor-control circuit board at J3. To do this, cut the camera's cable down to about 3". Strip 1" of insulation and slide on a 2" piece of heat-shrink tubing. Then solder the black and red wires to the 2-pin connector, slide the heat-shrink tubing over the connection, and shrink it with a heat gun or hair dryer.

4e. Using your double-sided tape, attach the wireless camera to the camera servo temporarily (you'll adjust it later).

5. INSTALL THE MOTOR BATTERY AND PROGRAM THE PIC MICROCONTROLLER

5a. For the motor's power source, you need something that will supply enough voltage and last a while between charges. I used a Thunder Power lithium polymer (Li-Poly) 1,320mAh/7.4V 2-cell rechargeable battery. I don't recommend using less than a 480mAh/7.4V battery, as it would need to be recharged more often.

> ⚠ **CAUTION: The battery will have a charge even if you buy it brand-new, so keep the 2 wires from touching!**

5b. Strip the black and red wires coming off the battery. Solder the wires to another 2-pin male connector, and heat-shrink. Attach the battery to the 9V battery holder with double-sided tape, and connect the rechargeable's wires to J5 (VBAT) on the circuit board.

5c. To program the PIC MCU, I used a PICkit 2 Starter Kit (#DV164120) that I bought from my company, Microchip Technology (microchipdirect.com) for $50. This kit has everything you need to write, debug, and program your source code directly into the MCU via header J2 on the circuit board. (If you're a student or educator, visit microchip.com/academic to learn how you can get a discount on development tools through Microchip's Academic Program.) Go to makezine.com/projects/living-room-baja-buggies. All you have to do is to import the *carprogorig.hex* program file using the PICkit 2 Starter Kit interface, and then click the Program button.

FINISH ❌

NOW GO USE IT »

USE IT.

CHARGE IT UP

CHARGE IT

Now that your buggy is assembled, you're almost ready to turn it on. However, you might want to charge the battery first. I used an Apache Smart Charger 2020 designed specifically for charging 2-cell Li-Poly batteries.

> ⚠ CAUTION: Trust me when I say that if you use lithium polymer batteries, you *must* use a charger designed to recharge Li-Poly batteries. I have a coworker who no longer has a garage because her husband didn't use the appropriate battery charger. Always follow the manufacturer's instructions on proper use of batteries and chargers.

ZERO THE SERVOS AND SET THE RADIO CONTROLS

You're now ready to fire up your buggy. Once you turn on S1 on the circuit board, the servos will power up and find their neutral positions. Hence, you'll need to detach the camera and reattach it so it faces toward the front of the car. (Good thing you used double-sided tape!) You may also need to make a slight adjustment to the steering servo. Simply remove the screw holding the lever arm to the servomotor, pull it off, and reattach it, pointing straight up. Now, when you turn S1 off and then turn it back on, the servos will be aligned to the correct starting points.

With the circuit board and hand-held transmitter both turned off, adjust the transmitter toggle switches in the lower right-hand corner to set the left control stick (forward and back) as the throttle; the left control stick (left and right) for steering; and the right control stick (left and right) for panning the camera. The radio system kit's instruction manual will help you do this.

When finished, turn the transmitter on, then turn the circuit board power on at S1. Take the throttle control stick on the transmitter and pull it all the way back toward you. Then push it all the way forward, then back to center. This will establish the maximum duty cycles for forward and reverse, and arm the speed control circuit. At this point, the buggy is ready to roll.

USING THE VIRTUAL REALITY GOGGLES

Plug the VR goggles (or a TV monitor) into the receiver that came with the wireless camera. Plug the receiver into an AC outlet. Adjust the tuning knob until a picture of what the camera sees comes up. I found that the wireless camera range is more than 500' straight line-of-sight, with no major obstacles. If you use it outside in a large open area, the range is even better.

RESOURCES

➕ Download the Baja buggy motor control program for your PIC microcontroller, *carprogorig.hex*, as well as the Gerber files and circuit schematic from makezine.com/projects/living-room-baja-buggies. You can also download an Excel file listing all the Digi-Key part numbers for components that fit this circuit board. And finally, there's a list of lower-cost build alternatives provided.

Microchip's Online Motor Control Design Center (microchip.com/motor) has a wealth of resources for programming your PIC microcontroller to control motors and more.

BOOM STICK

By Edwin Wise

Photograph by Garry McLeod

DECOMPRESSION THERAPY

The super-loud Boom Stick is a PVC air cannon that delivers maximum bang for the buck. It assaults the startle reflex of any nearby victim, adding an instant rush of physical terror to haunted houses, art pieces, pranks, and performances.

I work in haunted houses during the Halloween season, as an actor, guide, technician, makeup artist, and effects creator. Some of my effects instill fear through foreshadowing or complex storylines, but the most effective way to scare people is often just a simple, brute-force startle.

The air cannon is a great and safe device for such scares. In its simplest form, it consists of an air reservoir, a quick-exhaust valve (QEV), and sometimes a resonating chamber. Haunted house suppliers and special-effects houses sell commercial models with large-gauge QEVs, but these cost hundreds of dollars. Home projects that rely on a standard air compressor typically use smaller, cheaper water valves from washing machines or sprinklers, but for me, these designs have yielded only a disappointing "poof-hiss."

Inspired by the PVC-based designs of spud gun enthusiasts (but leaving out the potato), I've found a better approach: a two-stage, chamber-sealing, quick-exhaust, piston-valve air cannon that you can build out of common plumbing components for about $100. I call it the Boom Stick.

Edwin Wise is a software engineer with more than 25 years professional experience developing software during the day and exploring the edges of mad science at night. He can be found at simreal.com.

BUILDING BOOM

The Boom Stick creates a pressurized volume of air and releases it very quickly, generating a loud shockwave.

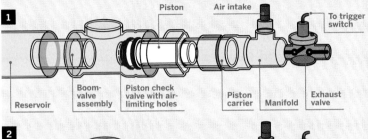

1

Piston Air intake To trigger switch

Reservoir Boom-valve assembly Piston check valve with air-limiting holes Piston carrier Manifold Exhaust valve

HOW IT WORKS

1. The piston rests in the piston carrier, and the entire system is at ambient air pressure. Small holes in the piston allow limited airflow through from behind; a rubber washer inside the piston acts as a check valve, passing air in only one direction and increasing efficiency.

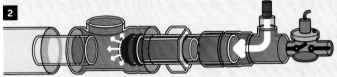

2

2. When air flows in behind the piston faster than it leaks out of the holes, pressure builds up in the manifold.

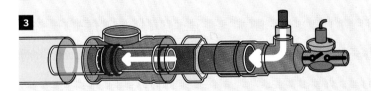

3

3. Pressure behind the piston pushes it into the boom valve tube, sealing the pathway between the air supply and reservoir.

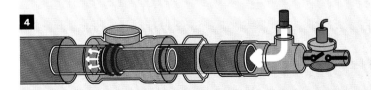

4

4. Pressurized air slowly fills the reservoir through the holes in the piston. The cannon is loaded once the pressures between manifold and reservoir balance.

5

5. To fire the cannon, a small exhaust valve opens and releases air pressure behind the piston, drawing the piston back into the manifold. As soon as the piston clears the boom valve tube, the pressurized contents of the reservoir release into the atmosphere with an impressive bang.

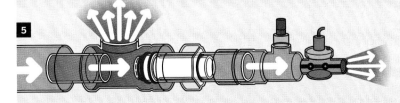

Illustration by Damien Scogin

SET UP.

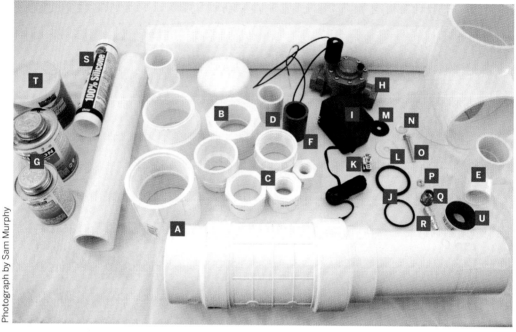

MATERIALS

[A] 3" diameter PVC parts:
» 18"–24" long pipe
» Tee fitting with female
   pipe thread (FPT) stem
» Repair coupling
» FPT to slip adapter
» Male pipe thread (MPT)
   to slip adapter
» End cap

[B] 2" diameter PVC parts:
» 15" long pipe
» FPT to slip adapter
» MPT to slip adapter
» Insert coupling
**ProPlumber model
#PPFC200. Only carried
at Lowe's.**

[C] PVC reducing bushings:
» 3" to 2" slip (2)
» 2" to 1½" slip
» 1½" slip to 1" FPT
» 1" slip to ½" FPT

[D] 1½" PVC pipe, 2" long

[E] PVC reducing tee
fitting, 1½"×1"×1½"

[F] 1" repair coupling,
Schedule 80 PVC **must fit
inside 2" insert coupling**

[G] PVC primer and
medium-thickness glue

[H] Irrigation valve, 24V
solenoid with 1" MPT ends
A cheap one ($10) is fine.

[I] 24V power supply **Look
in the irrigation aisle next
to the valves.**

[J] 2" O-rings, ³⁄₁₆" thick (4)
» 1¾" O-rings, ⅛" thick (4)

[K] On-off switch, SPST

[L] Fender washers, 1⅝"
with ⁵⁄₁₆" hole (2)

[M] Neoprene washer,
1½" with ⁵⁄₁₆" hole

[N] ⁵⁄₁₆" washers (2)

[O] ⁵⁄₁₆" bolt, 2" long

[P] ⁵⁄₁₆" lock nut with
nylon insert

[Q] Brass pipe adapter,
½" MPT to ¼" FPT

[R] Quick-release pneu-
matic coupling, ¼" MPT

[S] Silicone caulk

[T] Lithium grease

[U] Teflon pipe tape

TOOLS
[NOT SHOWN]

Plumber's epoxy putty
(optional)

Air compressor that can
produce 40–60psi **An air
tool compressor is best.**

Tape measure

Hacksaw or pipe cutter

Crescent wrenches
at least 2

Vise **capable of clamping
the flange of 3" bushing**

Gloves and goggles

Sandpaper **coarse and
medium grits**

Electric drill with grinding
stone bit

Drill press or lathe

Popsicle stick

File (optional)

MAKE IT.

BUILD YOUR OWN BOOM STICK

START ⋰⋰ Time: **A Day** Complexity: **Medium**

1. MODIFY THE PVC FITTINGS

1a. Take two 3" to 2" bushings and grind out the ridge inside them with a drill using a grinding stone bit, so that the 2" pipes can slide firmly through.

1b. Sand down both sleeves of the 2" insert coupling so they slide easily into 2" pipes. You can bolt the part between 2 fender washers, chucking it into a drill press, and sanding carefully on both sides to keep it from flying off. This will be our moving piston body.

NOTE: Don't let the PVC get hot or it will melt and deform. Use light pressure and moisten it occasionally to keep it cool. Start with coarse-grit sandpaper and finish with medium-grit.

Photography by Edwin Wise

⚠ **CAUTION: DANGEROUS PROJECT** At normal temperatures, standard Schedule 40 PVC has a working pressure of around 150psi, but heat, sunlight, solvents, scratches, and time make the material lose strength, and even at the 40–60psi used for this project, it will eventually fail. When it does, it will break into fragments that will be thrown with great force by the compressed air. Always operate your Boom Stick inside a solidly built plywood box or wall, so that shrapnel cannot reach anyone's tender flesh.

ABS plastic does not shrapnel like PVC, but the common type used for DWV (drain/waste/vent) applications is not pressure-rated, so it may or may not hold up. Foam-core PVC or ABS is even more lightweight and MUST BE AVOIDED AT ALL COSTS.

Pressure-rated ABS such as Duraplus from Ipex ( ) is the perfect material for this project, but it costs 10 times as much as Schedule 40 PVC. Copper and other metal pipes are similarly expensive.

1c. Cut the pipe pieces down to size. For the reservoir, cut an 18"–24" length of 3" pipe. For the boom valve and piston carrier tubes, cut 2 lengths of 2" pipe, one 6" and the other 8". For the air fittings, cut a 2" stub of 1½" pipe.

1d. Cut the 1" Schedule 80 repair coupling into a 1¼" section and a 1" section. There may be ¼" or so of scrap left over.

1e. Cut the sanded insert coupling into 3 pieces by trimming a ⅜" ring off one end and chopping enough of a sleeve off the other end to leave a ¹¹/₁₆" stub.

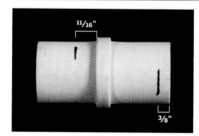

NOTE: The measurements need not be exact, but look at the photos for Steps 2c–2e on the next page to see how this piece is used.

1f. File or sand the ends of the pipes smooth. File or sand a bevel on one side of the ⅜" coupling ring and the boom valve (2"×6") pipe. These bevels will correct minor alignment errors during operation.

2. ASSEMBLE THE PISTON

2a. Drill four ⅛" or smaller holes in one fender washer, just outside the radius of the regular ⁵/₁₆" washers that will be mounted over them. Drill four ¼" holes in the other fender washer, also outside the radius of the smaller washers.

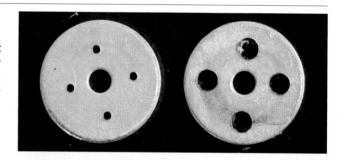

2b. Run the bolt through, in order: a small washer, the neoprene washer, the small-drilled fender washer, the 1¼" Schedule 80 PVC segment, the large-holed fender washer, another small washer, and the lock nut. Tighten the lock nut just enough to hold the assembly firm, but not so much so that the neoprene distorts.

NOTE: The neoprene washer limits air-flow, which lets pressure push the piston forward into position. When the pressure balance reverses, it seals the piston as it travels back in order to put all the air into the boom.

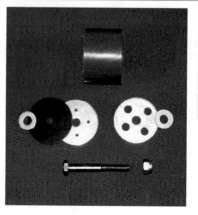

2c. Fit two 2" O-rings onto each end of the main piece of the sanded insert coupling. Then test-assemble the entire piston. Using the other 2 pieces of the insert coupling and two 1¾" O-rings, enclose the 1" Schedule 80 PVC segment and the piston valve assembly as shown.

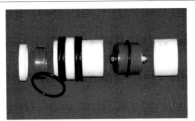

NOTE: Pipe fittings are tapered, which makes it harder to get the parts to fit together nicely when they're cut. Use epoxy putty or gel to reinforce the construction as needed.

2d. Use PVC glue or epoxy to glue the piston together. First, glue the ¾" Schedule 80 segment into the ⅜" ring, on the side opposite the bevel. Place a small O-ring as a spacer on the segment, and glue this subassembly into the body of the insert coupler.

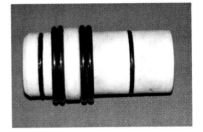

2e. Glue the piston valve halfway into the assembly, with the neoprene washer facing in. Then glue the remaining insert coupler piece around the outside half, with another small O-ring spacer in between. Reinforce the connections with epoxy.

3. FIT THE O-RINGS

The large O-rings are bumpers that protect the PVC during operation. The small O-rings form the piston's seal inside the piston carrier and boom valve cylinders. The long end of the piston must mate with the 2"×8" pipe, and the short end with the 2"×6" pipe. The goal is for the piston to be able to fit into the pipe and seal via the O-rings. These can be tricky to get right. With the PVC parts I bought, the perfect-sized small O-ring would be ⁵⁄₃₂" thick. But I only found them available in ⅛" and ³⁄₁₆", so I used the ⅛" size and pushed them out with a layer of silicone caulk underneath.

3a. Glue flowed onto the small O-rings in previous steps. This isn't good for them, so once the glue dries, cut or pry them off and discard. Replace with the remaining 2 small O-rings.

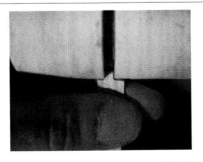

3b. Glue the large O-ring bumper pairs together with silicone caulk. This keeps them from jumping off. (You can also try one thicker O-ring on each side.)

3c. Out of a popsicle stick, make a small tool that fits a groove ⅛" deep and ⅛" wide. Fill the piston grooves with silicone caulk, and use the tool to remove all but the thin layer that it can't reach. This will help the small O-rings make a seal.

NOTE: You may need to do this several times before the O-ring seals and the piston slides. Even with everything fitting and well greased, the difference between success and jamming or leaking is subtle.

4. ASSEMBLE THE BOOM VALVE AND PISTON CARRIER

4a. For the boom valve assembly, use PVC glue to weld one of the slide-through-modified 3" to 2" bushings into one end of the 3" tee fitting, and the 3" female adapter into the opposite end. Weld the unmodified 3" to 2" bushing into the 3" repair coupling. Wait for the glue to dry, and then dry-fit each end of the 2"×6" pipe into the 2 bushings, with the beveled end in the tee assembly.

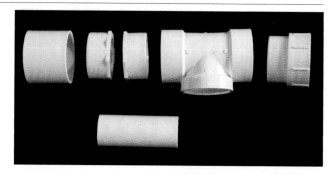

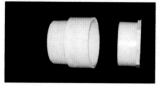

4b. For the piston carrier, PVC-weld the other modified bushing into the 3" male adapter. Let the glue dry, and then test-fit the 8" piston carrier pipe into the bushing.

4c. Slide the long end of the piston into the carrier pipe and screw the piston carrier male adapter into the boom valve assembly female adapter.

4d. Adjust the 2 pipes in their bushings until the piston (without its O-rings) travels freely between carrier pipe and boom valve assembly. The small O-rings should tuck inside both tubes when the piston is extended, and you can see a gap between the piston and the valve pipe when the piston is retracted. Proper alignment is key. Mark the position of the 2 tubes, remove them (and the piston) from the bushings, and then weld the tubes back into place at the marks.

4e. Mark the positions of everything. Then remove the piston, unscrew the 2 assemblies, and glue in the tubes. Apply primer and glue only to the 2" pipes inside the marks and not the bushings, or else you'll foul the ends of the pipes.

4f. Weld the unmodified bushing and repair coupling assembly to the other end of the valve pipe, and the 2" male adapter to the free end of the carrier pipe. Alignment isn't so important with these.

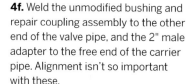

5. ASSEMBLE THE RESERVOIR AND MANIFOLD

Unlike the valve/piston system, the reservoir and manifold are low-precision designs that will tolerate variation.

5a. For the reservoir, weld the 3" end cap onto one end of the 3"×18"–24" pipe, and weld the other end into the repair coupling on the boom valve assembly.

5b. For the manifold, weld the short piece of 1½" pipe into one side of the 1½" tee fitting and weld its other end into the 2" to 1½" reducing bushing. Weld the 2" female adapter to the bushing.

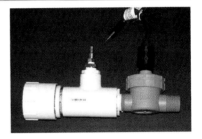

5c. Screw the inlet port of the 1" irrigation valve into the 1½" slip to 1" female threaded bushing, using Teflon tape to seal the threads. The flow arrows should point away from the bushing. Tighten firmly. This is the exhaust valve.

5d. Weld the exhaust valve subassembly into the other side of the manifold tee fitting, orienting the wiring connections as desired. For the trigger, connect the on-off switch to either of the 2 wires.

5e. Weld the 1" slip to ½" female threaded bushing to the center port of the tee fitting. Wrap Teflon tape around the brass adapter and thread it into the bushing, and then Teflon-tape the quick-release coupling into the adapter. This is the air intake.

6. FINAL ASSEMBLY

6a. Allow 24 hours for all of the PVC solvents and glues to cure. Remember, patience is a virtue.

6b. Lather up the piston and all the O-rings with lithium grease. Insert the piston into the piston carrier with its valve aiming toward the reservoir.

6c. Screw the piston carrier back into the boom valve assembly. This connection does not need to be airtight, but the carrier pipe and valve pipe must be aligned.

6d. Wrap several layers of Teflon tape around the 2" male adapter on the piston carrier, and screw it into the air manifold. You're done!

FINISH ☒

NOW GO USE IT »

USE IT.

GO BOOM

BOOM STICK OPERATION

1. Attach an unpressurized air hose to the quick-release fitting on the air intake.
2. Attach the 24V power supply to the exhaust valve and trigger switch.
3. Test the exhaust valve to make sure it works.
4. Put the entire system in a sturdy box or solid wall, or at least behind a blast shield.
5. Pressurize the manifold to about 40psi. The piston should snap into the valve tube and the reservoir should fill with a hiss. If the piston doesn't fit into the valve tube, the small O-rings may be pushed out too far. If the air leaks around the O-rings, they are not out far enough. If the tubes are misaligned, you may have to rebuild the piston carrier.
6. Activate the trigger switch for about half a second. The pipe behind the piston will lose pressure and the piston will slam back into the carrier pipe, exhausting the reservoir.
7. Jump for joy at the loud bang!
8. Repeat.

FIXES

The hard part in this design is getting the O-rings to seal firmly without jamming the piston's motion. If you just can't get them to seal, never fear; add a second irrigation valve to the air inlet, and only let air in just before you want to set off the device. The effect won't be as clean, but you'll lose less air during operation.

The piston carrier is modular for a reason: you can remove it easily and experiment with different piston designs (of which there are many), and you can replace the piston if it breaks. Also, if you glue the piston carrier into place with bad alignment to the valve tube, you only have to throw away a few inexpensive pieces to try again.

To keep stuff from falling into the Boom Stick, cover all openings in its box with hardware cloth.

RESONATING CHAMBERS AND CONFETTI

Once you get the basic Boom Stick working, create a resonating chamber by gluing a 3" male adapter onto some 3" pipe, and screwing it into the cleanout port on the boom valve's tee fitting. Try constricting the exhaust, putting a 3"×2" or even smaller bushing into the base of this chamber. Try long ones and short ones. Stuff confetti into the chamber and make a mess of your workshop. But never, ever launch anything directly at anyone!

ACTION VIDEO

See Wise's Boom Stick in action at makezine.com/go/boomstick. But note that the boom sound is mostly lost. Microphones can only do so much.

RESOURCES

Huge list of haunter how-tos:
halloweenmonsterlist.info

O-Ring Handbook:
dichtomatik.us/products/o-ring-handbook

Generally useful site with size and pressure specs for PVC: engineeringtoolbox.com

OSHA warning memo on pressurized PVC:
makezine.com/go/oshapvc

Wooden Mini Yacht

AUTHENTICALLY RIGGED MODEL BOAT SAILS ACROSS POOLS AND PONDS.

BY THOMAS MARTIN

Photography by Thomas Martin

When my son was 3 years old, I made a small bathtub boat with him, using scrap wood and a piece of dowel. It lasted much longer and got more of his attention than any dollar-store bath toy, and about six years later we decided to try building a larger boat for the pool and local ponds we fished.

Here's the result of our experimentation: a simple and worthy pond sailer that's rigged and scaled like a real yacht. You can build it in a weekend using readily available materials and tools.

Build Your Boat
Time: A Weekend Complexity: Easy

1. Prepare the sailcloth.

It's hard to find waterproof fabric that's easy to cut and won't fray. You can make your own by stretching ripstop nylon loosely over a frame or 2 hangers, and spraying it lightly (in a well-ventilated area) with polyurethane. First spray up and down, and then back and forth, until the fabric is well coated but not saturated. Let dry overnight.

2. Mark and cut the parts.

Download the project plan at makezine.com/projects/make-20/wooden-mini-yacht and print it at full size. Following the plan, measure and mark the mast, jib boom, and mainsail boom lengths on the ¼" dowel. Trace the hull from the printed pattern onto the top and 2 ends of the cedar block; cut templates or use carbon paper. Draw the keel and masthead crane patterns on the brass strips, and draw the bowser (rigging clip) pattern 8 times on the thin plastic.

A

B

C

Cut and drill all the parts. Any fine-tooth saw will cut the dowel, or you can roll it under an X-Acto blade and snap the score. Heavy-duty shears or a hacksaw will cut the brass; be sure to file away the sharp edges afterward. You can saw or file down the hull's shape, then use a hobby knife or thin chisel to excavate the slot for the keel. Drill all holes, plus pilot holes for the screw eyes (in the hull, just poke pilot holes in by hand with a thumbtack).

Finally, file, sand, and smooth all parts. The more time you spend here, the better — especially if you plan to use a clear finish over the wood.

3. Mount the keel.

On the underside of the hull, mask both sides of the keel's slot with tape. Wearing gloves, and in a well-ventilated location, mix and spread some 5-minute epoxy into the slot using a scrap stick or wooden match. Slide the keel into position and hold it there while the epoxy cures (Figure A). You can square it up using a business card on each side. Use a gloved finger to smooth the epoxy along the joint line, and fill any voids with more epoxy.

4. Finish the wood.

Finish the hull uniformly, or for a big-boat look, paint the outside of the hull and stain the deck (Figure B).

Sand the hull with 100-grit paper over a sanding block, and again with 150-grit. Apply a first coat of paint or varnish, and re-sand with 180-grit before each subsequent coat.

For a stained deck, first paint the hull upside down, then re-sand the top perimeter to remove any overspray. Rub stain into the deck and edge, let dry, and coat with varnish or polyurethane.

For the mast and boom pieces, bevel the cut edges for a more finished look, then sand with fine grit to remove any fuzz. Stain if desired, and cover with at least 2 coats of varnish or polyurethane sealer, sanding lightly between coats.

MATERIALS

Western red cedar wood, 2"×4"×18" for the hull. From a lumber yard, or better yet, see if a local fencing contractor can sell or give you an offcut.
¼" hardwood dowel, 36" length
Brass wire brad, #16×1¼" for the clew hook
Screw eyes, zinc-plated steel, ½" (11)
Brass strip, .032"×½"×12" for the masthead crane
Brass strip, .093"×2"×12" for the keel
¹⁄₁₆" sheet plastic, about 2" square cut from the bottom of a milk jug
Metal eyelets (grommets), ⁵⁄₃₂" (4mm) diameter (8)
Ripstop nylon, 2'×2' for sails
Braided dacron line, 30lb–80lb test, 6' for rigging, from a fishing supply store
Spray paint and/or stain (1+ cans)
Spray polyurethane
5-minute epoxy
Cyanoacrylate ("super") glue or wood glue

TOOLS

Scissors or carbon paper to avoid cutting the plan
Needlenose pliers, masking tape, gloves
Small hammer or mallet
Hacksaw
Jigsaw or coping saw
Bastard file or 4" hand rasp and file
Hand drill and bits: ¹⁄₃₂", ¹⁄₁₆", ⅛", and ¼"
Eyelet tool such as Dritz #104T
Sanding block and sandpaper: 100, 150, and 180 grit

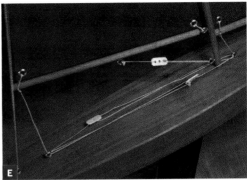

E

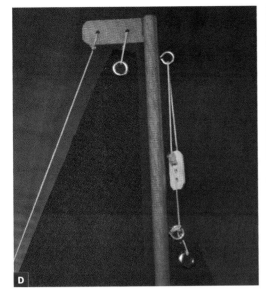

D

5. Assemble the mast and booms.

Cut a slot in the top of the mast and glue in the masthead crane. Once that's secure, follow the plan to install all screw eyes: 4 to the mast, 1 on the fore end of each boom, 1 more on the mainsail boom (for the boom vang), and 4 to the deck. Screw these in until the shank of the screw is completely into the wood. Insert the brass brad down through the hole in the jib boom and bend it into a clew hook (Figure C, previous page).

Use needlenose pliers to open the mainsail boom eye, hook it onto the eye on the mast, and close it. This forms the gooseneck, the joint that lets the boom swing from side to side (Figure C, far left).

Press the mast down into the hole in the deck with the masthead crane centered aftward, and tap it gently down into its hole with a hammer.

6. Add the sails.

After the sail material is dry, trace and cut it to the plan patterns. Lay the boat on its side on a hard surface with the masts and booms in place and fit the sails to the areas for rigging. For the grommets, cut a small X at each sail corner, insert a grommet up through the hole, press the cloth down around it, and tap the grommet flat with the eyelet tool until it firmly grips the cloth.

NOTE: It's a good idea to practice setting grommets first with a couple of sailcloth scraps and extra grommets.

It's time for rigging. Knot and cut a short length of dacron line, thread it through a bowser, and

string the boom vang. For these and all other knots, add a *tiny* drop of cyanoacrylate glue immediately after tying; the line is slippery and won't hold knots otherwise.

Use 5" lengths of line to tie each sail grommet to its corresponding screw eyelet or drilled hole with a square knot (Figures C–E). You'll need about 10" for the top of the jib sail, which threads through 2 eyelets before tying off to the uphaul bowser.

Referring to the plans, tie the 4 lower connections on the booms first, and then add the upper lines for tension, so there are no wrinkles in the sails along the booms. Thread a bowser onto the jib uphaul as indicated: for their final tensions, you'll adjust the jib using the uphaul at the top (Figure D), and the mainsail using the boom vang (Figure E).

For the backstay, tie in a long length of line at the masthead crane and install a bowser, routing the line through the eyelet at the stern.

Tighten the backstay and the sails so that they're fairly tight but the mast is not bowed forward or aft.

Finally, add the 2 lines called sheets. For these, cut two 15" lines. Tie each one through the hole in the aft end of a boom, thread it through the sheet eyelet on the deck just underneath, then through 2 holes in a bowser, through the other sheet's eyelet, and finally through the last hole in the bowser, double-knotting the line (Figure E).

NOTE: It's important to tie the bowsers exactly as shown on the plan to make them work.

The sheets let you adjust the angle (trim) of the sails — slack for downwind sailing or tight for crosswind — letting you cross a pond or pool in any direction that isn't too close to directly upwind.

Now it's time to go sailing!

Thomas J. Martin is an artist, illustrator, writer, and blogger at tmrcsailplanes.com and aerosente.com.

Crafts
AND Wearables

Making encompasses a broad range of activities—from ancient crafts to modern electronics. Each issue of *Make:* features projects and skill builders aimed at a variety of levels and interests. Making is often a blend of science, technology, and art. The technology may be as old as humanity or as cutting-edge as the latest developments, but the goal is always to inspire you to take it into your own hands, try it yourself, and make something in the world.

Making helps us understand how the products we consume are made. We all use soap every day, but how many of us have made our own? In "Making Bar Soap," journalist Alastair Bland tells you a bit about the history of soap and then shows you how to make an inexpensive, long-lasting supply for yourself, free of the toxins often found in commercial products.

Wendy Jehanara Tremayne is committed to the off-the-grid DIY lifestyle, which she chronicles in her book *The Good Life Lab: Radical Experiments in Hands-On Living* and her blog Holy Scrap. Wendy provides a quick introduction to one of the world's oldest food technologies, fermentation, in her article "Three-day Kimchi."

Larry Cotton and Phil Bowie team up to present a medley of projects that use humble PVC pipe from your local hardware or home improvement store to create attractive furniture and décor. Sean Michael Ragan contributes a piece on how to stain PVC pipe if you prefer that to painting it.

Art critic Robert Hughes once described the effect of modern art as "the shock of the new." *Make:* magazine editor Jordan Bunker shows you how to make conductive ink for your next art project that may not necessarily shock but will certainly carry an electrical current.

Scott Heimendinger was fascinated by sous vide, a technique in which you vacuum-seal food and cook it for long periods at low temperatures. Commercial sous-vide cookers were expensive, though; so Scott decided to make his own economically from scratch—and now you can, too. Scott took the concept he developed for this article and Kickstarted it into a commercial product called the Sansaire. He's successfully gone from maker to maker pro!

If you like to make an impression with your outfits, we have a trio of wearable electronics projects for you.

Clayton Ritcher wondered why kids should have all the fun when it comes to light-up shoes. The adult-sized footwear he hacks in "Luminous Lowtops" contains force-sensitive resistors and an Arduino mini that let you control the color gradient of the LED strips. Kathryn McElroy is a user-experience designer for IBM. In her article "The Chameleon Bag," she builds a messenger bag flap that responds to different RFID tags with colorful displays and animations. Becky Stern and Tyler Cooper combine an Adafruit Flora microcontroller with the NeoPixel LED ring, a GPS module, and an accelerometer to create a seriously cool watch with navigation and compass modes.

SOUS VIDE IMMERSION COOKER

By Scott Heimendinger

Photography by Scott Heimendinger

PRECISION GOURMET

I'm fascinated by sous vide cooking, in which foods vacuum-sealed in plastic are immersed in a precisely temperature-controlled hot water bath to achieve optimal doneness.

But most sous vide (soo-veed) cooking machines are commercial models that cost north of $2,000, and the first "home" version, the countertop SousVide Supreme, is priced in the neighborhood of $450 (not including vacuum sealer), which is still a steep investment for something that essentially keeps water warm.

I decided to build a better device on the cheap. Behold, the $75 DIY Sous Vide Heating Immersion Circulator! By scrapping together parts from eBay and Amazon, I created a portable device that heats and circulates water while maintaining a temperature accurate within 0.1°C. And unlike the SousVide Supreme, it mounts easily onto larger containers, up to about 15 gallons, for greater cooking capacity.

The water is heated by three small immersion heaters and circulated by an aquarium pump to keep the temperature uniform. An industrial process temperature module controls the heaters, and an eye bolt lets you clamp the entire apparatus to the rim of a plastic tub or other container.

To cook sous vide, you also need a vacuum sealer, which this project does not include. I bought a good one new for about $112.

Scott Heimendinger runs the blog SeattleFoodGeek.com where he offers tips for geeky recipes and projects.

Slow Food

Cooking sous vide ("under vacuum" in French) means vacuum-sealing ingredients in plastic and immersing them in a relatively low-temperature bath (typically around 140°F/60°C) for long periods of time. The process retains cell structure and juices that foods lose through other cooking methods.

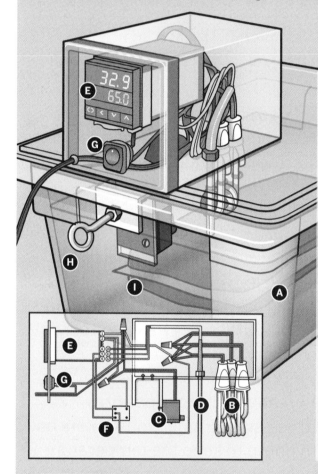

Ⓐ A container holds the hot water bath that cooks the food.

Ⓑ Three immersion heaters heat the water.

Ⓒ An aquarium pump circulates the water while the heaters are on, to keep the temperature distribution uniform.

Ⓓ A thermocouple probe reads the water temperature.

Ⓔ The PID controller runs a proportional-integral-derivative (PID) algorithm that monitors the water temperature and switches the immersion heaters on and off to keep it constant.

Ⓕ A relay converts 12V DC switching from the PID controller into 120V AC switching for the heaters.

Ⓖ A rocker switch turns the entire unit on and off.

Ⓗ An eye bolt clamps the unit to the side of the container.

Ⓘ A vacuum bag acts as a thermally conductive but waterproof barrier between the food and the water bath.

PID Control

A PID temperature controller is like an advanced thermostat. Regular thermostats click on and off when the measured temperature passes fixed thresholds, and the resulting temperature oscillates around the thresholds. This creates an opportunity for temperature "carry-over" in the food, like the way meat's internal temperature can rise after you take it out of the oven.

In contrast, PID controllers use the PID algorithm, common in industrial control applications, to track temperature changes and calculate how much ON time will raise the measured temperature (the process variable) to the target (the set point) asymptotically, rather than overshooting it.

By predicting, mathematically, the impact that turning on the heaters for 1 second will have on overall water temperature, the PID controller enables extremely precise temperature control and stability over long periods of time.

NOTE: Be aware of a minor technical nuance — the model CD101 PID controller used in this project is accurate when used in Celsius mode. For unknown reasons, people have experienced inaccurate readings and fluctuating temperatures when the controller is set to Fahrenheit.

Illustration by Damien Scogin

SET UP.

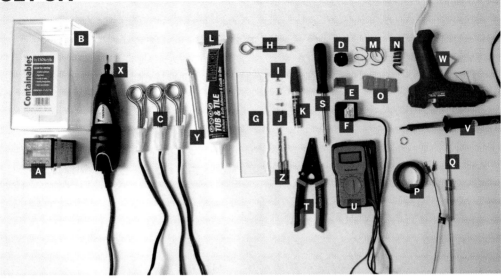

MATERIALS

[A] Digital PID temperature controller, model CD101 with PT100 support and voltage pulse output
Some model CD101 units will not work; you need to make sure that the 5th character of its code (on a sticker on the side) is V, for voltage pulse output. I bought mine (code FK02-VM*AN-NN) for $39 from eBay, or see Sure Electronics #RDC-TE11113 (sureelectronics.com).
   This controller has 12 screw terminals. Posts 1 and 2 connect to AC power, 5 and 6 control a relay for the heater, and 10–12 connect to a thermocouple. If you use a different controller, refer to its datasheet for the functionally equivalent connection locations.

[B] Clear acrylic storage container, 7"×4"×4"
item #B000NE80GO from Amazon (amazon.com)

[C] Immersion heaters, Norpro 559 (3) Amazon #B000I8VE68

[D] SPST rocker switch, heavy duty rated 10 amps at 125VAC or 250VAC, mounts in ¾"-diam. hole

[E] SPDT mini relay, 7V–9V DC, 12A RadioShack #275-005. You may want to upgrade to a solid-state relay for a more robust, reliable build. The wiring is the same for an SSR as it is for the relay included in the build instructions. A good option is Lightobject #ESSR-25DAC.

[F] Aquarium pump with suction cup "feet" Catalina Aquarium #A801 (catalinaaquarium.com)

[G] Clear acrylic sheet (plexiglass), 7"×2"×¼" thick From hardware stores or online vendors; you may have to cut it to size with a band saw, table saw, fine-toothed hacksaw, or jigsaw.

[H] Eye bolt, ¼"×2", with nut

[I] Machine screws, #4-40, ½" long, stainless steel, with matching nuts (2)

[J] Sheet metal screws, #6-40, ⅜" long, stainless steel (2)

[K] Super glue

[L] Waterproof silicone caulk aka tub and tile caulk

[M] Wire, 22 gauge, stranded, 1'

[N] Wire, 14 gauge, solid core, 4'

[O] Wire nuts, large (4)

[P] Electrical tape

[Q] Thermocouple temperature sensor probe with 3 leads #PT100 from Virtual Village (virtualvillage.com), or from eBay.

[NOT SHOWN]

Large, straight-sided container to hold the water. Because the bath doesn't get very hot or touch the food, plastic is OK; I use a 17qt plastic storage bin I bought at Bed Bath & Beyond (bedbathandbeyond.com).

TOOLS

[S] Phillips screwdriver

[T] Wire cutters and strippers

[U] Multimeter

[V] Soldering iron and solder

[W] Hot glue gun

[X] Rotary cutting tool with router bit such as a Dremel

[Y] Hobby knife

[Z] Drill and drill bits: ⅛", 5/32"

[NOT SHOWN]

CNC laser cutter (optional)

Marker

Stove and oven mitt

Bowl of cold water

MAKE IT.

BUILD YOUR SOUS VIDE IMMERSION COOKER

START ⋙   **Time: 6 Hours Over 2 Days   Complexity: Easy**

1. CUT THE ACRYLIC ENCLOSURE

This is the most difficult part of the project. For your cooker to be sturdy, water-resistant, and decent looking, the mounting holes must be cut precisely. I used a CNC laser cutter I have access to at work, but with a steady hand you can achieve the same results using a rotary tool like a Dremel.

1a. Download and print the cutting template in *Cutouts.pdf* from makezine.com/projects/sous-vide-immersion-cooker. This template matches the heaters, controller, and switches I used, so you'll need to adjust the shapes and sizes if you use different parts.

1b. Following the template, cut out the 3 holes for the immersion heaters on one side of the acrylic container, near its base.

1c. Mark and cut the small oval-shaped hole for the water pump power cord and a circular hole for the thermocouple.

1d. Follow page 2 of the template to mark and cut openings in the lid of the container for the controller, switch, and power cord.

2. MOUNT THE IMMERSION HEATERS

2a. Cut the power cord off each heater, leaving an 8" tail of wires from each coil end.

2b. Using your hobby knife, scrape the flat sides of the heater handles to remove the lettering and flatten the circular rim where it meets the cord.

2c. Fit all 3 heaters into their holes in the container, such that the coils stick out and line up. Your finger should be able to fit through all 3 coils at once. Trim the holes if needed to make the heater handles fit snugly.

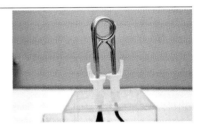

2d. Caulk a fully waterproof seal around the heater handles on the outside of the enclosure and let dry overnight.

For added strength, add hot glue over the caulk after it's dried. The hot glue only needs 5 minutes to dry before continuing.

3. ATTACH THE MOUNTING BRACKET

3a. Cut a 7"×2" rectangle out of ¼" acrylic. The cut sides don't have to be perfect as long as the rectangle dimensions are approximate. Use a rotary cutter to rout a shallow recess that's the same shape and size as the nut for your eye bolt, centered at one end, as shown on page 3 of the template. Drill a ¼" hole through the center of the recess for the eye bolt to pass through.

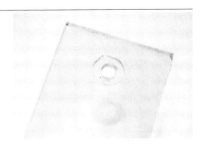

3b. It's time to bend the acrylic sheet. Clear a countertop near your stove, put a bowl of cold water within reach, and mark the bend point lines from the template onto the acrylic.

NOTE: If your countertop is too thick or thin for this process, you'll need to bend the sheet around a hard, stable object, approximately 2" thick, such as a 2×4.

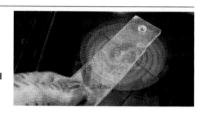

Turn one of the front burners on high, hold the undrilled end of the acrylic with an oven mitt, and place the drilled end with bend lines a few inches above the burner, moving the acrylic around and turning it to heat both sides. If it begins to form small bubbles, move it away from the heat a little. When the acrylic starts to curl away from the heat, it's ready to bend.

3c. With the recess for the nut facing outward, bend the acrylic around the edge of the counter (or other object) approximately along the marked lines to form a "J." While in place, press the middle segment of the J against the edge of the countertop to make it flat. Immediately drop the acrylic into cool water so that it holds its form.

3d. Position the middle segment of the J bracket against the heater side of the enclosure, with the bracket's nut side aligned along the enclosure's rim. This is how you'll bolt these 2 pieces together. Mark and drill two ⅛" holes through the bracket and 2 matching holes in the enclosure. Check that the 4-40 machine screws fit through both pairs of holes.

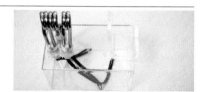

3e. Liberally apply super glue to the underside of the bracket, and bolt it to the enclosure with the machine screws and nuts.

3f. Liberally apply more super glue to the recessed area for the eye bolt nut, and set the nut in place. Reinforce the bond with hot glue, but don't obstruct the hole in the nut.

For more support, you can cut a 1½" acrylic round with a hole fitting the nut, and glue it to the J-clamp overlapping the opening. This will provide extra gluing surface and take stress off the nut.

4. ATTACH THE PUMP

4a. Cut the water pump's power cord about 8" from the pump. Reserve the severed cord, which will become the cooker's power cord.

4b. Position the pump on the long side of the bracket, with its cord pointing toward the enclosure and its top aligned with the top of the heater coils. The pump should be placed so its outlet is more or less pointing through the coils. Mark the locations of the suction cup feet on the bracket.

4c. Drill two ⁵⁄₃₂" holes through the bracket at the exact centers of the suction cup locations. Check that the holes can accommodate the sheet metal screws, and enlarge them if needed by wiggling the drill bit.

Remove the suction cups from the pump, and attach the pump to the bracket with the sheet metal screws in the suction cup holes.

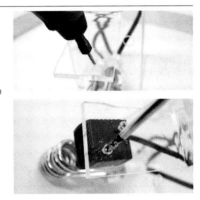

5. COMPLETE THE WIRING AND ASSEMBLY

For the big picture, refer to this wiring diagram.

5a. Split the heater cords apart and strip about 2" of shielding away from each end. Twist and solder together 1 lead from each of the heaters, then repeat for the other set of leads to wire the 3 coils in parallel.

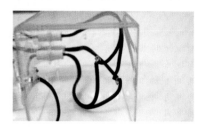

5b. Cut two 18" 14-gauge wires and strip ⅜" of shielding from each end. Use wire nuts to bundle one end of each wire with a set of heater leads. One wire will connect to AC power, and the other will connect to the relay.

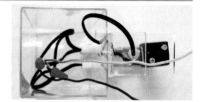

5c. Remove the mounting ring from the PID controller, fit its front panel out the opening you cut for it in the acrylic lid, and secure the controller from the back by replacing the mounting ring.

5d. Fit the power switch into its hole in the acrylic lid and secure it with its included mounting nut. Run the power cord you cut from the water pump through its hole nearby. Separate its leads, strip ¼" of shielding, and solder 1 lead to 1 leg of the power switch.

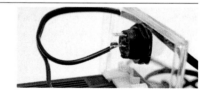

5e. Cut and strip 3 short (about 4") lengths of 14-gauge wire. Twist-bundle 1 wire with the free power cord lead, 1 lead from the water pump, and 1 of the 14-gauge leads from the heaters. Secure the bundle with a wire nut, and screw the short wire to terminal #1 on the controller.

5f. Solder the other 4" piece of 14-gauge wire to the unconnected leg of the power switch, and then screw the other end to the #2 terminal on the controller.

5g. Make a new bundle connecting the free lead from the water pump, the wire from the #2 terminal, and the third short length of 14-gauge wire.

5h. Orient the relay so that its underside (with the pins) faces you, and the row with 3 pins is facing down. Solder the free lead from the previous step to the upper left pin of the relay.

5i. Solder the free lead from the heater bundle to the middle bottom pin of the relay.

5j. Cut and strip two 6" lengths of 22-gauge wire. Solder one to the bottom right pin of the relay and connect it to the #5 terminal on the controller. Solder the other piece to the relay's bottom left pin and connect it to the #6 terminal.

5k. Insulate and strengthen the relay connections with a liberal application of hot glue. Wrap the relay and wires with electrical tape.

5l. Apply hot glue around the holes for the power cord and water pump cord, to prevent steam or water from leaking into the enclosure.

5m. Insert the PT100 thermocouple into its hole in the enclosure and secure with hot glue. Connect its red lead to the controller's #10 pin, its blue lead to the #11 pin, and the yellow lead to the #12 pin.

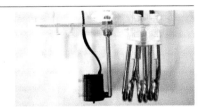

5n. Put the acrylic lid on the open end of the container, and apply a neat strip of electrical tape around the edges to hold it on. (Once you've verified that the cooker works, you can then glue the seam.) Insert the eye bolt through the nut.

5o. Everything is wired up and assembled. Now you probably want to see if it works. BUT WAIT! Don't turn the machine on (ever!) unless the coils are submerged in water or else they'll burn out in about 5 seconds. I learned this the hard way.

5p. Fill a container with enough water to cover the top of the circular part of the heater coils by ¼"–½". Mount the cooker inside with its bracket hanging over the lip, and tighten the eye bolt to secure.

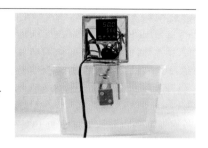

Plug in the cord and flip on the power switch. If the PID controller turns on and the pump starts pumping, that's a good sign. Note that the heaters may not warm up yet, depending on what the controller's default target temperature is.

> ⚠ **CAUTION: 120V AC current is dangerous and misuse can injure or kill. Seek the assistance of a qualified electrician if needed. Take care to keep the control box and wires out of the water or other liquids. And once you've tested your cooker, seal the lid on the control box.**

6. PROGRAM THE CONTROLLER AND TEST

For the model CD101 PID controller manual and a full sequence of photos showing how to program it for this project, visit makezine.com/projects/sous-vide-immersion-cooker.

6a. The PID controller defaults to a different thermocouple than the PT100 used here, so you need to reconfigure it for a PT100 sensor. Press and hold the SET button until **AL1** appears, then keep pressing SET until you see **LCK** (lock) with **1000** underneath. To unlock the settings, we need to change this value.

6b. Use the left-arrow key to move the cursor and the up-arrow key to change **LCK** to **1100**. Then press SET again to exit the menu.

6c. Push and hold the SET and left-arrow keys together until you see **COD** (code). Use the arrow keys as before to set this value to **0000**, then press SET again to see **SL1**, the sensor type symbol.

6d. Change SL1 to **1100**, for the PT100 sensor. Press and hold the SET and left-arrow keys again to exit. Then re-lock the settings by setting the value of **LCK** back to **1000**, as in Steps 6a–6b.

6e. It's time to heat things up. The top line of the display (green) shows the current temperature, and the second line (orange) shows the target. Set a target temp by tapping the SET button, using the up and down arrows to specify the target, and pressing SET again to confirm. 50°C is good for testing.

If all's well, the OUT1 light will turn on and you'll hear a soft clicking as the relay turns on the heaters. The temperature reading will increase, and you'll hear more frequent clicking as it approaches the target.

FINISH ☒ _____ **NOW GO USE IT ››**

MAKE THE ULTIMATE STEAK 'N' EGGS

VACUUM SEALING

A good vacuum seal is essential to sous vide. Without air pockets, the heat transfers into the food evenly and efficiently. A good seal also dramatically decreases the risk of bacterial contamination from long incubation at low temperatures.

I use a FoodSaver V3835, which I bought for just over $100 with a 20% coupon at Bed Bath & Beyond. It uses rolls of plastic, so you can make the bags any size you need. I recommend against using cheaper handheld sealers. I've tried a few and had problems creating a reliable seal.

With any sealer, always wipe clean and dry the end you're going to seal, or the plastic layers might not melt together properly.

Using a zip-lock bag instead of a vacuum sealer is actually OK for cooking times up to 4 hours, but also riskier for leaking and contamination.

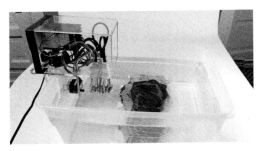

KITCHEN TESTS

To reveal the power of sous vide, cook an egg in the shell (no plastic needed) at 64.5°C for 1 hour. This yields an amazing transformation: perfectly soft whites, not runny or rubbery, and a yolk with the consistency of a rich pudding. It's impossible to achieve this through any other cooking method, and it's spectacular the first time you experience it.

I expanded on this amazing transformation by breading the yolks and quickly deep-frying them to add a crunchy shell. See my recipe at makezine.com/projects/sous-vide-immersion-cooker.

Next, try cooking a good, thick steak. Unlike conventional cooking methods, sous vide gives you a perfect medium-rare steak all the way through. To finish the steak, pat all sides dry and sear like crazy with a propane torch (the kind plumbers use).

Some things that don't work well sous vide, in my experience, are broccoli, kiwi, and strong aromatics. Note also that if you sous vide with too much garlic, your whole house will smell like feet.

GENERAL-PURPOSE TEMP CONTROLLER MOD

Several people have modified this project so that the relay switches a power outlet that the heaters and pump plug into, and the probe plugs into an added jack. Then, if a heater burns out, you plug in another one rather than having to mess with glue and solder. Modularizing the control and power components this way also lets the machine double as a precise temperature controller for a smoker box, such as Alton Brown's DIY flower-pot smoker (see makezine.com/projects/sous-vide-immersion-cooker). You just plug in the hotplate filled with wood chips.

Pick up a few steaks and break out your blow torch — it's time to start cooking like a geek!

➕ For hole templates, the CD101 PID controller programming instructions and manual, Scott's Deep-Fried Sous Vide Egg Yolks recipe, and other resources, visit makezine.com/projects/sous-vide-immersion-cooker.

Gregory Hayes

Making Bar Soap

Wash your hands of the toxins found in commercial soaps.

By Alastair Bland

JUST WHERE AND WHEN HUMANS FIRST observed the chemical reaction between oil and potash is unknown. One legend says it was at "Mount Sapo" in ancient Rome, where a creek flowed over a deposit of wood ash and animal fats created by sacrificial fires for the gods. There's evidence that the ancient Babylonians, Egyptians, and Celts all developed soap — but regardless of who was first, it's clear that observing nature led people to re-create the chemistry of water, potash, and oil to produce the first liquid soaps.

Bar soaps were innovated in the 19th century, but soon after, the soap-making industry introduced problematic chemicals to the equation, such as propyl alcohol, limonene, benzaldehyde, and methylene chloride. Today,

some soap ingredients are derived from animals rendered in factories, while others are known to be toxic, even carcinogenic, and contain byproducts of petroleum.

All natural, plant-oil-based soaps provide an eco-conscious alternative, but 2 downsides remain: the disposable packaging and the collateral damage of transportation. Fortunately, making soap at home is easy, and not much more complex than baking bread.

In its most basic form, soap consists of just 3 components — a strong base such as potash or lye, oil, and water. Potash (potassium hydroxide) is harder to find and is more conducive to liquid soap making, so we'll use lye (sodium hydroxide).

Blended at the right proportions and

temperatures, these ingredients produce a chemical reaction called *saponification* which renders the lye, normally caustic and dangerous when mixed with water, entirely benign while breaking apart the oils and eliminating their cloying greasiness.

And although curing (allowing the soap to dry and harden) takes about a month, the first and most demanding steps take just 30 minutes for practiced soap makers.

Now consider that a year's supply can be made in a single batch; each bar will cost about a dollar; you won't risk exposure to any lingering poisons — and one by one they mount: the good and healthy reasons to free oneself from chemical industries and never buy soap again.

1. Set up a soap-making station in your kitchen.

Weigh your ingredients and place each in its own vessel. Water goes in the heat-resistant jar, and coconut oil goes in the saucepan.

2. Melt the oil.

Gently warm the coconut oil on the stove. The chunks of fragrant fat will melt into clear grease (Figure A).

3. Carefully add lye to water.

Open your doors and windows. Don your protective gloves and glasses, and slowly pour the lye into the jar of cold water. Stir with a steel spoon. A fast exothermic reaction will occur as the temperature shoots to nearly boiling and the mixture momentarily emits a plume of toxic fumes. Do not inhale over the jar. Insert a thermometer and watch the temperature as it slowly drops (Figure B). Your target reading is 80°F.

⚠ **CAUTION:** Lye is highly caustic and will burn skin and eyes. Wear protective gloves and goggles, and follow all directions on the container for safe handling of lye. Clearly Figure B outs us throwing caution to the wind.

4. Take the temperature.

Insert the other thermometer into the pan of melting coconut oil. When the oil is 90°F and

MATERIALS

To make larger batches, keep proportions exactly the same.

8 fl oz (1c) cold water
79g (0.176lb) sodium hydroxide (lye) Buy a jar of 100% lye drain cleaner at your local hardware store.
10½ fl oz (297.6g or 0.656lb) olive oil It needn't be extra virgin. The quality is irrelevant since you'll mix it into a lye-and-water chemical bath.
10½ fl oz (297.6g or 0.656lb) coconut oil found at natural foods stores in plastic jars or, preferably, in bulk. Buy organic or fair-trade if possible.
8g essential oils (optional) for fragrance and nutrients
½c steel-cut oats (optional) for abrasiveness

TOOLS

Kitchen stove
Digital scale capable of accurate measurements in grams
Safety goggles
Rubber gloves
Kitchen thermometers (2)
Glass jar, heat resistant, 40oz or larger
Saucepan, stainless steel for heating oils on stove
Spoon, stainless steel for stirring lye and water
Electric eggbeater, stick blender, or pedal-powered blender
Milk carton, or wooden box with parchment paper for a mold for curing
Knife for cutting and shaping bars

⚠ **CAUTION:** Do not use aluminum kitchenware. Lye and aluminum react to form flammable hydrogen gas and can cause fire or explosion.

liquefied, turn off the heat and add the olive oil (Figure C). The temperature of the blended oils should read 80°F.

5. Combine the liquids.

When both the lye water and oils reach 80°F, combine the two in the glass jar. The mixture will abruptly turn cloudy (Figure D).

6. Blend it all together.

Blend the liquid for roughly 15 minutes using either an electric eggbeater or stick blender (Figure E). You can also whisk by hand, though this will take about an hour.

However you approach it, your goal is saponification, which occurs visibly as the liquid thickens and turns opaque (Figure F). To test for it, lift the blender from the liquid and drizzle the soap across the surface. When droplets remain on the surface for a moment before sinking — known as *tracing* — it's done.

7. Add scents and scrubs.

If you're going to add fragrances, essential oils, or oats, now's the time. Add and mix — and do it fast, because the soap may be thickening more quickly than you realize.

Any additional ingredients must be all-natural to avoid fouling up the delicate chemistry of the saponification process.

8. Mold it.

Pour the soap into your mold (Figure G). Paper cartons should be thoroughly cleaned, and wooden molds should be lined with parchment paper. Cover the filled mold with a cutting board or coffee table book and set it aside for 1–3 days to harden.

9. Cut, cure, and wash up.

Lift the long soap block from the mold, peel away the parchment paper, and cut the brick into roughly 10 bars (Figure H).

Stand each bar on end to allow the most surface-to-air contact, and set them aside to cure undisturbed. Three weeks should do it, at which point the lye's causticity has fully neutralized, and your soap is ready.

D

E

F

G

H

Congratulations. You can now wash your hands of the chemicals and toxins used by the commercial soap-making industry. Learn more at makezine.com/go/makesoap. ◪

Alastair Bland is a freelance travel, food, and news writer in San Francisco. After graduating from UC Santa Barbara, he wandered through California by bicycle and Baja California by foot for two years before falling into journalism. Follow his latest travels at blogs.smithsonianmag.com/adventure.

Three-Day Kimchi

Piquant, fiery, and fast to make, this lacto-fermented version might just get you hooked.

Written by Wendy Jehanara Tremayne ■ Photographed by Gunther Kirsch

Time Required:
**30 Minutes Prep,
3 Days Ferment**

Cost:
$5-$10

Materials

» 1 head of your favorite cabbage
» 1lb carrots
» 8 cloves chopped garlic or chopped onion
» 2 Tbsp minced ginger
» 2 Tbsp sesame oil or fish oil
» 2 tsp salt
» ½ tsp dry red pepper flakes
» juice of 2 limes
» 1 pint whey from kefir, yogurt, or live cheese. Making cheese and other dairy products produces a byproduct called whey, a protein-rich liquid full of probiotics.
» Mason jars with lids
» Airlocks (optional)

WENDY JEHANARA TREMAYNE

(gaiatreehouse.com) pursues a post-consumer life (blog. holyscraphotsprings.com) in Truth or Consequences, New Mexico, where she and her partner make their own fuel, power, food, medicine, and domestic goods. She founded Swap-O-Rama-Rama, now a mainstay of Maker Faire. Her book, *The Good Life Lab: Radical Experiments In Hands-On Living* (Storey 2013), won the 2014 Nautilus Book Silver Award for Green Living/Sustainability.

KIMCHI IS THE FUNKY, SPICY, ANCIENT KOREAN FERMENTED CABBAGE DISH that's being rediscovered by chefs today. While traditional kimchi takes months to make, our lacto-fermented kimchi is ready in days because it cooks the vegetables enzymatically. (Microorganisms work with enzymes, a chemical catalyst, to "predigest" your food.) It's full of healthy probiotics and vitamin C. And did I mention it's delicious and inexpensive?

1. Shred cabbage and carrots and fill mason jars with the mix. Press them down with the back of a spoon to pack the jar tight.

2. Mince the garlic and ginger in a bowl with sesame oil, salt, red pepper flakes, lime juice, and kefir whey that has been inoculated with a spoonful of kefir culture. (Any whey that contains a live culture will do: Try yogurt, live cheese, or kefir.)

3. Pour the liquid mixture over the vegetables so that they're completely covered. Leave a little space at the top for expansion. Cap tightly and store at room temperature, away from sunlight.

4. After about 24 hours, the lid will pop up from pressure. Refrigerate to slow fermentation.

Finishing: After 3 days in the fridge, the culture has fermented the vegetables and spices in the jar, and your kimchi is ready to eat. Serve over rice. ◐

Get more appetizing information and recommended reading at makezine.com/projects/three-day-kimchi-2
Share it: *#threedaykimchi*

More Fermented Favorites

Make these three and more at makezine.com/projects

Kombucha
Brew the staggeringly popular fermented tea.

Yogurt
Turn a crockpot into a yogurt bot using an Arduino.

Cider
Juice, strain, and bottle your own hard apple cider.

Luminous Lowtops

Written by Clayton Ritcher

Take light-up sneakers to the next level with full-color LEDs that respond to your moves.

CLAYTON RITCHER (claytonritcher.com) is a double-major student (electrical and computer engineering, and robotics) at Carnegie Mellon University who enjoys coding, tinkering, and building things. He was born in San Antonio but calls Atlanta home. He's also the creator of the Raspberry Rover (makezine.com/rasprover).

Time Required:
A Weekend
Cost:
$200–$240
Stunning show of light and motion leaves other LED shoes in the dark ages.

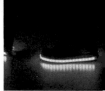

THE LIGHT-UP SHOES FROM YOUR CHILDHOOD ARE ALL GROWN UP
— these Luminous Lowtops are force-sensitive, full-color LED light-up shoes for adults. Each shoe has two embedded force-sensitive resistors (FSRs) — one under the heel and one under the ball of the foot — and up to 40 RGB LEDs that change color based on the force readings, giving brilliant visual effects when you walk, stomp, jump, or lean.

Not long ago I saw a little boy stomping around a store in his light-up shoes. Admittedly jealous, I searched online for adult light-up shoes. Disappointingly, none of them responded to *how* you moved, only to the fact *that* you moved. Also, most of them required a battery pack to be strapped to the leg or shoe, rather than putting it inside like the kids' shoes do. With those issues in mind, I decided to make the Luminous Lowtops.

The electronics are as simple as possible, to allow everything to fit within the shoe. The LEDs are individually addressable, so each one can be a different color at the same time, allowing the shoes to show shifts in weight and react to your movements. An Arduino Mini microcontroller reads an analog input from the front and rear FSRs, converts these values into colors and maps them to the front and rear LEDs, then calculates a color gradient for all the LEDs in between.

Each shoe is powered by 3 rechargeable AA batteries under the heel, and the components are embedded under the insole for a clean look. The LED strip is securely sewn to the exterior of the shoe, so you can jump, dance, or just gaze at the changing colors.

1. Prepare the shoes

There are 2 good mounting options for the batteries and Arduino. Mounting them on top of the tongue of the shoe is easiest, but they'll be more visible there.

To hide the components inside, choose shoes with a thick insole. This allows for some of the padding to be cut out and replaced with the 3×AA battery pack. Rip out the insoles and strip all extra padding off of them, leaving just a thin layer.

Save the extra padding you remove in this step, as you might want to replace some of it later for comfort.

2. Attach the LEDs

Using scissors, cut the LED strips to the proper length to wrap around the perimeter of each shoe. The SparkFun and NeoPixel strips can be cut between any 2 LEDs (**Figure 2a**); just avoid the copper contacts. Cut the Adafruit #306 strip along the lines that appear after every second LED.

Use a needle (with thread that matches your shoes) to sew the LED strip to the shoe. To do this, loop the thread from the inside of the shoe, out through the edge of the LED strip, and back through the opposite edge of the strip into the shoe (**Figure 2b**). Pull this loop tight and tie a knot.

Repeat this process about once every inch along the shoe's perimeter.

Materials

- » **Shoes** (2) ideally with thick insoles
- » **RGB LED strips, 1 meter, individually addressable** (2) SparkFun #COM-12027, 60 LEDs per meter. You can also use the inexpensive Adafruit NeoPixel strips, or the older Adafruit #306 (32 LEDs/ meter).
- » **Sewing kit**
- » **Arduino Pro Mini 328 microcontroller boards, 5V, 16MHz** (2)
- » **FTDI Basic Breakout board, 5V**
- » **Force-sensitive resistors, FlexiForce, 100lb rating** (4) from tekscan.com. Make sure to buy the 100 pound model. If your shoe size is too small to fit the 8" sensor, buy a shorter size for a bit more money.
- » **Resistors, 1MΩ** (4)
- » **Hookup wire, stranded insulated, 22 gauge**
- » **3-pin strip headers, female, 0.1" spacing** (2)
- » **Batteries, NiZn rechargeable, AA size** (6) Normal AA batteries provide 1.2V, but NiZn AAs provide 1.7V, so only 3 are needed to supply 5V, helping them fit in the shoe. You can also try AAAs; they fit into shoes easily, but won't last as long.
- » **Battery packs, 3×AA** (2)
- » **Battery chargers, NiZn AA** (2)
- » **Cotton padding or cotton balls**
- » **Battery packs, 2×AAA** (2) (optional) to hide the Arduino Minis

Tools

- » **Scissors**
- » **Drill or high-speed rotary tool** such as a Dremel
- » **Wire cutters / strippers**
- » **Soldering iron and solder**
- » **Duct tape** or packing tape
- » **Hot glue gun** (optional) to mount the Arduino
- » **Computer running Arduino IDE** free download from arduino.cc

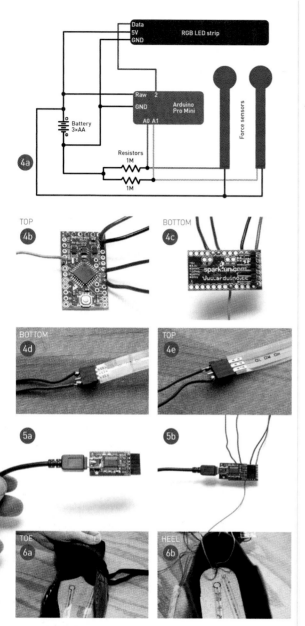

Data
5V
GND
RGB LED strip

Raw   2
GND   Arduino
       Pro Mini
A0 A1

Battery
3×AA

Force sensors

Resistors
1M
1M

4a

TOP
4b

BOTTOM
4c

BOTTOM
4d

TOP
4e

5a

5b

TOE
6a

HEEL
6b

3. Drill the shoes

Drill a small hole through the back of each shoe, so that it emerges under the insole. This will allow wires from the Arduino and battery pack inside to reach the LED strips outside.

4. Build the circuit

Following the wiring diagram (**Figure 4a**), solder together the circuit for each shoe.

The wires from the Arduino to the LED strips need to be about half the length of the shoe; the wires to the force sensors should be about three-

quarters of the length of the shoe. Cut them longer than you think you need; you can always shorten them later (**Figures 4b** and **4c**).

Don't solder directly to the force sensors, as they are plastic and could melt. Instead, solder to a 3-pin female strip header, and then plug the force sensor into the header (**Figures 4d** and **4e**). The middle pin of the force sensors isn't used.

While you're at it, solder the included headers to the FTDI breakout board.

5. Program the Arduino

Download the project code from makezine.com/projects/luminous-lowtops and open it in the Arduino IDE:

» If you're using SparkFun or NeoPixel LEDs, use the *Neopixel.ino* sketch and download Adafruit's Arduino library for NeoPixel LED strips from github.com/adafruit/Adafruit_NeoPixel.

» If you're using Adafruit #306 LEDs, use the *8806. ino* sketch and Adafruit's library for LPD8806 LED strips from github.com/adafruit/LPD8806.

Under the Tools → Board menu, choose Arduino Mini w/ATmega328. Also, under Tools → Serial Port, select the serial port that your board is plugged into.

Plug the FTDI breakout board into your computer and plug its header pins into the corresponding 6 pins on the end of the Arduino Pro Mini (**Figures 5a** and **5b**).

Count the LEDs on your shoe and update the nLEDs variable in the sketch with that number. (The default is int nLEDs = 40.) Click Upload in the Arduino IDE. Unplug the Arduino board.

Repeat for the second Arduino Mini.

6. Mount the force sensors

Use duct tape or packing tape to mount 2 force sensors inside each shoe, on top of the sole, so that the circular pad of one force sensor is under the ball of your foot, and the pad of the other sensor is under your heel (**Figures 6a** and **6b**).

Run the wires and resistors flat along the bottom of the inside of the shoe.

7. Mount the battery pack and Arduino

Tape the battery pack on top of the heel force sensor so that it fits comfortably under your heel (**Figure 7a**). I rip out any extra padding under the heel first. Most shoes have a hard pad under the heel to lift it; the battery pack essentially replaces this.

The Arduino should lie flat, just forward of the battery pack but tucked close to it, so that the batteries take your weight, not the Arduino (**Figure 7b**). Cover it with a bit of cotton padding or a cotton ball to protect it and to prevent it from poking you. Once you're sure the shoes are working great, you can seal the Arduino in with hot glue. Or you can stash it in an empty 2×AAA battery pack for extra protection (**Figure 7c**).

Replace the insole to cover the electronics and battery pack. Though you've thinned it out, it still offers a bit of a cushion.

If your shoes don't have room for these components under the heel, you can mount them on top of the tongue, above the area where your foot flexes (**Figure 7d**). Again, you can protect the Arduino by hiding it in an 2×AAA battery pack.

8. Power up the shoes

Charge the 6 NiZn AA batteries and place them in the battery pack of each shoe.

Put the shoes on, lace them up, and watch as they react when you walk, run, jump, and dance (**Figures 8a, 8b, and 8c**)!

The basic code loop reads an analog input from the front and rear force sensors (their resistance changes linearly with the amount of force, and they're connected to the Arduino with pull-down resistors). It then takes those force values and scales them to the color space of the LEDs based on some general estimates of the maximum and minimum resistances of the force sensors.

Once the code has calculated the corresponding colors for the front and rear LEDs, a **for** loop produces a color gradient of sorts for all the LEDs in between.

Finally, the code sends these color values to the individual LEDs.

Going Further

To improve the fit with the components mounted inside the shoe, try excavating a cavity in the top of the sole to accept the battery pack and Arduino.

Try modifying the `getColorFromForce` function in the code so that the default color (with no weight on the sensors) is your favorite color. You might also save power (and extend run time) by switching the default from blue to red.

You could easily apply the techniques and code from this project to modify other kinds of shoes, like these light-up high heels (**Figure 9**), or other garments altogether. What about light-up elbow or knee pads? Light-up gloves? Pants? ◐

Get the code, instructions, photos, and video at makezine.com/projects/luminous-lowtops

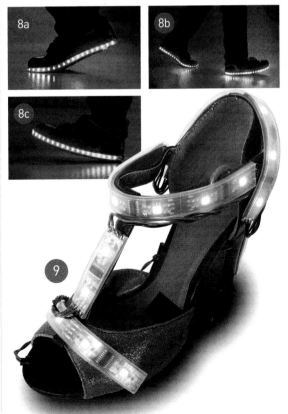

Flora NeoGeo Watch

Make a styling LED timepiece with GPS navigation and compass modes built in. Written by Becky Stern and Tyler Cooper

**Time Required:
A Weekend
Cost:
$140**

John de Cristofaro for Adafruit

**USE THE FLORA WEARABLE MICROCONTROL-
LER AND ITS GPS MODULE TO TELL TIME WITH
A STUNNING RING OF PIXELS.** A leather cuff
holds the circuit and hides the battery. The watch
is a bit chunky, but it still looks and feels great on
tiny wrists!

The circuit sandwich becomes the face of the
watch, and you'll use a tactile switch to select
its three modes: timekeeping, compass, and
GPS navigation. Customize your waypoint in the
provided Arduino sketch, and the LEDs will point
the way and then tell you when you're arriving at
your destination.

This is an intermediate-level project requiring

soldering and some precision crafting. Before
you begin, read the following guides at learn.
adafruit.com: *Getting Started with Flora*, *Flora
Wearable GPS*, *Flora Accelerometer*, and *Adafruit
NeoPixel Überguide*.

This project is adapted from our book *Getting
Started with the Adafruit Flora*, Maker Shed item
#9781457183225-P at makershed.com. Project
help and modeling from Risa Rose.

1. Assemble the circuit

1a. Prepare the components
Start by soldering small stranded wires to your
electronics components, about 2" long each.

WHY FLORA?

Flora is Adafruit's Arduino-compatible wearable electronics platform. Measuring only 1¾" in diameter, it's small enough to embed into any wearable project, and it has large pads for sewing with conductive thread. The round shape means there are no sharp corners to poke through your garment, and the 14 pads are laid out to make it easy to connect a variety of sensors and modules, such as Flora NeoPixels: addressable, color-changing LED pixels.

With its powerful ATmega32u4 microprocessor, Flora has built-in USB support. It's programmed via Mac or PC with free software you download online. Adafruit publishes hundreds of tutorials and dozens of free code libraries for Arduino-compatible boards, so you'll never be lacking project ideas or sample code to get you started.

CIRCUIT DIAGRAM
Each component connects to the Flora main board as follows:

GPS:
3.3V → 3.3V
TX → Flora RX
RX → Flora TX
GND → GND

Accelerometer:
GND → GND
SCL → SCL
SDA → SDA
3.3V → 3.3V

NeoPixel ring:
Vcc (power) → Flora VBATT
IN (Data Input) → Flora D6
Gnd (ground) → GND

Tactile switch:
Any 2 diagonal pins → Flora D10 and GND

Battery:
Plugs in via the onboard JST port

Becky Stern for Adafruit

BECKY STERN (sternlab.org) is a DIY guru and director of wearable electronics at Adafruit. She publishes a new project video every week and hosts a live show on YouTube. She lives in Brooklyn, NY, and belongs to art groups Free Art & Technology and Madagascar Institute.

TYLER COOPER
Tyler Cooper is a creative engineer at Adafruit Industries, where he helps develop the Adafruit Learning System. In 2010, he co-founded the open-source hardware company Coobro Labs. He's also co-owner of the Minneapolis/St. Paul, Minn., makerspace Nordeast Makers.

Materials

- » Adafruit Flora microcontroller board Maker Shed #MKAD58, makershed.com or Adafruit #659, adafruit.com
- » Flora Wearable Ultimate GPS Module Adafruit #1059, or get both the Flora and its GPS module in the Flora GPS Starter Pack, Maker Shed #MKAD51 or Adafruit #1090
- » NeoPixel RGB LED ring, 16 LEDs Maker Shed #MKAD75 or Adafruit #1463
- » Flora Accelerometer/Compass Sensor Adafruit #1247
- » Tactile switch Adafruit #1119
- » Battery, LiPo, 3.7V 150mAh, with charger Adafruit #1317 and #1304
- » Leather watch cuff Check out Labyrinth Leather on Etsy.
- » Scrap of fabric
- » E6000 craft adhesive
- » Binder clips
- » Thin-gauge stranded wire
- » Double-stick foam tape
- » Black gaffer's tape

Tools

- » Soldering iron
- » Solder, rosin core, 60/40
- » Multimeter
- » Scissors
- » Flush snips
- » Wire cutters / strippers
- » Pliers
- » Tweezers (optional)
- » Helping hand tool (optional)

Strip the wire ends and twirl the stranded core to make it pass easily through the circuit boards' holes. Solder wires to the NeoPixel ring's IN, Vcc, and Gnd pads on the back of the board (**Figure A** on the following page).

Also solder wires to the GPS module (3.3V, RX, TX, and GND) and the accelerometer/compass module (3V, SDA, SCL, and GND) (**Figure B**).

Trim off any wire ends with flush snips.

1b. Mount the GPS and accelerometer
Use double-stick tape or E6000 glue to mount the GPS in the center back of Flora. Wire the GPS connections to Flora according to the circuit diagram on the right, keeping the wires short and flush to the board (**Figure C**).

Flip the circuit over and solder the wires to the Flora main board (**Figure D**). Trim any excess wire ends.

Use double-sided foam tape to mount the accelerometer in the center of the Flora, on top of the 32u4 microcontroller chip. The foam helps distance the boards from one another to avoid short circuits (**Figure E** on page 357).

Trim, strip, and solder the wire connections for the accelerometer/compass (**Figure F**).

A

B

C

D

1c. Mount the switch

To prepare the tactile switch, flatten and snip off any 2 diagonal pins (**Figure G** on the next page).

Insert the switch into D10 and GND on the component side of the Flora board (**Figure H**). Bend out the leads to hold it in position and solder the joints. This big tactile button makes it easy to switch watch modes by holding down the whole face of the watch for a few seconds.

1d. Test your work

Use a multimeter to verify that your solder connections are secure.

Before you proceed, test the GPS and the accelerometer

with the sample sketches provided in their respective code libraries.

1e. Mount the NeoPixel ring
Trim, strip, and solder the NeoPixel ring's wires to the Flora according to the circuit diagram, routing them inside the ring. Load the NeoPixel test code to be sure the ring is connected and functioning properly.

Glue the NeoPixel to Flora, lining up the PCB edges exactly. Don't pinch the boards together too much — there should be a cushion of glue between (**Figure I**).

Allow the glue to set for at least 1 hour. The circuit is finished!

2. Assemble the watch
The circuit is held in place by the small strap on a leather watch cuff. The USB and JST ports line up perpendicular to the band for easy access, and the switch is on the "top" of the watch.

Unthread the small strap from one half of the cuff and lay it over the component side of the Flora board. Then glue 2 small strips of fabric onto the circuit board to make "belt loops" (**Figure J**). Clamp with binder clips until dry.

Glide your watch face toward the buckle side and thread the free end back through the cuff (**Figure K**).

3. Upload the code
Grab the NeoGeo watch sketch from makezine.com/neogeo and upload it to Flora. The Flora NeoGeo Watch is very easy to use. The watch fetches the time of day from GPS satellites, so when it first powers on, it needs to get a GPS fix by directly seeing the sky. Once set, the watch automatically displays the current time, with one pixel lit for the hour and another for the minute (**Figure L** on the following page).

To change modes after calibration, you'll press and hold the tactile switch near the top of the watch. Compass mode lights up blue in the direction of north (**Figure M** and **Figure N**), no matter which way you turn.

GPS navigation mode points toward the coordinates you configure in the Arduino sketch (**Figure O** on page 359), and grows redder the closer you get to your destination.

The code for the Flora NeoGeo Watch is straightforward. We're using the standard

NOTE: The watch code will allow you to adjust which LED is at 12 o'clock, so the orientation of the ring doesn't matter.

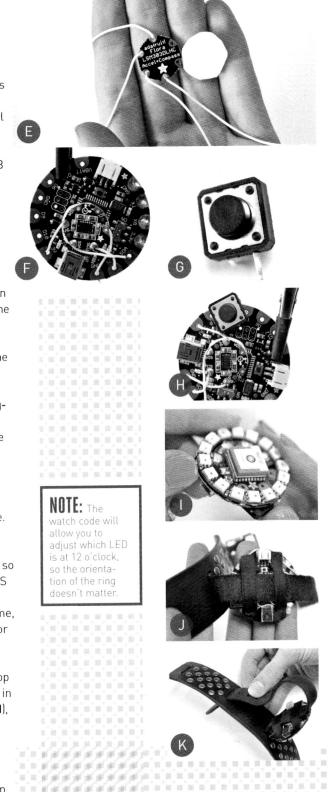

Adafruit GPS Library, Time Library, Pololu's LSM303 accelerometer/compass library, and the Adafruit NeoPixel Library. You'll find links to all required libraries on the NeoGeo Watch Github page. Follow the NeoPixel tutorial to install the library and run the *strandtest.ino* sample code.

Then follow the Flora GPS tutorial to test your GPS module. Make sure your GPS has a direct view of the sky.

Next, calibrate the compass module. Follow the steps on the Pololu LSM303 Github page on how to use the calibration sketch (github.com/pololu/LSM303). Then take the numbers from the calibration sketch and dump them into the NeoGeo Watch sketch in the **Calibration values** section of the code.

In the **WAYPOINTS** section of the code, dump in a location, such as your home, so you'll always be able to find your way home. We like iTouchMap (itouchmap.com) for finding latitudes and longitudes online.

Now you'll calibrate your watch. First calibrate your NeoPixel Ring. Simply light up each pixel using the NeoPixel code to determine which one is your **TOP_LED** pixel (from 0–15).

Then make sure your watch knows which way north is. Select the compass mode. Then, using a compass on your smartphone, or the old-fashioned kind, point your **TOP_LED** north. Count clockwise how many pixels away the lit LED is away from the **TOP_LED**. So, if you aim your **TOP_LED** north, and the LED 4 spots over is lit up, you would change the 0 in the **LED_OFFSET** code above and replace it with **4**.

That's it. Upload the code to your Flora, and start using your NeoGeo Watch!

O

4. Wear it

Plug in a tiny LiPo battery and tuck it into one of the "side pockets" where the cuff overlaps the strap. A bit of gaffer's tape holds it nicely.

Flora's onboard power switch is hidden under the part of the watch closest to you. Use a fingernail or other pointy, nonconductive object to flip the switch.

Place the watch on a windowsill or go for a walk outdoors to get a GPS fix — the clock will then set automatically, and you'll be ready to navigate through space and time. Enjoy your new Neo Geo Watch! ◉

John de Cristofaro for Adafruit

NOTE: It can take several minutes to acquire a GPS fix, but it only needs to do this once on power-up. An optional backup battery will allow the GPS to keep its fix at all times.

CAUTION: This watch is not waterproof! ⚡ Take it off and power it down if you get stuck in the rain. Don't wear it while doing the dishes, etc.

MODES:

WATCH MODE — Shows the "hour hand" as an orange LED, the "minute hand" as a yellow LED, and if both hands are on top of each other, the LED will glow purple. Once the GPS locks on, it will automatically update the time for you. If you lose GPS signal, it will remember the time.

NAVIGATION MODE — Points an LED in the direction you need to go to reach the coordinates you entered in the sketch. When you get close to your destination, the LED will change from yellow to red. Requires a constant GPS lock.

COMPASS MODE — A blue LED always points to the north.

To change between modes, press and hold the button for 2–3 seconds. Hold down the button longer and it will cycle through all 3 modes.

The Chameleon Bag

Make an interactive messenger bag that reacts to your RFID-tagged objects.

Written by Kathryn McElroy

Time Required:
6-8 Hours
Cost:
$180-$200
Solder up the electronics, then brush up on your sewing to make this spectacular bag.

Jeffrey Braverman

THE CHAMELEON BAG IS AN INTERACTIVE MESSENGER BAG WITH A REACTIVE FRONT PANEL. I wanted the bag to display animations and patterns across its front flap as the user places different RFID-tagged objects into it, and I accomplished this by combining a Boarduino microcontroller and an RFID reader inside the bag, along with 49 RGB LEDs on its front flap.

Designed into the coding are three specific uses, with potential for more. First, the bag keeps track of the RFID-tagged items placed inside and warns the user through light displays if an important item, such as keys or cellphone, is missing. Alternately, the lights can change color to match clothing or accessories embedded with RFID tags. And finally, it can be programmed to display cheery animations

when a favorite totem with an RFID tag is placed in the bag, enabling the user to share her good mood with the people around her. Since the user has access to the microcontroller, she can code additional uses to change the colors and animations of the display.

1. Set up the electronics

1a. First, solder your Boarduino following the directions at Adafruit (makezine.com/adafruit-boarduino). Connect it to the computer using the FTDI cable; the colored wires with female headers go onto the back wall of pins (black goes to ground, for orientation).

Open the project code in the Arduino IDE software, select Tools → Board, and select Duemilanove with ATmega328. Upload the code to the board and launch the Serial Monitor by clicking the icon in the upper right-hand corner. It should tell you that no RFID reader is attached.

KATHRYN McELROY
is a Design Lead and UX Designer for IBM Watson in Austin, TX. She is an award-winning designer and photographer and is passionate about near-future technology, smart objects, and open hardware. Kathryn is a talented maker in many mediums, including: sewing, electronics, baking, papercraft, and graphic design.

NOTE:
If you're having trouble programming the Boarduino, try moving the jumper over one pin.

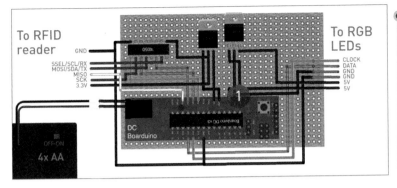

```
To RFID          To RGB
reader    GND    LEDs
                 CLOCK
SSEL/SCL/RX      DATA
MOSI/SDA/TX      GND
MISO             GND
SCK              5V
3.3V             5V
          050k
OFF-ON
4x AA
              DC
              Boarduino
              1
```

1b. Next, set up the RFID reader using the Adafruit tutorial (makezine.com/adafruit-rfid-tutorial). Solder the two 3-pin strips, but don't solder the 8-pin strip. Follow the directions to connect the RFID reader to the included 4050 level-shifter pin. Mount the 4050 onto the small breadboard, and temporarily connect it to the RFID reader and to the Boarduino with jumper wires, using **Figure 1** for reference.

1c. Now you'll test the RFID reader with some RFID tags. First, reload the code and open the Serial Monitor. It should now recognize that the reader is attached; if not, check your connections.

1d. Each RFID tag has a unique number, and you'll see these numbers flash onto the Serial Monitor when you place a tag within 4" of the reader. Copy and paste these tag numbers into a text document for reference. In the project code, about halfway down, change the `cardidentifier` variables to your unique tag numbers to choose what animation or colors you'd like to see when the reader reads that number. Then reload the code to your Boarduino. Now you can label your RFID tags to keep track of which tag launches which display.

1e. You're ready to set up the RGB LEDs. Refer to the Adafruit tutorial (makezine.com/adafruit-rgb-led) for the specific wires. You'll need to power the LEDs with the 5V plug-in power supply while testing so they don't try to pull 5V through the Boarduino. Connect the other wires temporarily to the Boarduino as outputs. With the Serial Monitor open, test the RGB LEDs by trying the tags you've assigned colors to. You may need to troubleshoot if something is not properly connected.

1f. Once you've gotten all the components to work together, you're ready to make these connections permanent! Arrange all your components onto a 2"×4" protoboard. Use the 16-prong socket for the RFID reader's 4050, and two 15-pin strips of female headers to make a strip for each side of the Boarduino. Solder the socket and female headers in place. Solder all the connections between the 4050 and the Boarduino.

1g. Cut extra long wires for the connection between the 4050 and the RFID reader, and solder the wires to the protoboard, waiting to attach the reader until we determine the length we need. Cut extra long wires for the RGB LEDs as well, about twice as long as the RFID wires.

1h. Now you can solder the power connections for the RGB LEDs and the RFID reader. We'll be using 4 AA batteries for our power supply to the Boarduino. The RFID reader needs 3.3V, so connect one of the 5V pins to the 3.3V voltage

Materials

» Boarduino microcontroller Adafruit Industries item #91, adafruit.com
» FTDI cable, USB, 3.3V
» RFID reader Adafruit #364
» Breadboard, small for testing
» RGB LED strips (2) Adafruit #738
» Power supply, 5V DC for testing
» RFID tags Adafruit #365
» Protoboard, 2"×4"
» Female headers, 0.1"
» DIP socket, 16-pin, 0.3"
» Voltage regulator, 3.3V, LD1117V33
» Voltage regulator, 5V, 3A, 3-pin
» Battery holder, 5V, 4xAA
» Jumper wires for testing
» Luan plywood, ⅛"×11"×14"
» Electrical tape
» Upholstery foam, 1"× 26"×20" non-yellowing

FOR THE BAG:
Alternately, you can repurpose an existing messenger bag.
» Canvas, gray, 2 yards
» Canvas, white, 1 yard
» Thread, gray and white
» Magnetic snaps (2)
» Metal loops, 2" (2)
» Metal slide, 2"
» Zipper

Tools

» X-Acto knife
» Computer with Arduino IDE free download from arduino.cc
» Laser cutter or drill
» Foam cutter
» Soldering iron and solder
» Sewing machine

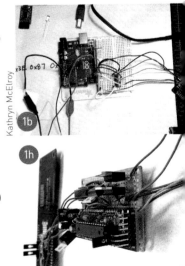

Kathryn McElroy

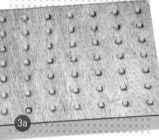

existing bag — following the instructions and templates at makezine.com/projects/chameleon-bag.

3. Assemble

3a. Using the downloadable template (at the URL above), laser-cut or drill holes in the thin plywood to fit the RGB LEDs and hold them in place in a grid pattern. (If you're repurposing a bag, adjust the template to fit into your bag's front flap.) Then push each LED into a hole, and use electrical tape on the backs to hold them in place.

3b. Cut your upholstery foam into two 13"×10" rectangles (or the size of your repurposed bag), one for the back panel and one for the front panel that will cover the LEDs. On the back foam panel, cut out cavities (not all the way through) to hold the electronics. On the front foam panel, use an X-Acto knife to cut Xs where the LEDs will push into the foam. The foam will diffuse the LEDs' colors and make them flow together.

3c. Insert the electronics in the spaces you cut out in the back foam panel, then trim and solder the wires to the RFID reader. Cover the LED board with the front foam panel and insert it into the white front flap of the bag. Measure where the other half of the magnetic snaps should go on the bottom of the front flap and attach them. Measure and finalize the wire lengths from the Boarduino to the LEDs, take the electronics out of the bag, and solder the final wires in place.

3d. Put all the electronics back into their respective bag parts. Insert the front flap's extra fabric down into the back of the bag.

Fold the raw edges of the back panel into itself and pin everything together. Sew the panel together, taking care not to damage the electronics or break your machine's needle on the LED wires.

3e. Unzip the back zipper, insert the battery holder, plug it into the Boarduino, and test. Now your Chameleon Bag can tell what's missing, express how you're feeling, or just match your outfit. ◉

regulator, then connect it to the 4050 and the reader. The LEDs need 5V connected to 2 different wires; since there is only one 5V pin left, connect that to the RGB LEDs, then connect the Vin (total voltage from the batteries) to a 5V regulator before connecting it to the other power wire.

2. Sew a bag or repurpose one

When you want a break from the electronics, start sewing the messenger bag — or repurpose an

Get busy with the project code, downloadable templates, sewing instructions, and video at makezine.com/projects/chameleon-bag

PIPE DREAMS
Build sturdy furnishings
with PVC pipe and a few tricks.
By Larry Cotton and Phil Bowie

Humble PVC drain pipe is cheap, widely available, easy to work with, and almost endlessly useful for making everything from patio furniture to elegant sculptures.

Here are four family-friendly projects that use 3"- or 4"-ID (inside diameter) PVC pipe. In a weekend you can easily make all four: a kids' table with a dry-erase top and matching stool, a two-faced clock to help you remember friends in another time zone, a hanging planter, and an accent lamp that seems to float on light.

You can make them with handheld tools, but bench tools such as a band saw or table saw with a fine-toothed blade work best for making square and accurate cuts. PVC also bends easily when heated in boiling water, which opens up all kinds of new shapes and design possibilities.

If cutting pipe from a 10' length, ask a friend to help support it. Use a face mask and ear protection for cutting and sanding.

Fill any dings with automotive body filler and/or glaze. Then sand the pipe parts with 180-grit sandpaper, prime, and paint. If you want to skip the primer, there are new spray paints that adhere directly to plastic.

⚠ **WARNING:** PVC pipe tends to roll while cutting on a table saw, so hold it firmly and cut slowly. Gripper gloves help. For cutting off sections on a table saw, set the blade just slightly higher than the pipe wall thickness. Don't use a ruler or tape to set blade height; instead, make trial cuts in a scrap of wood and measure the cuts. Always wear eye protection when using power saws.

Larry Cotton is a semi-retired power-tool designer and part-time community college math instructor. He loves music and musical instruments, computers, birds, electronics, furniture design, and his wife — not necessarily in that order.

Phil Bowie is a lifelong freelancer with 300 magazine articles published. He has three acclaimed suspense novels and a short story collection available on Amazon. Visit him at philbowie.com.

Gregory Hayes

CUT IT, DRILL IT, BEND IT, PAINT IT

Polyvinyl chloride (PVC) plastic pipe is strong, works like wood, and accepts various fillers and finishes. It's also thermoplastic, so it can do something wood can't — bend into new shapes with the application of heat. Here are techniques for making your own furnishings from PVC.

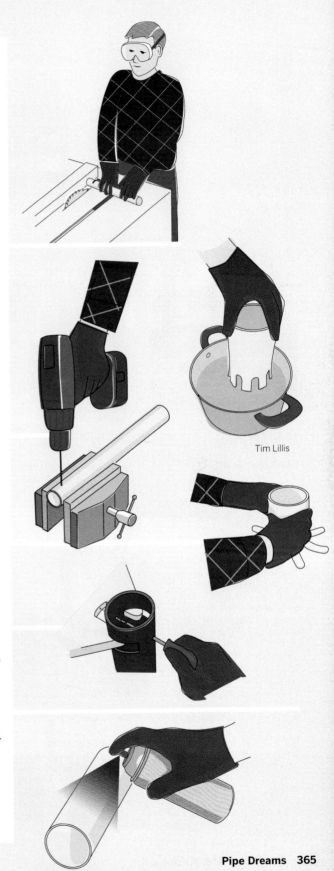

Tim Lillis

CUTTING

PVC pipe is easily cut with small-toothed hand-saws. A hacksaw is slow but relatively accurate.

For straighter cuts, you can use a table saw, band saw, or miter saw (aka chop saw). To avoid chip-outs and minimize the danger of kickback, use a sharp, fine-toothed blade and move the work piece into the blade very slowly. On a table saw, use a rip fence and miter gauge to keep cuts square. On a band saw, set the upper guard properly.

To keep pipe from rolling while you're cutting, clamp it to a workbench or in a vise (with a handsaw), or to the fence (with a miter saw), and wear gripper gloves with any power saw.

DRILLING

When drilling PVC, the bit will grab the plastic firmly — so commit to your drilling motion and follow through. Stopping halfway can leave chunky waste material in the hole, or cause chip-outs. Grip the pipe in a vise or clamp it in a V-groove in a board to prevent rolling.

BENDING

PVC is thermoplastic; heat it up and bend it into any shape you like. In this project you'll use boiling water; we also like the PVC Bendit heating tool (see box, following page).

FASTENING

Wood screws or self-tapping sheet metal screws provide a strong hold in PVC pipe. The kids' table in this project also makes use of interlocking slots for a stronger connection.

FINISHING

Flaws in PVC are easily filled using automotive epoxy putties and glazing and filling compounds.

PVC sands like wood and accepts automotive primer before painting. A new generation of spray paints will bond directly to PVC without primer.

PVC is also easily stained — see "Stain PVC Any Color" on page 375. Acetone will clean off most factory markings.

SET UP.

A

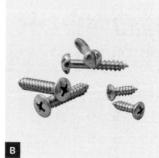

B

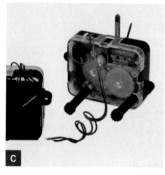

C

D

Gregory Hayes

COOL NEW TOOL

The **PVC Bendit** wasn't used in these projects but it does a nice job creating PVC furniture and structures. This new gadget gently heats plastic pipe up to 4" in diameter, making it pliable so you can bend it to any shape. MAKE Labs tested it and it works great.

MATERIALS

TABLE AND STOOL
A. PVC pipe:
» White, 4" ID × ¼" wall, 10½" length for the stool. Buy a 10' length, Lowe's part #23838, and use it for the other projects too. We used cheap, foam-core Schedule 40 sewer-and-drain PVC but other types of PVC will work.
» White, 3" ID × ¼" wall, 10' length Lowe's #23834, for table legs

B. Screws:
» Sheet metal screws, #10×1" Phillips head (4)
» Wood screws, #8×1" flathead (12)

» Plywood, ¾" thick: 2'×2' (1); 12" diameter rounds (2); 4" diameter rounds (2)
» Dry-erase board, about 48"×32" Lowe's #61082, for table and stool tops
» Contact cement, Weldwood ("The Original"), 14oz
» Wood dowel, ¼" diameter, 6" length
» Scrap wood for making a miter gauge extension

CLOCK
C. Clock movements, 2⅛"×2⅛"×⅝", with hands (2) for 12- or 24-hour time.

» PVC fitting for 4" ID pipe, white, 45° elbow with ⅛" wall Lowe's #24124. This has a ⅛" wall at the openings.
» Plywood, ¾", 4½" diameter (2)
» Hardboard, ⅛", 4½" diameter (2) or dry-erase board
» Photo paper, white, heavyweight
» Sheet metal screws, #6×½" (2)
» Batteries, AA (2)

LAMP
» PVC pipe, white, 4" ID × ¼" wall, 14" length
» Switch, SPST, with round bushing and nut Lowe's #71393
» Plywood, ½", 4" diameter round
» Threaded nipple, 1"×⅜" diameter Lowe's #46816
» Socket, for standard lamp bulb Lowe's #70826
» Wire nut
» Lamp cord, 120V rated
» Lamp bulb, 40W maximum, frosted incandescent or CFL
» Acrylic rod, ¼" diameter, 10" length for the feet. You can also use a tilt wand for Venetian blinds.
» Cyanoacrylate glue aka super glue or crazy glue

PLANTER
» PVC pipe, white, 4" ID × ¼" wall, 8" length
» Nylon monofilament fishing line
» Clay or plastic pot, 4" diameter

For all projects:
D. Paint and primer:
» Spray paint various colors. If you're using primer, you can use most any paint. To skip the primer, use Krylon Fusion or Rust-Oleum Paint for Plastic; they're formulated to bond directly to plastic.
» Spray automotive primer (optional) Rust-Oleum or equivalent

» Automotive body filler putty and/ or automotive glaze such as Bondo filler or DuPont 315 glaze
» Sandpaper, 180 grit
» Masking tape
» Hot glue and/or epoxy

TOOLS

» Saw for cutting plastic pipe. Handheld saws will work, but we recommend a band saw or a table saw with a miter gauge and a fine-toothed blade.
» Drill and drill bits: ¹⁄₁₆", ⅛", ¼", ⅜", ¾" spade, countersink
» Combination square (optional)
» Marking compass or ¾" dowel.
» Jigsaw with fine-toothed blade for cutting plywood
» File or high-speed rotary tool (optional) such as a Dremel
» Screwdriver or long driver bit, Phillips head
» Measuring tape
» Hot glue gun (optional)

MAKE IT.

BUILD YOUR PVC FURNITURE

Time: A Few Hours
Complexity: Easy

KIDS' TABLE AND STOOL

This small table fits young kids perfectly — and they can scribble to their hearts' content on the dry-erase tabletop.

1. MAKE THE TABLE LEGS

1a. Cut them from a 10' length of 3"-ID pipe. It's best to use a table saw with a rip fence and a miter gauge to keep the slots and pipe ends square and parallel.

1b. Wrap a measuring tape around one end of a leg and put marks at the starting point and at exactly half the distance around the leg. Drill — from the outside, not straight across — ⅛" holes through both sides of the legs. Then drill straight across to enlarge both holes to ¼".

1c. Using this technique and the first leg as a guide, drill the holes in the other 3 legs. All holes are 19½" from the bottom ends.

1d. Insert a 6"-long, ¼"-diameter wood dowel in the first leg, held by tape inside.

NOTE: One leg is longer than the others (for storing markers and an eraser).

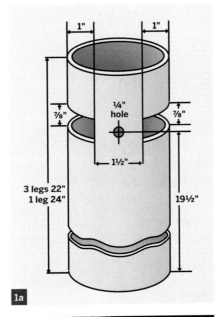

1" · 1" · ¼" hole · ⅞" · ⅞" · 1½" · 3 legs 22" 1 leg 24" · 19½"

1a

1b

1b

1c

1d

Larry Cotton and Phil Bowie

Pipe Dreams 367

1e. Slot one leg to mate with the tabletop. The dowel in the pipe will ride on its top surface so the finished leg slots will be perfectly aligned.

Set the table-saw blade depth to exactly 1". Using the miter-gauge extension, cut the slots in the leg. The slot dimensions and positions are critical to ensure that your table is sturdy and all 4 legs are perpendicular to the top.

1f. Remove the dowel. (The holes will be used to attach the legs to the table.)

1g. Repeat Steps 1d–1f for the other legs. When slotting the longer one, the rip fence (parallel to the blade) must be moved to accommodate that leg's extra length.

Paint the legs a bright color. Hanging the legs horizontally while painting helps.

2. MAKE THE TABLETOP

2a. For the tabletop core, we used smooth ¾" plywood, 2'×2'. For more durability, use exterior plywood. The top surface is ⅛" dry-erase board.

2b. Cut a slightly oversized piece of dry-erase board and laminate it to the plywood using Weldwood ("The Original") contact cement. Follow the container directions exactly. After pressing the 2 pieces tightly together, trim all sides and sand the edges smooth. Avoid scratching the dry-erase surface.

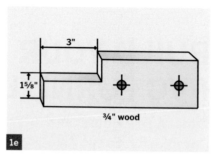

1e

3"

1⅝"

¾" wood

1e

1e

1g

NOTES: If you use a table saw, first make a miter-gauge extension from ¾"-thick scrap wood as shown in this diagram.

While cutting, make multiple small passes and be sure the side of the dowel always stays in contact with the miter-gauge extension.

2c. Lay out the 4 identical corners. A combination square is helpful. Draw the radius at the back of each slot with a thin ring of 3"-ID pipe, then use a compass or the end of a ¾" dowel to mark the 8 radii at the slot ends.

2d. To facilitate cutting, drill ⅜"-diameter holes in the corners of each slot, then cut with a handheld jigsaw with a fine-toothed blade. Sand, file, or Dremel the slots until the legs fit snugly into them. This will ensure the table doesn't wobble.

2e. Fill any imperfections in the legs and the top's edges with Bondo and/or glazing putty. Mask the top and paint its edges white. Paint its bottom for more durability.

3. ASSEMBLE THE TABLE

3a. Use four #10×1" Phillips-head sheet metal screws to screw the legs into the tabletop.

3b. To cap the legs, cut 3" disks from the dry-erase board and drop them into the tops of the legs.

4. MAKE THE STOOL

4a. For the column, cut a 10½" length of 4"-ID pipe and drill 6 clearance holes in it as shown in the diagram. Countersink them so the #8 screws will sit flush.

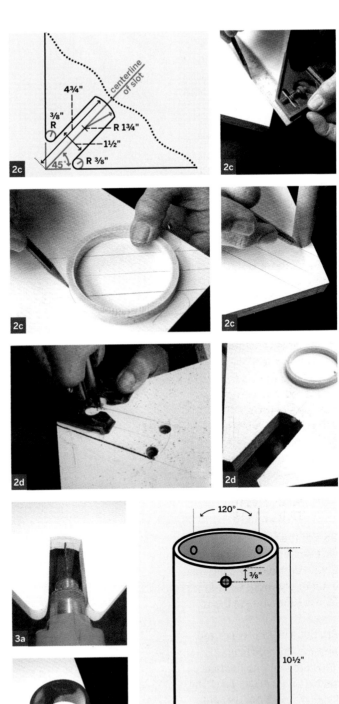

4b. Cut 2 disks of ¾" plywood to fit snugly into the 4" pipe. Cut 2 more ¾" plywood disks 12" in diameter for the stool's top and bottom.

4c. Attach the small disks to the centers of the large disks using three #8×1" screws each.

Laminate an oversized disk of dry-erase board to one of the disks for the top.

4d. Trim, sand, and finish the top and bottom to match the tabletop.

Assemble the stool with six #8×1" flathead wood screws.

4b

4c

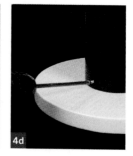

4d

▶ TIP: For speed and accuracy, cut plywood disks on a band saw and pivot the plywood on a brad with its head snipped off.

TWO-FACED CLOCK

If you and a friend or relative live in different time zones, keep track with this two-faced clock. The housing is a 45° PVC elbow for 4" ID pipe. Be sure to get the lighter one with ⅛" walls. The clock movements are the ubiquitous AA-battery plastic boxes. Remove them from old clocks or buy them online.

1. MAKE SPACERS

Cut two ¾" plywood spacers to slip-fit into the elbow ends and core them out with a jigsaw or band saw to clear the clock movements.

2. MAKE FACE BACKING PLATES

Cut 2 disks from dry-erase board to slip-fit on top of the

4d

TWO-FACED CLOCK

2

2

spacers. Sand the edges.

To drill the exact center of a disk, you can draw 2 chords, bisect them with perpendicular lines, and drill where the lines intersect.

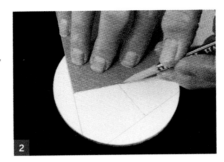

3. MAKE CLOCK FACES

Use your imagination to design 2 clock faces, then print them on thick paper. Add names to the faces to match whichever time zone your friend or relative lives in.

4. ASSEMBLE THE CLOCKS

Glue the faces, backing plates, and spacers together. Add the movements, mounting nuts, and AA batteries.

Clip the hands to fit (if necessary), push them onto their respective spindles, and set their times.

How about a NASA photo of the world? Or make 12-hour and 24-hour faces. Or a galaxy, with hands painted to resemble meteors? Here are a few that we designed.

5. INSTALL THE CLOCKS

Orient one face carefully (the seam on the elbow can help), then insert the assembly into the elbow. Affix it with one #6×½" sheet metal screw. Repeat for the other opening.

6. GO FURTHER

We left our elbow unmodified (including bar code label), but here are a few suggestions for other clocks and bases using PVC pipe.

FLOATING ACCENT LAMP

This lamp adds a romantic glow to any room. Designed to provide good airflow around the bulb, it seems to float magically on a soft ring of light around its base.

1. MAKE THE HOUSING

Cut and drill the bulb support from ½" plywood, following the diagram.

One 45°-angled cut in the 4" PVC easily allows you to make 2 lamp housings if you wish. Drill the hole in back to fit your switch, 2" from the pipe bottom. Thoroughly sand the angle and the bottom surfaces.

2. ASSEMBLE THE LAMP PARTS

Thread the nipple through the bulb support, add the socket, and wire up the switch. Ensure that all electrical connections are tight and insulated.

Screw in a 40W (maximum) frosted or CFL bulb and test.

3. PAINT

Mask the inside of the PVC housing, then sand and paint the outside a bright color.

Reverse the masking and paint the elliptical rim, the bottom rim, and the inside white.

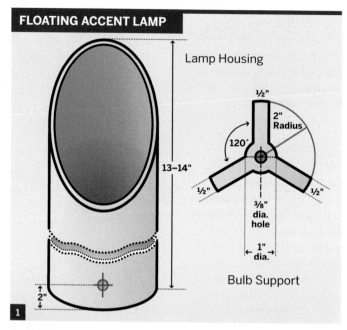

FLOATING ACCENT LAMP

Lamp Housing

13–14"

½"

2" Radius

120°

½" ½"

⅜" dia. hole

1" dia.

Bulb Support

2"

1

1

1

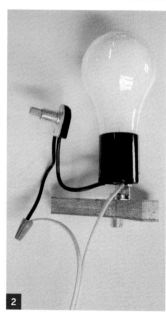

2

4. INSTALL THE LAMP ASSEMBLY

Mount the switch, and then hot-glue the bulb support assembly into the lamp about 1" from the bottom.

5. GLUE ON THE FEET

Super-glue at least six ¼"-diameter, 1½"-long clear feet inside the bottom rim. We used acrylic rod. Sand the mating surfaces thoroughly before gluing. Notch a scrap of wood and use it as a jig to ensure that the feet protrude the same amount.

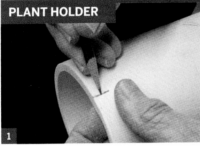

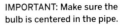

IMPORTANT: Make sure the bulb is centered in the pipe.

PLANT HOLDER

This versatile and attractive plant holder holds a standard 4" flowerpot in a variety of ways. You can set the pot in either end and use the holder upside down or right side up. In either position, you can hang it, or just place it on any surface.

1. CUT AND DRILL THE PIPE

Cut an 8" length of 4"-ID pipe and mark 8 evenly spaced spots 3" from one end to drill holes.

To space the holes, wrap a strip of paper around the pipe and mark a line across both ends. Hold the strip in front of a light to align the marks and fold the strip in half. Then fold in half twice more. Mark the pipe at the fold lines.

Using a spade bit, drill eight ¾" holes.

PLANT HOLDER

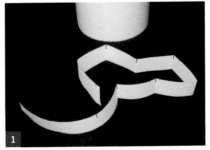

2. MARK AND CUT SLOTS

Draw lines from each hole to the pipe end. Then cut the slots with a handheld jigsaw.

3. BEND THE LEGS

Boil 3½" of water in a large cooking pot. Using gloves, immerse the slotted pipe end until the legs become very pliable. Bend the legs out away from the body a bit.

Remove the pipe from the water, then while keeping the pipe perpendicular, push the legs down onto a cookie sheet. The legs will splay out. Allow to cool for a minute. Repeat if you need to correct any faults.

4. FINISH THE LEGS

Round the leg ends with a disc sander, then file and sand the legs smooth. To hang your plant holder, drill a 1/16" hole in the center of each leg end.

5. FIT A FLOWER-POT AND PAINT

Some pots fit the holder better if you chamfer the inside top edge. Use the pot itself as a backing surface for coarse sandpaper.

Spray it a bright color or leave it white. Attach mono-filament line to each leg for hanging. ◪

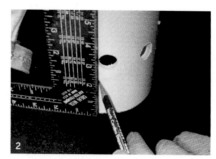

TEST BUILDER:
Max Eliaser,
MAKE Labs

1+2+3 Stain PVC Any Color
BY SEAN MICHAEL RAGAN

PVC PIPE IS GREAT, BUT IT'S KINDA UGLY
— it only comes in white, gray, sometimes black, and clear. Sure, you can paint it, but paint can flake and can screw up dimensional tolerances. Stain doesn't flake or add thickness, so the pieces will still fit together.

⚠ **CAUTION:** Work in a well-ventilated workspace and wear gloves and goggles when handling the solvent or dye.

1. Mix the stain.
Visit makezine.com/projects/stain-pvc-any-color-you-like for a list of dye-to-PVC-cleaner ratios for red, orange, yellow, green, blue, indigo, violet, brown, and black.

Using your pipette, draw up the required volume of each dye and transfer it to the PVC cleaner container. Be careful not to cross-contaminate the dyes. Note that solvent dyes are very strong; 1 ounce goes a long way.

Close the can lid tightly. Wipe off any stray liquid on the outside of the can. Gently shake for about 15 seconds to mix.

2. Apply the stain.
You can use a holder for the PVC, such as a piece of bent wire hanger. Generously slather the stain onto the pipe using the cleaner can's built-in applicator. Work quickly, smoothing out streaks before they have time to dry.

3. Dry and test.
The solvent will dry quickly — an hour will be more than enough. Once dry, the stained PVC should be able to pass a "white glove test" and not transfer even a small amount of color to anything that touches it.

NOTE: Dyes can fade over time; try using light-fast dyes or adding UV stabilizers. ◪

Sean Michael Ragan (smragan.com) is descended from 5,000 generations of tool-using hominids. Also he went to college and stuff.

YOU WILL NEED
Nitrile gloves
Safety goggles
PVC cleaner Check the label and make sure it contains tetrahydrofuran. I used Oatey Clear Cleaner, a product used to prepare PVC pipe for gluing.
Volumetric pipette, measuring 1mL
Solvent dye I found Rekhaoil Red HF (Solvent Red 164), Rekhaoil Yellow HF (Solvent Yellow 126), and Rekhaoil Blue (Solvent Blue 98) on eBay by searching "petroleum dye."
Paper towels
Bent wire hanger (optional)

Damien Scogin

DIY

CONDUCTIVE INK

TIME REQUIRED:
2 DAYS
COST:
$45–$130

MATERIALS
» **Silver acetate (99%), 1g**
» **Ammonium hydrox-ide (28%-30%), 3.0mL**
» **Formic acid (88% or higher), 0.5 mL**
» **Wood, ½"×3"×3"** or larger
» **Bolt, 2"**

TOOLS
» **Band saw**
» **Bench vise**
» **Hacksaw**
» **Hot glue gun**
» **Test tube with stopper**
» **Small glass vial**
» **Beaker (2)**
» **Dispensing syringe, 100mL (3)**
» **Syringe filter, 0.2μm**
» **Weight boat**
» **Digital scale**
» **Drill**
» **Neoprene Gloves**
» **Safety goggles** rated for chemical splashes

JORDAN BUNKER
is a polymathic jack-of-all-trades who enjoys manipulating ideas, atoms, and bits. When he's not braving the daylight, he can be found in his basement workshop in Seattle.

Mix up a homemade batch using some basic chemistry skills

WRITTEN AND PHOTOGRAPHED BY JORDAN BUNKER

A

B

C

D

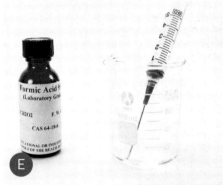

E

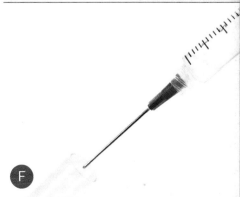

F

THANKS TO RECENT SCIENTIFIC ADVANCES, YOU CAN BUY CONDUCTIVE INKS IN THE FORM OF PENS, PAINTS, AND EVEN PRINTER CARTRIDGES, but have you ever wondered if you could make your own?

You can, and following a simple process developed by the University of Illinois Urbana-Champaign Materials Research Laboratory, it's actually quite easy to produce the conductive ink at home.

The following steps have been adapted from the UIUC paper titled "Reactive Silver Inks for Patterning High-Conductivity Features at Mild Temperatures," and have been simplified for the amateur chemist.

PREPARATION

Clean all glassware and tools and lay them out on your work surface. It is important to read through all of the steps, and make sure that you understand them thoroughly before beginning .

1. MAKE A VORTEX MIXER

Rather than using an expensive laboratory vortex mixer, you can make your own using a 2" bolt and a circular piece of wood.

Cut a circle approximately 2½" in diameter from a piece of ½" thick wood. Drill a centered hole large enough to fit the shaft of the 2" bolt. Drill a second ½" hole halfway into the wood, slightly off-center and overlapping the centered hole (Figure A).

Place the 2" bolt in a bench vise, and then use a hacksaw to remove the bolt head. Insert the bolt shaft into the center hole until it is flush with the bottom of the ½" hole, and then secure it with hot glue.

2. MAKE THE INK

Pour roughly 3mL of ammonium hydroxide into a glass beaker. Using a dispensing syringe, draw exactly 2.5mL out of the beaker and deposit it into the test tube (Figure B).

Place the weight boat on a digital scale and tare the scale. Measure out exactly 1g of silver acetate powder (Figure C) and pour it into the test tube.

Insert the bolt of your homemade vortex mixer into the electric drill. Holding the top of the test

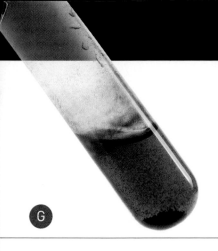
G

H

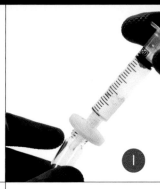
I

tube firmly, place the base of the test tube against the hole in the vortex mixer (Figure **D**). Slowly increase the speed of the drill until a vortex appears in the test tube, mix for 15 seconds, then set the test tube aside.

Pour approximately 0.5mL of formic acid into a second glass beaker. Using a new dispensing syringe, draw exactly 0.2mL of formic acid out of the beaker (Figure **E**).

Drip one drop of formic acid into the mixed solution in the test tube (Figure **F**), then vortex-mix using the same method from step 2. Repeat this process until all 0.2mL of formic acid has been mixed.

After mixing, the solution will be a gray or black color. Place a stopper in the top of the test tube, and set the test tube aside to react for at least 12 hours (Figure **G**).

3. FILTER THE INK

After 12 hours, the solution should look clear with gray sediment of silver particles in the bottom (Figure **H**). In order to use the ink in an inkjet printer or airbrush, these particles must be filtered out to avoid clogging.

Remove the plunger from a new dispensing syringe, and place a 0.2µm syringe filter onto the syringe tip. Fill the syringe with the prefiltered solution, and replace the syringe plunger (Figure **I**).

Pressing slowly but firmly on the plunger, force the solution through the filter and into a small glass vial for storage (Figure **J**).

4. USING THE INK

Before using the ink, it's important to choose a suitable material to deposit it onto. In order to boost the conductivity of the ink, it must be heated to 90°C (192°F), so any material you choose must be able to withstand at least that much heat. Using this ink on porous materials such as paper or fabric will not result in a conductive coating, so it is recommended that the material have a smooth surface.

Using a paintbrush, apply the ink to your material of choice; a stencil can be used to create complex patterns. Allow the ink to dry until it turns to a dull gray color.

Heat the material to 90°C (192°F) for at least 15 minutes. You can use a toaster oven or hot plate as a heat source.

TIPS FOR USE

The ink is quite fragile and will easily scratch off most materials. To increase adherence, scuff the target surface before ink application. Once the ink is dry, apply clear nail polish to protect the traces.

Unfortunately, you can't solder onto the dry ink, as molten solder will leach the silver coating away. If you have a CircuitWriter silver-based ink pen, you can use it to create pads on the traces, which can then be carefully soldered to. ⊘

J

For complete step-by-step instructions and photos see makezine. com/projects/diy-conductive-ink.

CAUTION: Heating in a toaster oven will make the oven unsafe for food preparation. ⚡